Lecture Notes in Physics

Managing Editor

W. Beiglböck
Assisted by Mrs. Sabine Landgraf
c/o Springer-Verlag, Physics Editorial Department II
Tiergartenstrasse 17, D-69121 Heidelberg, Germany

The Editorial Policy for Proceedings

The series Lecture Notes in Physics reports new developments in physical research and teaching – quickly, informally, and at a high level. The proceedings to be considered for publication in this series should be limited to only a few areas of research, and these should be closely related to each other. The contributions should be of a high standard and should avoid lengthy redraftings of papers already published or about to be published elsewhere. As a whole, the proceedings should aim for a balanced presentation of the theme of the conference including a description of the techniques used and enough motivation for a broad readership. It should not be assumed that the published proceedings must reflect the conference in its entirety. (A listing or abstracts of papers presented at the meeting but not included in the proceedings could be added as an appendix.)

When applying for publication in the series Lecture Notes in Physics the volume's editor(s) should submit sufficient material to enable the series editors and their referees to make a fairly accurate evaluation (e.g. a complete list of speakers and titles of papers to be presented and abstracts). If, based on this information, the proceedings are (tentatively) accepted, the volume's editor(s), whose name(s) will appear on the title pages, should select the papers suitable for publication and have them refereed (as for a journal) when appropriate. As a rule discussions will not be accepted. The series editors and Springer-Verlag will normally not interfere with the detailed editing except in fairly obvious cases or on technical matters.

Final acceptance is expressed by the series editor in charge, in consultation with Springer-Verlag only after receiving the complete manuscript. It might help to send a copy of the authors' manuscripts in advance to the editor in charge to discuss possible revisions with him. As a general rule, the series editor will confirm his tentative acceptance if the final manuscript corresponds to the original concept discussed, if the quality of the contribution meets the requirements of the series, and if the final size of the manuscript does not greatly exceed the number of pages originally agreed upon. The manuscript should be forwarded to Springer-Verlag shortly after the meeting. In cases of extreme delay (more than six months after the conference) the series editors will check once more the timeliness of the papers. Therefore, the volume's editor(s) should establish strict deadlines, or collect the articles during the conference and have them revised on the spot. If a delay is unavoidable, one should encourage the authors to update their contributions if appropriate. The editors of proceedings are strongly advised to inform contributors about these points at an early stage.

The final manuscript should contain a table of contents and an informative introduction accessible also to readers not particularly familiar with the topic of the conference. The contributions should be in English. The volume's editor(s) should check the contributions for the correct use of language. At Springer-Verlag only the prefaces will be checked by a copy-editor for language and style. Grave linguistic or technical shortcomings may lead to the rejection of contributions by the series editors. A conference report should not exceed a total of 500 pages. Keeping the size within this bound should be achieved by a stricter selection of articles and not by imposing an upper limit to the length of the individual papers. Editors receive jointly 30 complimentary copies of their book. They are entitled to purchase further copies of their book at a reduced rate. As a rule no reprints of individual contributions can be supplied. No royalty is paid on Lecture Notes in Physics volumes. Commitment to publish is made by letter of interest rather than by signing a formal contract. Springer-Verlag secures the copyright for each volume.

The Production Process

The books are hardbound, and the publisher will select quality paper appropriate to the needs of the author(s). Publication time is about ten weeks. More than twenty years of experience guarantee authors the best possible service. To reach the goal of rapid publication at a low price the technique of photographic reproduction from a camera-ready manuscript was chosen. This process shifts the main responsibility for the technical quality considerably from the publisher to the authors. We therefore urge all authors and editors of proceedings to observe very carefully the essentials for the preparation of camera-ready manuscripts, which we will supply on request. This applies especially to the quality of figures and halftones submitted for publication. In addition, it might be useful to look at some of the volumes already published. As a special service, we offer free of charge LaTeX and TeX macro packages to format the text according to Springer-Verlag's quality requirements. We strongly recommend that you make use of this offer, since the result will be a book of considerably improved technical quality. To avoid mistakes and time-consuming correspondence during the production period the conference editors should request special instructions from the publisher well before the beginning of the conference. Manuscripts not meeting the technical standard of the series will have to be returned for improvement.

For further information please contact Springer-Verlag, Physics Editorial Department II, Tiergartenstrasse 17, D-69121 Heidelberg, Germany

Detlev Koester   Klaus Werner (Eds.)

White Dwarfs

Proceedings of the 9th European Workshop
on White Dwarfs
Held at Kiel, Germany,
29 August – 1 September 1994

 Springer

Editors

Detlev Koester
Klaus Werner
Institut für Astronomie und Astrophysik
Christian-Albrechts-Universität
Olshausenstrasse 40
D-24098 Kiel, Germany

ISBN 978-3-662-14059-8   ISBN 978-3-540-49202-3 (eBook)
DOI 10.1007/978-3-540-49202-3

CIP data applied for

© Springer-Verlag Berlin Heidelberg 1995
Originally published by Springer-Verlag Berlin Heidelberg New York in 1995
Softcover reprint of the hardcover 1st edition 1995

Typesetting: Camera-ready by the editors
SPIN: 10481046   55/3142-543210 - Printed on acid-free paper

PREFACE

Twenty years after the 1st European Workshop on White Dwarfs in 1974, and ten years after the 5th in 1984, this series of meetings returned again to Kiel. In the years since the beginning this workshop has been held almost every two years: Frascati (1976), Tel Aviv (1978), Paris (1981), Frascati (1986), Toulouse (1990), and Leicester (1992). Together with the two IAU Colloquia in Rochester (1979) and Hanover (1988) these have been the milestones in white dwarf research of the last two decades.

The last workshop in Kiel was, with 78 registered participants, the largest ever, documenting the importance of this field for our understanding of stellar evolution and its growth in recent years. This growth is at least partly due to the prominent role white dwarfs play as targets for the EUV and X-ray space experiments of the eighties and nineties, starting from EINSTEIN to the most recent EUVE.

We thank our colleagues of the Scientific Organising Committee (Franca d'Antona, Martin Barstow, Irmela Bues, Gilles Fontaine, Uli Heber, Jordi Isern, James Liebert, Gerard Vauclair, Dayal Wickramasinghe) for their help and suggestions. We also thank our colleagues in Kiel for their great support with the local organisation. We gratefully acknowledge financial support for the meeting from the Deutsche Forschungsgemeinschaft and from the Land Schleswig-Holstein.

The series of European Workshops was founded in 1974 by Volker Weidemann and owes much of its spirit and success to his work. Through Weidemanns research and teaching, white dwarf research in Germany has spread from Kiel to a number of other places, as can be seen in these proceedings. On October 3, 1994 he celebrated his seventieth birthday; on this occasion we want to dedicate these proceedings to Prof. Volker Weidemann in recognition of his outstanding contributions to our field.

Detlev Koester and Klaus Werner
Kiel, December 1994

Contents

Thirty Years of White Dwarf Research in Retrospect
V. Weidemann 1

I   Luminosity Function of White Dwarfs

**The Mass Distribution and Luminosity Function
of Hot DA White Dwarfs**
J. Liebert and P. Bergeron 12

**The Stellar Formation Rate and the White Dwarf
Luminosity Function**
*J. Isern, E. García-Berro, M. Hernanz, R. Mochkovitch, and
A. Burkert* 19

**On the Luminosity Function of White Dwarfs
in Wide Binaries**
T.D. Oswalt and J. Allyn Smith 24

**A New Determination of the Luminosity Function
of Hot White Dwarfs**
A.G. Gemmo, S. Cristiani, F. La Franca, and P. Andreani 31

The Luminosity Function of Dim White Dwarfs
M. Hernanz, E. García-Berro, J. Isern, and R. Mochkovitch 36

Theoretical White Dwarf Luminosity Functions: DA Models
M.A. Wood 41

Cool White Dwarfs in a Southern Proper Motions Survey
M.T. Ruiz 46

II   Evolution and Interior Structure

The Born Again AGB Phenomenon
I. Iben Jr. and J. MacDonald 48

What is Wrong With Stellar Modeling
I. Mazzitelli 58

Transformation of AGB Stars into White Dwarfs
T. Blöcker 68

The Role of G in the Cooling of White Dwarfs
E. García-Berro, M. Hernanz, J. Isern, and R. Mochkovitch 73

Calibrating White Dwarf Cosmochronology
C.F. Claver and D.E. Winget 78

Thermodynamic Quantities for Evolutionary
and Pulsational Calculations
  W. Stolzmann and T. Blöcker 83

Numerical Simulations of Convection and Overshoot
in the Envelope of DA White Dwarfs
  B. Freytag, M. Steffen, and H.-G. Ludwig 88

The Formation of Massive White Dwarfs
and the Progenitor Mass of Sirius B
  F. D'Antona and I. Mazzitelli 93

HST Observations of White Dwarf Cooling Sequences
in Open Clusters
  T. von Hippel, G. Gilmore, and D.H.P. Jones 95

Effects of Updated Neutrino Rates in Evolutionary
White Dwarf Models
  P.A. Bradley 96

III   Atmospheres I

Non-LTE Line Blanketed Model Atmospheres
of Hot, Metal-Rich White Dwarfs
  I. Hubeny and T. Lanz 98

New He I Line Profiles for Synthetic Spectra
of DB White Dwarfs
  *A. Beauchamp, F. Wesemael, P. Bergeron, R.A. Saffer,
  and J. Liebert* 108

Line Broadening in Hot Stellar Atmospheres
  T. Schöning 113

Diffusion in PG 1159 Stars
  K. Unglaub and I. Bues 118

Pure Hydrogen in the Spectrum of GD229
  D. Engelhardt and I. Bues 123

Spectroscopic Effects of T-Inhomogeneities
in the Atmospheres of DA White Dwarfs
  H.-G. Ludwig and M. Steffen 128

The π Line Polarization in Magnetic White Dwarfs
  N. Achilleos and D.T. Wickramasinghe 129

Compton Scattering in Stellar Atmospheres at T_{eff}
Below 100 000 K
  J. Madej 130

Hydrogen Absorption in High Field Magnetic White Dwarfs
N. Merani, J. Main, and G. Wunner 131

LTE for the Analysis of White Dwarfs?
R. Napiwotzki 132

Spherically Symmetric and Plane Parallel NLTE
Model Atmospheres for Hot High Gravity Stars
J. Kubát 133

The Influence of the Bound-Free Opacity on the Radiation
from Magnetic DA White Dwarfs
S. Jordan and N. Merani 134

Models of 3 Magnetic White Dwarf Stars in Flux
and Polarization
A. Putney and S. Jordan 135

Cyclotron Absorption Coefficients in the Stokes Formalism
H. Väth 136

IV   Atmospheres II

IUE Echelle Observations of the Photospheres
and Circumstellar Environments of Hot White Dwarfs
J.B. Holberg 138

The Mass Distribution of DA White Dwarfs
D.S. Finley 150

PG 1159 Stars and Their Evolutionary Link
to DO White Dwarfs
S. Dreizler, K. Werner, and U. Heber 160

Detection of Ultra-Hot Pre-White Dwarfs?
K. Werner, T. Rauch, S. Dreizler, and U. Heber 171

White Dwarfs in Old Planetary Nebulae
R. Napiwotzki 176

New Analyses of Helium-rich Pre-White Dwarfs
T. Rauch and K. Werner 186

The High Mass White Dwarf in HR 8210
W. Landsman, T. Simon, and P. Bergeron 191

The Blue Edge of the ZZ Ceti Instability Strip
D. Koester, N. Allard, and G. Vauclair 196

L745-46A: Lyman Alpha Broadening and Metal Abundances
D. Koester and N. Allard 197

A Model for the WC-type Central Star WR 72
L. Koesterke and W.-R. Hamann 198

**White Dwarf Observations with the Ultraviolet
Imaging Telescope**
W. Landsman, P. Hintzen, and T. Stecher 199

LB 8827: A DB Star With Peculiar Line Profiles
F. Wesemael, A. Beauchamp, J. Liebert, and P. Bergeron 200

**Magnetic or Pressure Shifts of Carbon Bands
in Very Cool Helium-rich White Dwarfs?**
I. Bues and L. Karl-Dietze 201

IUE and Optical Observations of DAO White Dwarfs
J.B. Holberg, R.W. Tweedy, and J. Collins 202

CD -38° 10980 Revisited
J.B. Holberg, F.C. Bruhweiler, and J. Andersen 203

**Analysis of the DO White Dwarf PG 1034+001: Solution
of the He II 4686Å Line Problem**
K. Werner, S. Dreizler, and B. Wolff 204

Upper Limits for Mass–Loss Rates of PG 1159 Stars
U. Leuenhagen 205

V   Binaries, Cataclysmic Variables, Subdwarfs

Exposed White Dwarfs in Dwarf Novae
E.M. Sion 208

**The Spatial and Kinematic Distributions
of Field Hot Subdwarfs**
R.A. Saffer and J. Liebert 221

Fundamental Properties of Magnetic White Dwarfs in CVs
D.T. Wickramasinghe 232

Searching for Binary White Dwarfs: A Progress Report
A. Bragaglia 238

Iron– and Nickel Abundances of sdO Stars
S. Haas, S. Dreizler, U. Heber, T. Meier, and K. Werner 243

The HST White Dwarf Project: V 471 Tauri and Procyon B
H.L. Shipman 248

The Interacting Binary White Dwarfs
J.L. Provencal 254

GD1401 and GD984: X-ray Binaries
with Degenerate Components
  I. Bues and T. Aslan 259

The White Dwarf in AM Her
  B.T. Gänsicke, K. Beuermann, and D. de Martino 263

The Final Results of: A NLTE Analysis of the Helium-rich sdO
Stars in the Palomar Green Survey of High Galactic Latitude
Blue Objects
  P. Thejll, R. Saffer, F. Bauer, J. Liebert, D. Kunze, and H. Shipman 264

^{3}He- and Metal Anomalies in Subluminous B Stars
  U. Heber and L. Kügler 266

HST Observations of the White Dwarf in V471 Tauri
  H.E. Bond, E.M. Sion, K.G. Schaefer, R.A. Saffer, and J.R. Stauffer 267

Hot Stars in Globular Clusters
  S. Moehler, U. Heber, and K.S. de Boer 268

From Interacting Binary Systems to DB White Dwarfs
  J.-E. Solheim 269

White Dwarfs in AM CVn Systems
  J.-E. Solheim and C. Massacand 270

Hot Subdwarf Stars in Binary Systems
  A. Theissen, S. Moehler, T. Bauer, U. Heber, K.S. de Boer, and
  J.H.K. Schmidt 271

CCD Imaging, Optical Spectroscopy and UBVRIJHK
Photometry of Cool Companions to Hot Subdwarfs
  A. Ulla, P. Thejll, C.S. Hansen-Ruiz, J.L. Rasilla, A. Theissen, and
  J. MacDonald 272

VI   Pulsating White Dwarfs

The Multi-Periodic Pulsating PG1159
White Dwarf PG0122+200
  G. Vauclair, B. Pfeiffer, A.D. Grauer, J.A. Belmonte, A. Jimenez,
  M. Chevreton, N. Dolez, and I. Vidal 274

Asteroseismology of DA White Dwarf Stars
  P.A. Bradley 284

The Hydrogen Layer Mass of the ZZ Ceti Stars
  J.C. Clemens 294

Grasping at the Hot End of ZZ Ceti Variability
A. Gautschy and H.-G. Ludwig 295

WET Experiences in Central Asian Observatories
E.G. Meištas 296

The Vista for Seismological Thermometry
of the DBV White Dwarfs
D. O'Donoghue 297

Ionized Helium and Carbon in the DBV GD358
J.L. Provencal and H.L. Shipman 298

Stabilized Restoration of Pulsating White Dwarf Spectra
B. Serre, S. Roques, B. Pfeiffer, N. Dolez, G. Vauclair, P. Maréchal,
and A. Lannes. 299

VII EUV and X-Ray Observations

Extreme Ultraviolet Spectroscopy of White Dwarfs
M.A. Barstow, J.B. Holberg, D. Koester, J.A. Nousek,
and K. Werner 302

White Dwarfs in Close Binaries in the Extreme
Ultraviolet Explorer All-Sky Survey
S. Vennes and J.R. Thorstensen 313

Constraints on DAO White Dwarf Composition
from the ROSAT EUV Survey
M.R. Burleigh and M.A. Barstow 318

Spectroscopy of the Lyman Continuum of Hot DAs Using
the Extreme Ultraviolet Explorer
J. Dupuis and S. Vennes 323

An EUV Selected Sample of DA White Dwarfs
M.C. Marsh, M.A. Barstow, J.B. Holberg, D. O'Donoghue,
D.A. Buckley, T.A. Fleming, D. Koester, and M.R. Burleigh 328

The EUVE Spectrum
of the Hot DA White Dwarf PG 1234+482
S. Jordan, D. Koester, D. Finley, K. Werner, and S. Dreizler 332

ORFEUS and EUVE Observations
  of the Cool DO HD 149499 B
  R. Napiwotzki, S. Jordan, D. Koester, V. Weidemann, S. Bowyer,
  and M. Hurwitz 337

ROSAT Studies of DA White Dwarfs and the Calibration
  of the PSPC Detector
  B. Wolff, S. Jordan, and D. Koester 338

List of Participants 339
Author Index 347

Thirty Years of White Dwarf Research in Retrospect

V. Weidemann

Institut für Astronomie und Astrophysik der Universität, D-24098 Kiel, Germany

Abstract: A review of past milestones is followed by a more personal account of white dwarf research during the past three decades. Especially covered are cool He-rich white dwarfs with metal lines (spectral type DZ), the evaluation of color-color and color-magnitude diagrams, the determination of masses and the mass distribution for H-rich stars (spectral type DA), and the derivation of initial-to-final mass relations for stellar and galactic evolution.

1 Introduction

During the last thirty years our field has grown beyond any expectation. Whereas the First Conference on Faint Blue Stars in Strasbourg 1964 was a gathering of only 11 astronomers (Luyten 1965), and the IAU Colloquium No.42 on White Dwarfs in St.Andrews 1970 was restricted to 25 participants (Luyten 1971), the IAU Colloquium No.53 on White Dwarfs and Variable Degenerate Stars in Rochester 1979 had an attendance of 123 (Van Horn & Weidemann 1979) and the last IAU Conference on White Dwarfs in Dartmouth Hanover 1988 counted 112 participants (Wegner 1989). The series of European Workshops on White Dwarfs started in 1974 in Kiel with 13, continued through Rome (1976, 22), Tel Aviv (1978, 22) Paris (1981, 25), Kiel (1984, 31) Frascati (1986, 35), Toulouse (1990, 46), Leicester (1992, 72), and comes now back to Kiel for the 9.Workshop with 78 participants, not at all restricted to Europe any more.

Similar milestones in white dwarf research are monographies and reviews, appearing during the years, which may be listed here for easy reference: S.Chandrasekhar (1939), W.J.Luyten (1958, 1963), E.Schatzman (1958) — the only book specifically on white dwarfs up to now! —, J.L. Greenstein (1958, 1960), L.Mestel (1965), V.Weidemann (1968), J.Ostriker (1971), V.Weidemann (1975), J.Liebert (1980), E.Sion (1986), D.Koester/G.Chanmugam (1990), V.Weidemann (1990a), F.D'Antona/I.Mazzitelli (1990). As these reviews reflect the progress during the last decades sufficiently, I shall restrict myself in the following to a few topics in which I was especially involved: the DZ phenomenon (still unexplained), the interpretation of two-color and color-magnitude diagrams, the determination of birth rates and luminosity functions, the masses and mass distribution, initial-to-final mass relations, and the role of white dwarfs in galactic evolution.

2 Prehistory

Up to 1958 one can distinguish two main lines of white dwarf research.

On the observational side: W.J.Luyten (1922, 1952) mainly using proper motions and color classes, Humason and Zwicky (1947) photometry, and spectroscopy mainly by Greenstein, using the 200inch Palomar telescope (1958). From the theoretical side: the basic theory of degenerate configurations (Chandrasekhar 1939) and the consideration of physical processes in the interior and at the surface (Schatzman 1944-1958). A third line from which I entered the field 1957 in Pasadena was given by the theory of stellar atmospheres as developed in Kiel (Unsöld, 1938, 1955, Vitense 1951, 1953, Böhm 1954) Since my 1954 thesis had been on "Metal abundances, pressure stratification and pressure broadening in the solar atmosphere" (Weidemann 1955), it was natural that I accepted the challenge to apply my knowledge to an analysis of the atmosphere of the solar temperature white dwarf van Maanen 2 the spectrum of which was kindly provided to me by Jesse L.Greenstein. The astonishing result: a hydrogen deficiency of at least a factor of 50, and an extreme metal underabundance of 10000 compared to the sun (Weidemann 1960) proved that this star belonged to the class of white dwarfs with helium-rich atmospheres (DB) whereas the metal deficiency is not a signature of age (van Maanen being very old) but due to gravitational separation as suggested by Schatzman. In modern classification vMa2 is a DZ (Z indicating the presence of metal lines).

3 The DZ phenomenon

Van Maanen 2 has been the subject of more investigations in the next decade. After K.H.Böhm had investigated convection (1968) and H.Van Horn (1968) cooling, the thesis of G.Wegner (1972) and Grenfell (1974) covered aside from vMa2 the analysis of L745-46A, Ross 640, G47-18 and G99-37, demonstrating that the sum of $(H+C+N+O)/He$ is smaller than 1/10000. Although the lines of vMa2 appear strong they confirm the low abundances, $\log(Ca/He) = -10.7$, or $\log(Mg/He) \approx -9$, $\log(Fe/He) \approx -9.4$. A decade later, with better IDS spectra, Wehrse and Liebert (1980) confirm the analysis of vMa2 (Mg/He 1/1000 solar) but at the same time present the extreme case of G 165-7, which is by comparison with other cool DZ stars extremely Mg-rich (Mg/He 1/30 solar). More calcium-rich cases of hotter stars were also found: GD 40, DBZ (Shipman et al.1977, Shipman & Greenstein 1983) with $Ca/He \approx 3 \cdot 10^{-8}$, and later CBS78 with $Ca/He = 10^{-8}$ (Sion et al. 1986), similar to the DZ GD 401 (Cottrell & Greenstein 1980). Photon-counting and vidicon devices as well as multichannel spectrophotometry helped to improve the observational basis, and lead to the investigation by Liebert et al.(1987), which showed hydrogen traces to be present in at least two other DZ stars besides Ross 640. A comprehensive study, also evaluating IUE results, and discussing dredge-up vs. accretion mechanisms, observability and selection effects was made by Zeidler et al.(1986). As for the interpretation, the cloud accretion hypothesis by Wesemael (1979) and Wesemael and Truran (1982), also Dupuis et al.(1991) seems still to be the best explanation, however the differences of relative metal abundances in individual cases are hard to understand, even if grain accretion is postulated and different diffusion time scales (Vauclair et al.1979, Muchmore 1984) are considered. Synthetic spectra for DZ stars of a large variety of abundances have been published by Zeidler (1987). Recently traces of Ca have been found with solar abundances in several hotter DB stars (Kenyon et al.1988) which are also due to accretion, and some carbon had been found in IUE spectra of L 119-34 (Weidemann & Koester 1989). Furthermore Hammond et al.(1991, 1993) demonstrated that for several DZ stars hydrogen is present although in rather low abundances. They doubt if there are any hydrogen-rich cool white dwarfs - implying increased mixing of DA atmospheres with underlying helium. In this case trace hydrogen need not to be accreted. In summary, the DZ phenomenon with all its variety is far from being understood and invites further studies.

4 Two-color and color-magnitude diagrams

After the historical placement of the first white dwarfs in HR-diagrams (Russell 1914, Bottlinger 1923 – who pioneered the use of photoelectric color indices and concluded that for the explanation of the then three white dwarfs a new equation of state was necessary! –, and Kuiper 1941) comprehensive color-magnitude diagrams were presented by Luyten (1952) which demonstrated that the majority of the white dwarfs have between 1 and 4 earth radii. Using reduced proper motions he showed that for about 90 stars a kind of cooling sequence appeared which extends from very blue to far red in photographic color indices. With the advent of photoelectric photometry (Harris III, 1956 and following, see Greenstein 1958, 1960) Weidemann (1963) was able to give a theoretical explanation of two-color diagrams in which the location of the stars was interpreted by deviations from black body energy distributions. Comparing main sequence ($\log g \approx 4$) and white dwarfs ($\log g \approx 8$) it appeared that the DA stars follow a line of about constant $\log g = 8$, a result confirmed independently by evaluation of Balmer line spectra for 22 DA stars listed by Greenstein (1960).

After Eggen and Greenstein(1965 and following) published comprehensive lists of spectroscopic and photometric data for several hundred white dwarfs, I demonstrated that color-magnitude diagrams using parallax stars and two-color diagrams determining $\log g$ and using a theoretical mass-radius-(surface gravity) relation gave statistically the same distribution in a physical (luminosity/effective temperature) HR-diagram along lines of constant radii (Weidemann 1968) thereby implying the validity of the M-R relation. There was no indication of a division into an "upper and lower sequence" which was suggested by Eggen and Greenstein (1965) from their $M_V(U - V)$ diagram . Since in the following years the cooling theory of white dwarfs had been improved (Mestel and Ruderman 1967) Van Horn (1968) predicted crystallization sequences inclined against the lines of constant radii and parallel-shifted for different core compositions. He tried to explain the Eggen-Greenstein sequences as caused by descendants from different initial mass ranges with different chemical core composition, however retracted in view of my 1968 results and my suggestion that the cooler DA stars have lower luminosities in color-magnitude diagrams due to blanketing effects. I therefore suggested to J.Graham to observe more cooler DA stars with Strömgren photometry - which provides line-free positions in two-color-diagrams - for a crucial test. His results (Graham 1972) gave a convincing proof of absence of a second cooling sequence. One should think that the problem was settled, especially since also the red subluminous stars which had been used to fix the upper sequence, had been shown not to be degenerates (Liebert 1975) but the two-sequence crystallization hypothesis was brought up again by Tapia (1978)! However, Koester (1972) had shown that the luminosity function of white dwarfs should increase monotonically since crystallization starts at the center and spreads out slowly . Even today the luminosity function is not well enough determined that one could notice a cooling delay and corresponding pile-up of stars due to latent heat released at crystallization. Finally, a color-magnitude diagram due to Greenstein (1985), with M_V plotted against the multichannnel color index $(g - r)$, does not show a trace of a second sequence, the remaining overluminous point for the upper sequence being given by EG11 (L 870-02) which is now known to be a white dwarf binary (Saffer et al.1988).

5 Luminosity function, space density, birth rate

With the accumulating number of spectroscopically confirmed white dwarfs (Eggen & Greenstein 1965) a first estimate of the luminosity function became possible. According to my investigation (Weidemann 1967, 1969) it was confirmed that white dwarfs cool according to the Mestel-Ruderman (1967) theory. The space density down to a limiting absolute magnitude was then derived for different ensembles: the Luyten-Palomar material, the Eggen-Greenstein data, the Sandage-Luyten blue stars and white dwarfs within 10 parsec. I obtained space densities between 0.005 and 0.025 WD/pc^3 , down to $M_{bol} = 15.5$. With a cooling age of 5 Gyr this corresponds to an average birth rate of $(1 - 5) \cdot 10^{-12}$WD/pc^{3}yr. Today we know that the Sandage-Luyten blue star ensemble included many hot subdwarfs: if it is removed, the birth rate goes down to $(1-2) \cdot 10^{-12}$ WD/pc^{3}yr. The study also revealed that the observations are incomplete even within 10pc, mainly due to the fact that the redder degenerates are too faint. Although strong efforts were made to detect more cool red degenerates there appeared to be a true "red deficit" (Greenstein 1969) which could be caused either by accelerated Debye cooling, or by a finite age of the Galaxy, a highly interesting possibility (see Weidemann 1975). In the following years cooling theory and calculations were improved by Van Horn (1968), Koester (1972), Shaviv and Kovetz (1976), with phase transitions considered by Mochkovitch (1983 and following), and recently by M.Wood (1990, 1992),- a review is given by D'Antona and Mazzitelli (1990). The deficit of white dwarfs with very low luminosity, established by the luminosity functions of Liebert et al. (1988, 1989) was finally taken to be evidence for a finite age of the Galaxy by Winget et al. (1987), however the method and the accuracy of the estimates has been criticized by Yuan (1989). What can be determined, - provided the observations establish the sharp turndown of the luminosity function unambiguously, and the cooling theory is on certain ground -, is the age of the oldest white dwarfs in the local disk which is not at all identical to the age of the Galaxy. Chemo-dynamical evolutionary models of the Galaxy which are at present under investigation in Kiel (see Theis 1994) must clarify how the galactic disk is slowly formed.

We have also attempted to reproduce the white dwarf luminosity function (and mass distribution, see below) by synthetic evolutionary calculations for a closed local cylinder, starting from the main sequence luminosity function and deriving initial mass functions consistently under different assumptions about core-overshoot with correspondingly changing main sequence lifetimes (Yuan 1992). It appears that the calculated luminosity function predicts about a factor of two more white dwarfs: many of which can be hidden in binaries (see Weidemann 1994 for a review). The determination of the space density - which came out fairly low in the investigation of Fleming et al. (1986) - has been repeated according to the method of Liebert (1978, 1980) by Weidemann (1991) and yields somewhat higher values due to the fact that cooled down white dwarfs of smaller space motions are underrepresented in the observed (proper motion selected) ensembles. If a binary factor is included the birth rate now appears to be around $2 \cdot 10^{-12}$WD/pc^{3}yr in reasonable agreement with what is found for planetary nebulae (see Weidemann 1989).

The contribution of cooled-down degenerates to dark matter was of special interest as long as the dynamically determined local mass appeared to be twice as large as the visible mass. However the fraction of local dark baryonic mass in degenerates, neutron stars and black holes as predicted by Yuan models is small and amounts to only about 4 percent of the now revised local dynamical mass of 50 $M_\odot$/pc^2 (see Weidemann 1990b). These results depend heavily on the age of the local disk, on the history of star formation (initial burst or constant?), on the initial-to-final mass relation (see below) and - for comparison with locally determined space densities - on the evolution of scale heights (inflation effect) and will certainly be improved in the future.

6 White dwarf masses and mass distribution

Already with crude estimates Greenstein (1958) established that average masses for DA white dwarfs are 0.56 $M_\odot$ with a spread between 0.2 and 1.2 $M_\odot$. This has been confirmed by my analysis for 22 DA stars which yielded $\log g = 8.0\pm0.5$, corresponding to 0.6 ± 0.3 $M_\odot$ under the assumption of the validity of a mass-radius relation for hydrogen-free interiors (Weidemann 1963). For stars with known parallaxes it is also possible to determine masses via radii and the M-R relation, or if the surface gravity is known to test the validity of the M-R relation. The first investigations were inconclusive since spectroscopic g-determinations as well as temperatures and derived radii were too uncertain. However after

Strömgren photometry demonstrated that the surface gravity scatter of the DA stars is in reality very small (Weidemann 1971, based on Graham's observations, see above) the main deviations from the theoretical M-R relation were due to uncertain parallaxes. During the next decade the Palomar multichannel-spectrophotometry (MCSP) (Oke 1974, Greenstein 1976) provided valuable data which were evaluated together with UBV and $ubvy$- photometry and Hγ line profiles in a least square analysis by Koester, Schulz and Weidemann (KSW) (1979). KSW used pure hydrogen, line-blanketed model atmospheres, which were applied in the whole DA temperature range. Although for cooler DA stars under-mixing of helium is now under discussion,(see Bergeron et al. 1990, Hammond et al. 1993) the well established narrow sequence in the two-color diagrams (both for Strömgren as well as for MCSP based broad-band colors) all the way down to 7000 K seems to justify the KSW approach. Additional MCSP material (see Greenstein 1984) and restriction to the temperature range in which the gravity dependence of the energy distribution is largest (8000 to 16000 K) provided for 70 DA stars a reliable mass distribution, based on $M(g)$ for a zero-temperature Hamada-Salpeter M-R relation (Weidemann & Koester 1983). In the meantime it had become possible to evaluate redshift measurements based on high resolution Hα cores with greater reliability (Koester 1987, Wegner & Reid 1989) which confirmed the average values of 0.6 M$_\odot$ rather than the higher values of about 0.8 M$_\odot$ found by Trimble and Greenstein (1972). The situation around 1989 has been summarized in my review on "Masses and Evolutionary Status of White Dwarfs and their Progenitors" (1990a).

The next step ahead was made by Bergeron, Saffer and Liebert (1992) in an extensive analysis based on CCD observations of Balmer line profiles for 129 DA stars with temperatures above 13000K. In essence the KSW results were confirmed: a very narrow mass distribution with a maximum around 0.56 M$_\odot$, somewhat lower than KSW. However since this depended strongly on the way the higher Balmer lines overlap, and some revisions on the broadening theory have to be incorporated (Bergeron 1993) this is not final. Evaluation of another ensemble of 52 DA mainly from the southern sky by Bragaglia et al. (1993) reproduced the BSL results with somewhat larger average mass, 0.602 M$_\odot$. However it must be kept in mind that these masses were derived from surface gravity under the assumption of finite temperature M-R relations for white dwarf models with thin helium (10^{-4} M$_\odot$) and negligibly thin hydrogen surface layers, whereas recent seismological results confirm the evolutionary predicted thick layers (10^{-2}He, 10^{-4}H). As I have demonstrated (Weidemann 1993) for the Hyades white dwarfs this would increase the derived masses $M(g)$ about 0.04 M$_\odot$. In summary: although the narrow mass distribution with its slowly decreasing high mass tail (as predicted by population synthesis, see Weidemann & Yuan 1989, Yuan 1992) is well confirmed, the absolute values of the average and the maximum are still open to revision. The Yuan calculations were performed for single star evolution, however there can be no doubt that binary evolution is very important (see Weidemann 1994), and there is evidence in the recent CCD results that white dwarfs with masses below the predicted lower limit for single star evolution do exist, probably due to early mass exchange when the degenerate helium core has not yet grown to its helium flash mass of 0.45 $M_\odot$.

7 Initial-to-final mass relations

Since white dwarfs are present in the Hyades with a turn-off mass of about 2.5 $M_\odot$ it became clear fairly early (Auer & Woolf 1965) that mass loss is able to reduce the initial masses below the Chandrasekhar limit, even down to about 0.6 $M_\odot$. The detection and formulation of wind mass loss (Reimers 1975) and the presence of a white dwarf in the Pleiades with turnoff mass around 6 $M_\odot$ (Woolf 1974) lead me to consider mass loss towards the white dwarf stage in some detail (Weidemann 1977) and to discuss different initial-to-final mass relations (IFMR). With a simple model of galactic evolution WD mass distributions were predicted and shown to have roughly the observed shape. Koester and Weidemann (1980) proceeded to demonstrate how the white dwarf data: median mass, mass distribution, and ratio of WD to supernova birth rates constrain initial mass functions, star formation rates and IMFRs. Koester and Reimers started their program of spectroscopic observations of WD in open clusters (Koester & Reimers 1981, Reimers & Koester 1982), which established the upper mass limit for WD progenitors to be around 8 $M_\odot$ (Weidemann & Koester 1983). Although this result is now considered to be a basic fact of stellar evolution it must be pointed out that it depended on stellar models with intermediate main-sequence core-overshoot. A revision, including newer results from the ongoing Koester/Reimers program, new stellar evolution models, and WD masses, is overdue and at present about to be completed (Herwig 1994). The variety of IFMRs and their influence on models of galactic evolution has been studied extensively in Kiel during the last years (Yuan 1989, 1992). Evolutionary calculations for different initial masses all the way from the main- sequence to the WD region with mass loss on the AGB according to a modified Bowen law have been performed by Blöcker (1993, 1994). The resulting IFMR follows closely the one which was empirically determined under consideration not only from WD in clusters but also from AGB tip luminosities and AGB luminosity functions (Weidemann 1987), however it runs somewhat steeper between 2.5 and 5 $M_\odot$ (see Weidemann 1993, where also the IFMR dependence on metallicity is considered). Although for several reasons interest in IFMRs continues to grow (Bragaglia et al. 1994), I want to stress the fact that differential mass loss in pre-WD stages definitely occurs (Weidemann 1977, 1981, 1990a) and thus a single valued IFMR can only be a first approximation.

8 Closing remarks

Another topic of special interest, in which the Kiel group was strongly involved, was the interpretation of WD of spectral type DQ. IUE observations - which provided so many interesting new WD results (see Weidemann 1988 for a summary) - had revealed the presence of ultraviolet carbon lines. Analysis of the spectra was carried out by Koester et al. (1982) and the dredge-up mechanism by convection and diffusion first proposed. The quantitative study of this mechanism is due to Pelletier et al. (1986). However recent results show a large variety of atmospheric carbon abundances which indicate a different thickness of the surface helium-layer (Weidemann & Koester, 1994)

Since extreme carbon abundances have been derived for hot pre-white dwarfs (see Werner, 1993) the problem of the surface evolution of WD and the way they enter their cooling tracks has become a focus of recent research. The thirty years old question to understand DA vs. non-DA stars remains still unanswered!

I want to close with a word of thanks to all the colleagues who over the years became cooperative – sometimes competitive – friends and my best wishes for the future to those who work in our exciting field of white dwarf research.

References

Auer C.H., Woolf N.J., 1965, ApJ 142, 182

Bergeron P., 1993, in White Dwarfs: Advances in Observation and Theory, M.A.Barstow Ed., NATO ASI Ser.C 403, p.267

Bergeron P., Saffer R., Liebert J., 1992, ApJ 394, 228

Bergeron P., Wesemael F., Fontaine G., Liebert J., 1990, ApJ 351,L21

Blöcker T., 1993, in White Dwarfs: Advances in Observation and Theory, M.A.Barstow Ed., NATO ASI Ser.C.403, Kluwer, p.59

Blöcker T., 1994, A&A submitted

Böhm K.H., 1954, Z.f.Astrophysik 34, 182

Böhm K.H., 1968, Astrophys.Space Sci.2, 375

Bottlinger K.F., 1923, Veröff.Sternwarte Babelsberg 3, Heft 4

Bragaglia A., Renzini A., Bergeron P., 1993, in White Dwarfs: Advances in Observation and Theory, M.A.Barstow Ed., NATO ASI Ser.C 403, Kluwer, p.325

Bragaglia A., Renzini A., Bergeron P., 1994, ApJ in press

Chandrasekhar S., 1939, Introduction on Stellar Structure, Dover 1957

Cottrell P.L., Greenstein J.L., 1980, ApJ 242, 195

D'Antona F., Mazzitelli I., 1990, Ann.Rev.Astr.Ap.28,139

Dupuis J.,Fontaine G., Wesemael F., 1991, in White Dwarfs, G.Vauclair, E.Sion Eds., Kluwer, p.333

Eggen O., Greenstein J.L., 1965, ApJ 141,83; 142,925; 150,927

Fleming T.A., Liebert J., Green R.F., 1986, ApJ 308, 176

Graham J., 1972, AJ 77, 144

Greenstein J.L., 1958, Encycl.Physics 50,161. Springer

Greenstein J.L., 1960, Stars and Stellar Systems 6, 676

Greenstein J.L., 1969, Comments Astrophys.Space Sci.1, 62

Greenstein J.L., 1976, AJ 81, 323

Greenstein J.L., 1984, ApJ 276, 602

Greenstein J.L., 1985, PASP 97, 827

Grenfell T.C., 1974, A&A 31, 303

Hammond G.L., Sion E.M., Kenyon S.J., Aannestad P.A., 1991, in White Dwarfs, G.Vauclair, E.Sion Eds., Kluwer, p.317

Hammond G.L., Sion E.M., Aannestad P.A., Kenyon S.J., 1993, in White Dwarfs: Advances in Observation and Theory, M.Barstow Ed., Kluwer, p.253

Harris III D.L., 1956, ApJ 124, 665
Herwig F., 1994, Diplomarbeit Kiel
Humason M.L., Zwicky F. 1947, ApJ 105, 85
Kenyon S.J., Shipman H.L., Sion E.M., Aannestad P., 1988, ApJ 328, L65
Koester D., 1972, A&A 16, 459
Koester D., 1987, ApJ 322, 852
Koester D., Chanmugam G., 1990, Rep.Progr.in Physics 53, 837
Koester D., Reimers D., 1981, A&A 99, L8
Koester D., Schulz H., Weidemann V., 1979, A&A 76, 262
Koester D., Weidemann V., 1980, A&A 81, 145
Koester D., Weidemann V., Zeidler-K.T., E.-M., 1982, A&A 116, 147
Kuiper G.P., 1941, Coll. on Novae and White Dwarfs, Hermann, Paris, p.201 and PASP
    53, 248
Liebert J., 1975, ApJ 200, L95
Liebert J., 1978, A&A 70, 125
Liebert J., 1980, Ann.Rev.Astr.Ap.18, 363
Liebert J., Dahn C.C., Monet D.G., 1988, ApJ 332, 891 and 1989 in White Dwarfs,
    G.Wegner Ed., Springer, p.15
Liebert J., Wehrse R., Green R.F., 1987, A&A 175, 17
Luyten W.J., 1922, PASP 34, 132, 356
Luyten W.J., 1952, ApJ 116, 283
Luyten W.J., 1958, Vistas in Astronomy 2, 1048
Luyten W.J., 1963, Adv.Astron.Astrophys.2, 199
Luyten W.J., ed., 1965, First Conference on Faint Blue Stars. Obs.Univ.Minnesota
Luyten W.J., ed., 1971, IAU Coll.No.42 on White Dwarfs, Reidel
Mestel L., 1965, Stars and Stellar Systems 8, 297
Mestel L., Ruderman M.A., 1967, MNRAS 136, 27
Mochkovitch R., 1983, A&A 122, 212
Muchmore D., 1984, A&A 278, 769
Oke J.B., 1974, ApJS 27, 21
Ostriker J., 1971, Ann.Rev.Astr.Ap.9, 353
Pelletier C.,Fontaine G., Wesemael F., Michaud G., Wegner G., 1986, ApJ 307, 242
Reimers D., 1975, Mem.Soc.Roy.Sci.Liège, 6.ser.8, 369
Reimers D., Koester D., 1982, A&A 116, 341
Russell H.N., 1914, Popular Astronomy 22, 275
Saffer R.A., Liebert J., Olszewski E., 1988, ApJ 334, 947
Schatzman E., 1958, White Dwarfs. North Holland Publ.
Shaviv G., Kovetz A., 1976, A&A 51, 583
Shipman H.L., Greenstein J.L., 1983, ApJ 266, 761
Shipman H.L., Greenstein J.L., Boksenberg A., 1977, AJ 82, 480
Sion E., 1986, PASP 98, 821
Sion E.M., Shipman H.L., Wagner M., Liebert J., Starrfield G., 1986, ApJ 308, L67
Tapia S., 1978, Thesis and IAU Symp.80, The HR Diagram, A.G.Davis Philip, D.S.
    Hayes Eds., Reidel, p.133
Theis C., 1994, Thesis Univ.Kiel, Verlag Shaker, Aachen
Trimble V., Greenstein J.L., 1972, ApJ 177, 441
Unsöld A., 1938, 1955, Physik der Sternatmosphären. Springer
Van Horn H., 1968, ApJ 151, 227
Van Horn H., Weidemann V. Eds., 1979, IAU Coll.No.53 on White Dwarfs and Variable
    Degenerate Stars. Univ.Rochester

Vauclair G.,Vauclair S., Greenstein J.L., 1979, A&A 80, 79

Vitense E., 1951, Z.f.Astrophysik 28,81; 29, 13; 29, 73

Vitense E., 1953, Z.f.Astrophysik 32, 135

Wegner G., 1972, ApJ 172, 451

Wegner G., ed., 1989, IAU Coll.No.114 on White Dwarfs, Springer

Wegner G., Reid I.N. 1989, ApJ 375, 674

Wehrse R., Liebert J., 1980, A&A 86, 139

Weidemann V., 1955, Z.f.Astrophysik 36, 101

Weidemann V., 1960, ApJ 131, 638

Weidemann V., 1963, Z.f.Astrophysik 57, 87

Weidemann V., 1967, Z.F.Astrophysik 67, 286

Weidemann V., 1968, Ann.Rev.Astr.Ap.6, 351

Weidemann V., 1969, in Low Luminosity Stars, S.S.Kumar Ed., Gordon, p.311

Weidemann V., 1971, in White Dwarfs, ed.W.J.Luyten, Reidel, p.81

Weidemann V., 1975, in Problems in Stellar Atmospheres and Envelopes, eds.
    B.Baschek, W.H.Kegel, G.Traving, Springer, p.173

Weidemann V., 1977, A&A 59, 411

Weidemann V., 1981, in Effects of Mass Loss on Stellar Evolution, C.Chiosi, R.Stalio
    Eds., Reidel, p.339

Weidemann V., 1987, A&A 188, 74

Weidemann V., 1988, in A Decade of UV Astronomy with the IUE Satellite, ESA
    SP-281, Paris, 1, 18

Weidemann V., 1989, A&A 213, 155

Weidemann V., 1990a, Ann.Rev.Astr.Ap.28, 103

Weidemann V., 1990b, in Baryonic Dark Matter, G.Gilmore, D.Lynden-Bell Eds.,
    Kluwer, p.87

Weidemann V., 1991 in White Dwarfs, G.Vauclair, E.Sion Eds. p.67

Weidemann V., 1993, A&A 275, 158

Weidemann V., 1993, in Mass Loss on the AGB and Beyond. H.E.Schwarz Ed., ESO
    Conf. and Workshop Proceedings No.46, p.55

Weidemann V., 1994, ASP Conf. Ser. 57,209

Weidemann V., Koester D., 1983, A&A 121, 77

Weidemann V., Koester D., 1989, A&A 210, 311

Weidemann V., Koester D., 1994, A&A , in press

Weidemann V., Yuan J.W., 1989, in IAU Coll.No.114 White Dwarfs, G.Wegner Ed.,
    Springer, p.1

Wesemael F., 1979, A&A 72, 104

Wesemael F., Truran J.W., 1982, ApJ 260, 807

Werner K., 1993, in White Dwarfs: Advances in Observation and Theory. M.A.Barstow
    Ed., NATO ASI Ser.C 403, p.67

Winget D., Hansen C.J., Liebert J. et al., 1987, ApJ 315, L77

Wood M.A., 1990, Thesis and 1992, ApJ 386, 539

Woolf N.J., 1974, in IAU Symp.No.66, Late Stages of Stellar Evolution, R.J.Taylor
    Ed., Reidel, p.43

Yuan J.W., 1989, A&A 224, 108

Yuan J.W., 1992, A&A 261, 105

Zeidler-K.T. E.-M., 1987, A&AS 68, 469

Zeidler-K.T. E.-M., Weidemann V., Koester D., 1986, A&A 155, 356

Part I
Luminosity Function of White Dwarfs

The Mass Distribution and Luminosity Function of Hot DA White Dwarfs

J. Liebert[1] *and P. Bergeron*[2]

[1] Steward Observatory, University of Arizona, Tucson, AZ 85721, USA
[2] Département de Physique, Université de Montréal, C.P. 6128, Succ. Centre Ville, Montréal, Québec, Canada H3C 3J7

1 Introduction

The luminosity function (LF) of white dwarfs, whose cutoff at faint luminosities is believed to be caused by the finite age of the Galactic disk, is a fossil record of the star-formation history of this population; that is, the LF is closely related to the stellar birthrate function. The mass distribution (MD) of white dwarfs, likewise should be linked closely to the stellar initial mass function (IMF); in an ideal galactic population, there might be a one-to-one correspondence between the initial mass of the star and final mass of the remnant. In reality, the breadth of horizontal branches in globular clusters is proof that mass-loss in the red giant phases has a dispersion which depends on parameters other than the initial mass and chemical composition. Hence, there will also be dispersion in the final mass for stars of a given initial mass. Moreover, close binary stellar evolution and the likelihood that some white dwarfs can form from mergers will further complicate the relationship between the MD of white dwarfs and the stellar IMF.

In Galactic population studies, it is generally assumed that the stellar IMF $[\psi(M)]$ and birthrate functions $[b(\tau)]$ can be decoupled. That is, the IMF is assumed not to vary with time. Likewise, the related functions – the white dwarf LF $[\phi(L)]$ and MD $[\xi(M)]$ – are also discussed as separate functions. Yet it is quite clear from the physics of white dwarf cooling and other considerations that the two are functions of each other; that is we should begin thinking in terms of the white dwarf luminosity – mass distribution function – call it $\phi(L, M)$.

In the last several years, we have been reobserving the hot DA white dwarfs from the Palomar Green Survey for a variety of purposes – especially to redetermine the LF and current formation rate of these stars, after Fleming, Liebert & Green (1986). However, due to the development of the techniques of Bergeron, Saffer & Liebert (1992, hereafter BSL), we are obtaining detailed information on the mass distribution of this complete sample as well. In this review, which includes a preliminary report of this analysis, we link up the MD and LF of these stars. Coauthors in the overall effort include R.W. Tweedy, T.A. Fleming, J.B. Holberg and R.A. Saffer.

2 The PG Sample

At the time of this writing some 200 DA white dwarfs from the Palomar Green Survey, including some originally misclassified, have been reobserved and analyzed using essentially the methods of BSL. The new PG sample has been generally restricted to those stars with estimated $T_{eff} \gtrsim$ 15,000 K. The assumptions include a composition of pure hydrogen, unless helium appears in the spectrum, and the atmospheres are in local thermodynamical equilibrium. However, the BSL sample, drawn from heterogeneous literature sources, was restricted largely to stars with $T_{eff} \lesssim$ 40,000 K.

2.1 Concerns about analyzing hot DA stars

The reason for the upper temperature limit in BSL was that it was recognized already a few years ago that there are greater uncertainties associated with DA white dwarfs hotter than about that temperature. Concerns about the hotter stars now include the following:

(1) The presence of additional continuum opacity due to a trace helium abundance that is not readily detectable in a He II λ4686 line, ie. the star is not of DAO type. With our LTE atmospheres the determined T_{eff} is lower if a finite helium abundance is assumed (Bergeron et al. 1994); oddly enough the derived gravity is very insensitive to this effect.

(2) For very hot stars, especially planetary nebula nuclei with DAO spectra, a self-consistent fit cannot be achieved with all Balmer lines simultaneously (Napiwotzki 1992; Napiwotzki & Schönberner 1993; Bergeron et al. 1993, 1994). This is generally believed to be caused by the presence of additional opacity due to heavy elements, more so than helium. There is strong evidence for the presence of heavy elements in the form of additional opacity observed at the extreme ultraviolet (EUV) wavelengths.

(3) While LTE has been shown to be a good assumption for lower temperature DA stars, there is the question as to at what temperature and gravity the assumption breaks down (Napiwotzki 1995).

(4) It has been unclear whether to analyze the stars as homogeneous atmospheres, with some mixture of hydrogen and helium, or as stratified atmospheres, with a thin layer of pure hydrogen not fully opaque at all frequencies atop a deep helium zone. Bergeron et al. (1994) do find, however, that the majority of DAO stars show no evidence for stratified atmospheres, and the He II and H line spectra are well fit with homogeneous atmospheres.

Since the great majority of hot DA stars in the PG Survey are either cooler than 40,000 K or not very much hotter, these concerns may not seriously compromise the derived parameters. Hence, we proceeded with a similar analysis.

2.2 PG0948+534 and very hot DA planetary nuclei

In previous analyses of the hottest DA white dwarfs identified in the PG Survey, none were found to have T_{eff} hotter than 80,000 K (Holberg 1987; Holberg et al. 1989). The possibility that a sequence with hydrogen-rich atmospheres was discontinuous in T_{eff} from the very hot planetary nebula nuclei (PNN) to the hottest DA white dwarfs was disputed by Napiwotzki & Schönberner (1993), who found PNN in the alleged gap region. We confirm their claim that the nucleus of WDHS 1 is such an object (Liebert, Bergeron, & Tweedy 1994). However, since the PG included ∼ 10 helium-rich DO and PG 1159 stars, all with estimated T_{eff} of 80,000 K and much hotter, it was curious that no DA stars of similar T_{eff}, with or without nebulae, were found in that survey.

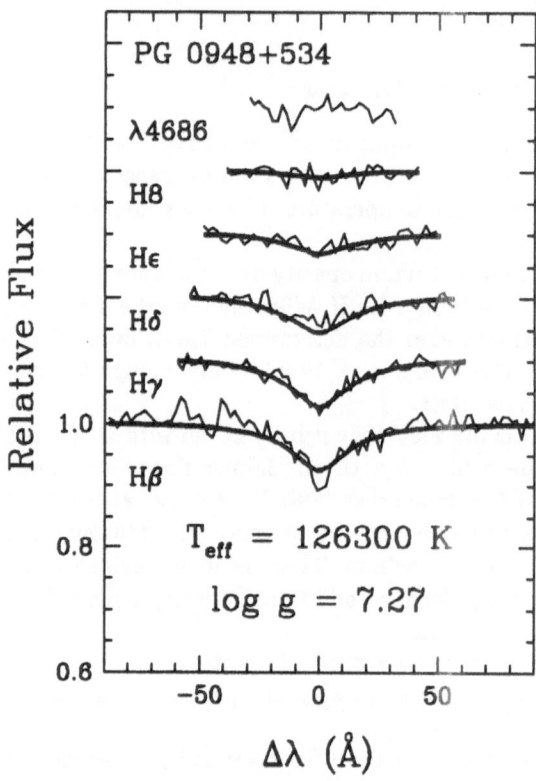

Fig. 1. Simultaneous Balmer line profile fits, following BSL.

In our reanalyzed sample, however, the DA star PG 0948+534 has $T_{\text{eff}}\sim$ 126,000 K, log g ∼ 7.3, assuming a pure H composition. PG 0948+534 had been misclassified as a hot subdwarf O star on the basis of an early image tube spectrum covering only the $H\gamma$ line (Green & Liebert 1979). In Fig. 1 are shown the simultaneous Balmer line fits to the CCD spectrum of this star. No trace of He II λ4686 is evident, though the T_{eff} estimate might be lower if some trace helium abundance is assumed in the modeling. Interestingly enough, PG 0948+534 represents the first DA star in which the so-called Balmer line problem is observed. So far, such a phenomenon had been reported in DAO stars only (see, e.g., Bergeron et al. 1994). This result reinforces the conclusions of Bergeron et al. (1994) that the presence of helium in DAO stars is not the source of the observed Balmer line discrepancy.

PG 0948+534 is not quite unique in the revised PG sample. We would now classify BE UMa (PG 1155+492) as a borderline DAO white dwarf, but with $T_{\mathrm{eff}} \sim 105,000$ K, log $g \sim 6.5$, based on a spectrum fit with a non-LTE model atmospheres analysis (Liebert et al. 1995). This object is a pre-cataclysmic binary with a low mass companion and an orbital period of 2.2 days. In the paper cited above, it is also found to have an old planetary nebula, and the parameters of the primary star are similar to some PNN analyzed by Napiwotzki & Schönberner (1993). Finally, we also derived a substantially higher T_{eff} estimate for the enigmatic PNN PG 0950+139 than did Liebert et al. (1989); however, the parameters (100,000 K, log $g \sim 7$) for this star are very uncertain due to the filling-in of the weak H line cores by nebular emission.

3 The Temperature (Luminosity) and Mass Distributions

In Fig. 2 we plot the distribution of surface gravities vs log T_{eff} for the 200 reanalyzed PG Stars. Also shown are the new 0.5, 0.6, 0.8, and 1.0 $M_{\odot}$ evolutionary models of Wood (1995) with thick hydrogen layers ($M_{\mathrm{H}} = 10^{-4} M_{\star}$), as well as the 0.5 and 0.6 $M_{\odot}$ models of Wood (1990) with no hydrogen layer; the latter are equivalent to models with thin hydrogen layers (see BSL). The three stars hotter than log $T_{\mathrm{eff}} = 5.0$ are PG 0948+534, PG 0950+139, and BE UMa, discussed above, though the number of stars increases rapidly only below 80,000 K.

The concentration of stars near the peak value of the mass ($M = 0.5 - 0.6$ $M_{\odot}$) is obvious at lower temperatures, where the number of stars is greater, and the spread in the log g distribution may appear to increase in the range log T_{eff} $\sim 4.3 - 4.6$. There appears to be a higher relative number of more massive stars in that range of T_{eff} as well. We return to this point later.

In Fig. 3 we show the cumulative log g and mass distribution functions. Masses have been obtained from Wood's evolutionary models with thick hydrogen layers discussed above. Note that the mean is shifted to 0.609 $M_{\odot}$, compared to 0.582 $M_{\odot}$ when the analysis is done with zero hydrogen. Similarly, the mean mass of the BSL sample increases from 0.562 $M_{\odot}$ to 0.590 $M_{\odot}$. The assumption of a thick hydrogen layer thus alleviates the concern of Bragaglia, Renzini & Bergeron (1995) about the low mass cutoff of the white dwarf distribution formed from single star evolution.

The mass distribution shown in Fig. 3 is strongly biased by the variability in radii, however. For a complete, magnitude-limited sample such as the PG, one wants to construct a volume-limited sample in order to calculate the space density of white dwarfs at each mass. Clearly the white dwarfs of high mass have a more limited search distance than their low mass counterparts, an effect which has long been discussed in the literature. The search distance is linearly proportional to the radius (R), so the volume depends on R^3. We can correct for this effect by weighting each star in proportion to R^{-3} (relative to some

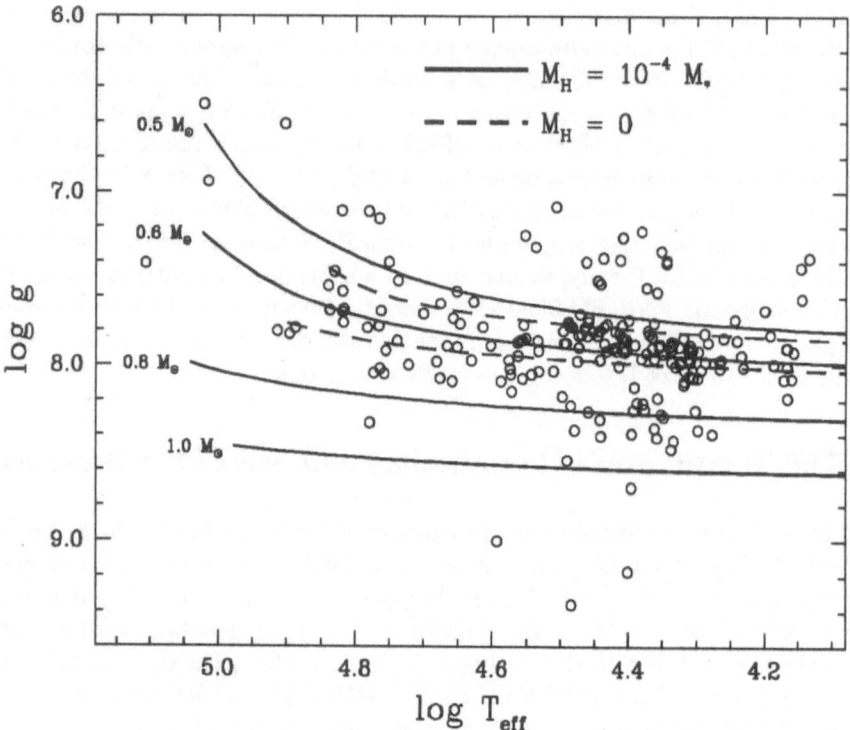

Fig. 2. Distribution of surface gravities vs. log T_{eff}. Thick curves are models with thick hydrogen layers, dashed curves no hydrogen.

radius near the peak value). This corrected mass distribution for the PG sample is shown in Figure 3 as well (thick line). The absolute vertical scale has no particular meaning here.

   The problem with the weighted distribution is that the few massive stars with small radii can have weights an order of magnitude larger than the average, but there are so few of them (one per mass bin) that the right end of the distribution is also consistent with zero! However, there is potential significance to the increased population of objects moderately more massive than the peak, near 0.8 $M_\odot$ or so.

   One more bias is inherent to this sample, however. In the Mestel part of the white dwarf cooling curve, the more massive stars cool more slowly than their low mass counterparts. This is precisely the T_{eff} range where the more massive stars appear to be overrepresented in Fig. 2. This effect can be substantial: for example, Wood's 1.2 $M_\odot$ model at log $T_{eff} = 4.5$ has a cooling age more than a factor of ten longer than a 0.6 $M_\odot$ model. Clearly, we would expect different

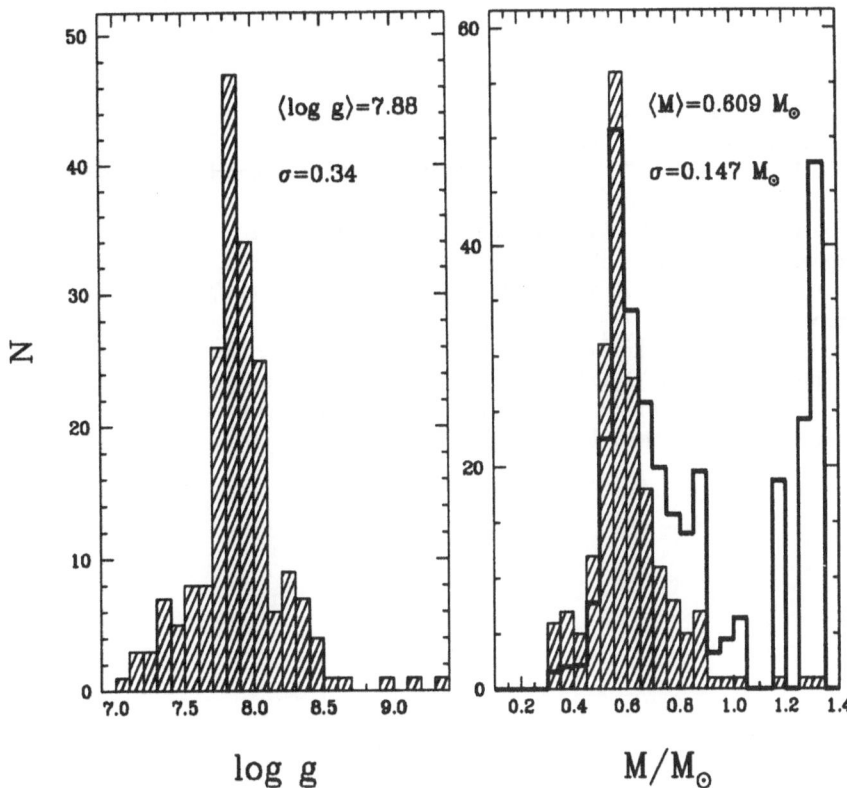

Fig. 3. The cumulative *logg* (left) and mass (right, shaded) distribution functions, using Wood models with thick hydrogen layers. The mass distribution corrected for the stellar radii is shown as the unshaded, thick histogram.

mass distributions for very hot and cool white dwarfs, where neutrino and Debye effects accelerate the evolution of the more massive stars.

This work was supported in part by the NSERC Canada, by the Fund FCAR (Québec), and by the NSF grant AST 92–17961.

References

Bergeron, P., Saffer, R.A., Liebert, J. 1992, ApJ, **394**, 228 (BSL)
Bergeron, P., Wesemael, F., Beauchamp, A., Wood, M.A., Lamontagne, R., Fontaine, G., Liebert, J. 1994, ApJ, **432**, 305
Bergeron, P., Wesemael, F., Lamontagne, R., & Chayer, P., 1993 ApJ, **407**, L85
Bragaglia, A., Renzini, A., & Bergeron, P. 1995, ApJ, in press
Green, R.F., & Liebert, J. 1979, in *White Dwarfs and Variable Degenerate Stars*, eds. H.M. Van Horn & V. Weidemann (Rochester NY: Univ. of Rochester Press), p. 118

Holberg, J.B. 1987, in *The Second Conference on Faint Blue Stars*, Proc. IAU Coll. 95, eds. A.G.D. Philip, D.S. Hayes, & J. Liebert (Schenectady NY: L. Davis Press), p. 285

Holberg, J.B., Kidder, K., Liebert, J., & Wesemael, F. 1989, in *White Dwarfs*, Proc. IAU Coll. 114, ed. G. Wegner (Berlin: Springer-Verlag), p. 188

Liebert, J., Bergeron, P., & Tweedy, R.W. 1994, ApJ, **424**, 817

Liebert, J., Green, R.F., Bond, H.E., Holberg, J.B., Wesemael, F., Fleming, T.A., & Kidder, K. 1989, ApJ, **346**, 251

Liebert, J., Tweedy, R.W., Napiwotzki, R., & Fulbright, M.S. 1995, ApJ, in press

Napiwotzki, R. 1992, in Proc. of Univ. Kiel/CCP7 Workshop, *Atmospheres of Early-Type Stars*, ed. U. Heber & C.S. Jeffery (Heidelberg: Springer), p. 310

Napiwotzki, R. 1995, these Proceedings

Napiwotzki, R., & Schönberner, D. 1993, in *White Dwarfs: Advances in Observations and Theory*, NATO ASI Ser., ed. M.A. Barstow (Dordrecht: Kluwer), p. 99

Wood, M. A. 1990, Ph. D. thesis, University of Texas at Austin

Wood, M. A. 1995, these Proceedings

The Stellar Formation Rate and the White Dwarf Luminosity Function

J. Isern[1,2], E. García-Berro[3], M. Hernanz[1], R. Mochkovitch[4] and A. Burkert[5,1]

[1] Centre d'Estudis Avançats (CSIC), Camí de Santa Bàrbara, 17300 Blanes, Spain
[2] Institut d'Estudis Espacials de Catalunya (FCR)
[3] Dept. de Física Aplicada, Universidad Politécnica de Cataluña, Jordi Girona Salgado s/n, Módul B-4, Campus Nord, 08034 Barcelona, Spain
[4] Institut d'Astrophysique de Paris (CNRS), 98bis bvd. Arago, 75014 Paris, France
[5] Max Planck Institut für Astrophysik, Karl Schwarzschild Straße 1, 85740 Garching bei München, Germany

1 Introduction

The evolution of white dwarfs is essentially a cooling process that lasts for about 10 Gyrs. The comparison between the theoretical and the observational luminosity functions allows to determine several aspects of galactic evolution such as the star formation rate as a function of time (Noh and Scalo, 1990; Díaz–Pinto et al., 1994), the age of the disk (Winget et al., 1987; García–Berro et al., 1988a,b) or the properties of the halo (Mochkovitch et al., 1990). Of course, due to the intrinsically low luminosity of the objects, the properties deduced in this way are only valid for the solar neighborhood.

The luminosity function can be written as:

$$n(l) = \int_{M_i}^{M_s} \Phi(M)\psi[T_{disk} - t_{cool}(l, M) - t_{MS}]\tau_{cool}dM \tag{1}$$

where l is the logarithm of the luminosity in solar units, $l = \log(L/L_\odot)$, M is the mass of the parent star (for convenience all the white dwarfs are labelled with the mass of their main sequence progenitor), $\tau_{cool} = (dl/dt)^{-1}$ is the cooling rate, M_s is the maximum mass of the main sequence stars able to produce a white dwarf, and M_i is the minimum mass of a main sequence star able to produce a white dwarf with luminosity l now. Therefore, it satisfies the condition:

$$t_{cool}(l, M) + t_{MS} = T_{disk} \tag{2}$$

where t_{cool} is the time necessary to cool down to a luminosity l, t_{MS} is the lifetime on the main sequence and T_{disk} is the age of the disk. $\Phi(M)$ is the initial mass function and $\Psi(t)$ is the star formation rate per unit volume. All these values can be obtained from Hernanz et al., (1994) and Ségretain et al., (1994). The adopted carbon oxygen distributions inside the white dwarfs are those of Mazzitelli and D'Antonna (1986).

There are two ways to obtain the star formation rate from this equation. One is to solve the inverse problem. The other is to restrict the luminosity function to massive white dwarfs only. In this case, the lifetime of the parent stars ($M \geq 2M_\odot$ in the main sequence) is much shorter than the age of the disk and can be neglected. Since t_{cool} is weakly dependent of the mass of the white dwarf, it is possible to remove Ψ from the integral. This method has been described in detail by Hernanz et al., (1993) and Diaz–Pinto et al., (1994) and it will not be considered here. Therefore, we are going to discuss the solution of the inverse problem.

2 Model and results

In order to solve the inverse problem it is convenient to change the variable of integration, M, in equation (1) by the time, t_b, at which the progenitors of the white dwarfs were born:

$$t_b = T_{disk} - t_{cool}(l, M) - t_{MS}(M) \tag{3}$$

This implies that it is possible to define the mass, $M = M(t_b)$, of the stars that, being born at the time t_b, provide a white dwarf of luminosity l at the present time. Equation (1) can be written as:

$$n(l) = \int_0^{t_b^{up}} K(l, t_b)\Psi(t_b)dt_b \tag{4}$$

with

$$K(l, t_b) = \Phi[M(t_b)]\tau_{cool}[l, M(t_b)]\frac{dM}{dt_b} \tag{5}$$

where $t_b^{up} = T_{disk} - t_{cool}(l, M_u)$ is the time at which stars with the maximum possible mass have to be born to produce a white dwarf of luminosity l. The function $K(l, t_b)$ is the kernel of the transformation. It is not symmetric in l and t_b and it has a quite complicated behavior. Therefore, according to Piccard's theorem, Ψ cannot be directly obtained and a unique solution is not guaranteed.

One way to tackle the problem is to define a trial function and to adjust its parameters using maximum likehood techniques. The last part can be achieved by using the subroutine MINUIT, from the CERN Library, while the trial function can be obtained in the following way: Many models of chemical evolution of the galaxy assume that the star formation rate per unit of disk surface, $\dot{\Sigma}$, is proportional to the surface density of gas. The solution thus takes the form $\dot{\Sigma} = \dot{\Sigma}_0 \exp(-t/\tau_s)$ in $M_\odot/Gyr/pc^2$. Since the gas has been gradually seetling onto the disk, it seems reasonable to assume that its scale height is of the form $h = h_1 + h_0 \exp(-t/\tau_h)$, where h_1 is the present value. Therefore, the star formation rate per unit volume can be written as:

$$\Psi(t) = \frac{\dot{\Sigma}(t)}{h(t)} \propto \frac{e^{-t/\tau_s}}{He^{-t/\tau_h} + 1} \tag{6}$$

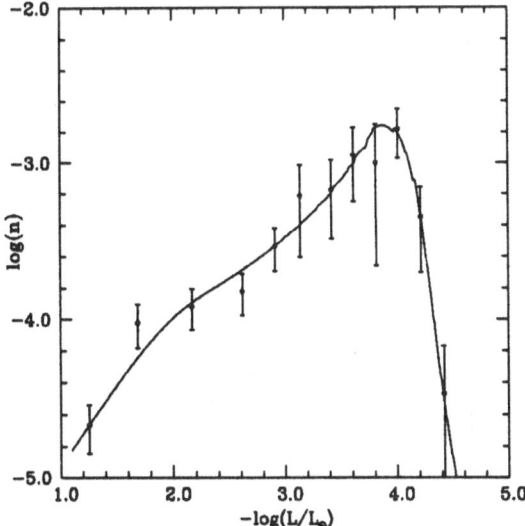

Fig. 1. White dwarf luminosity function obtained from a SFR given by equation 6. The observational data are binned in half magnitude bins. See details in text

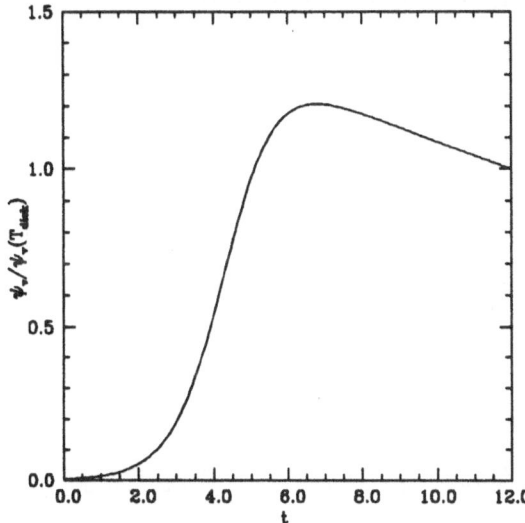

Fig. 2. SFR obtained by fitting the white dwarf luminosity function to the observational data. See details in text

where the constant of proportionality is obtained by normalizing the luminosity function to $l = -3$ in order to avoid the uncertainties linked to the total density of white dwarfs and $H = h_0/h_1$.

Figure 1 displays the best fit in the case in which the luminosity function is binned at intervals of half magnitude. Figure 2 displays the star formation rate responsible of such a fit. The values of the parameters are: $\tau_s = 24$ Gyr, $\tau_h = 0.7$ Gyr, $H = 485$ and $T_{disk} = 12$ Gyr. If this result is taken at face value

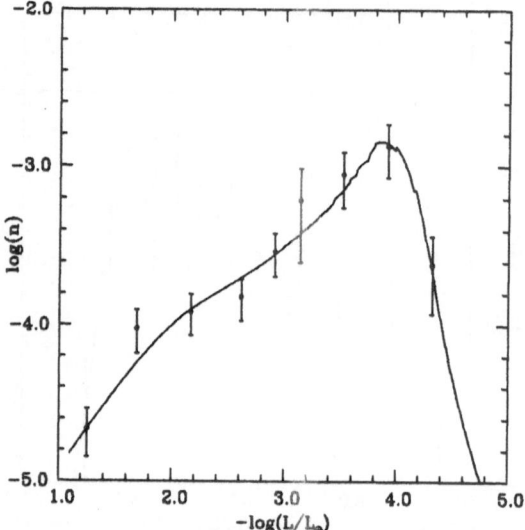

Fig. 3. Same as figure 1 but using the luminosity function binned in one magnitude bins. See details in text

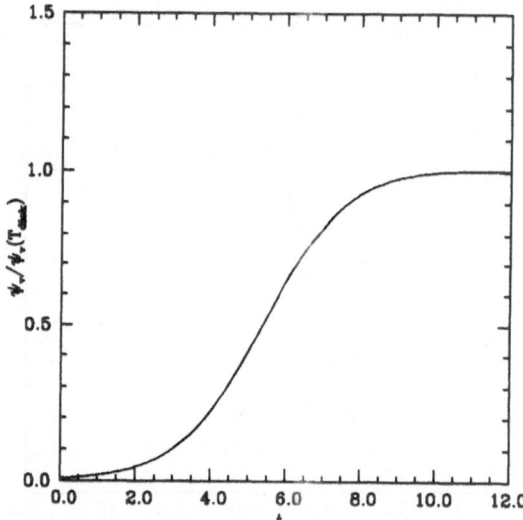

Fig. 4. Same as figure 2 but using the luminosity function binned in one magnitude bins. See details in text

it turns out that after 2 Gyrs of low activity (the coolest stars on the luminosity function were born at this epoch) the star formation rate suddenly increased reaching a maximum value 4 Gyrs later. Since then it remained nearly constant during 6 Gyrs (the star formation activity has merely decreased by 20% during that time). Therefore, the total age of the solar neighborhood is 12 Gyrs.

We must wonder to which extent the results are dependent on the binning of the luminosity function and on the shape of the trial function. For instance, it is evident that the adopted shape does not fit possible bursts of star formation.

For instance, there is no way to adjust the prominent bump at $l \approx -1.5$ and a possible solution is to assume that a burst was produced. Nevertheless, before obtaining conclusions it must be remembered here that at this particular value of luminosity, the characteristic cooling time changes because photon cooling starts to dominate neutrino cooling. The exact value and the total amount is controlled by the envelope, which properties are still very uncertain.

If the luminosity function is binned in intervals of one magnitude (Figures 3 and 4), the results are roughly the same except for the presence of a tail that is 1.1 Gyr longer than in the previous case. In this case, the age of the disk is 13.1 Gyr. This implies that the present distribution of white dwarfs is compatible with the existence of a long period of low activity, but due to the present uncertainties it is impossible to provide a precise shape for these tails.

3 Conclusions

The star formation rate that best fits the white dwarf luminosity function shows that in the solar neighborhood there was a period of low activity at the beginning of the life of the galaxy. After 2-3 Gyr, the activity suddenly increased and reached a maximum 4 Gyr later. Since then, the star formation rate has decreased very slowly, 20%, or remained more or less constant. The estimated age of the the solar neighborhood is 12-13 Gyr.

Acknowledgements. This work has been supported by the DGICYT grants PB91-060, PB93-0820-C02-02, by the CESCA consortium and by an AIHF. One of us (A.B.) gratefully acknowledges a grant from the EC Action "Access to Large Scale Supercomputing Facilities in Europe" within the Human Capital and Mobility Programme.

References

Díaz–Pinto, A., García–Berro, E., Hernanz M., Isern J., Mochkovitch R., 1994, *A&A* **282**, 86.
García–Berro E., Hernanz M., Isern J., Mochkovitch R., 1988a, *Nature*, **333**, 642.
García–Berro E., Hernanz M., Mochkovitch R., Isern J., 1988b, *A&A*, **193**, 141.
Hernanz M., Díaz–Pinto A., Isern J., García–Berro E., Mochkovitch R., 1993, in "White Dwarfs: Advances in Observation and Theory". Ed. M.A. Barstow (Kluwer), 15.
Hernanz M., García–Berro E., Isern J., Mochkovitch R., Ségretain L., Chabrier G., 1994, **434**, 652.
Mazzitelli I., D'Antonna F., 1986, *ApJ*, **308**, 706.
Mochkovitch R., García–Berro E., Hernanz M., Isern J., Panis J.F., 1990, *A&A*, **233**, 456.
Noh H.R., Scalo J., 1990, *ApJ*, **352**, 605
Ségretain L., Chabrier G., Hernanz M., García– Berro E., Isern J., Mochkovitch R., 1994, *ApJ*, **434**, 643.
Winget D.E., Hansen C.J., Liebert J., Van Horn H.M., Fontaine G., Nather R., Kepler S.O., Lamb D.K., 1987, *ApJ*, **315**, L77.

On the Luminosity Function of White Dwarfs in Wide Binaries*

Terry D. Oswalt and J. Allyn Smith

Department of Physics and Space Sciences
Florida Institute of Technology
Melbourne, Florida 32901-6988 USA

Abstract: Photometric parallaxes have been used to determine the luminosity function (LF) for a large sample of spectroscopically identified white dwarfs (WDs) found in the Luyten and Giclas common proper motion binaries (CPMBs). The LF derived for WD components of CPMBs has been corrected for incompleteness to $\mu \geq 0.2''/y$ and $m_{pg} \leq +18$. Comparison with the LF derived for single field WDs suggests that WDs are more commonly found in binaries.

1 Background

Luyten (1969, 1974, 1979) and Giclas et al. (1971, 1978) called attention to over 500 CPMBs which were suspected to contain WD components. Both identified WD candidates by using reduced proper motion, $H_{pg} \equiv m_{pg} + 5log\mu + 5$ and crude color estimates. For any particular color class, stars with H_{pg} larger than a certain limit are likely to be WDs (see Jones 1972).

Luyten (1970) measured the original Palomar Schmidt plates and second epoch red plates for all stars with $\mu > 0.1''/y$ down to near the plate limit of $m_{pg} \approx +21.2$, ultimately identifying 407 CPMBs with suspected WD components. Giclas found $\sim$100 additional CPMBs with suspected WDs within the Lowell survey limits of $m_{pg} \leq +17$ and $\mu \geq 0.2''/y$. Because Luyten specifically avoided listing these Giclas pairs, it is essential to include them in any CPMB sample where completeness is a desirable attribute.

The size and faint limiting magnitude ($m_{pg} \approx +21$) of the Luyten-Giclas (L-G) sample present a formidable observational challenge. We are conducting a long-term program to obtain optical (BVRI) and infrared (JHK) photometry of the entire CPMB WD sample (see Leggett et al. 1993). Using our most recent V,I data, Smith and Oswalt (1994) derived the LF for the main sequence (MS) components and showed that it compared favorably with that of single field MS

* This research was supported by observations made at Kitt Peak National Observatory & Cerro Tololo Inter-American Observatory (operated by the Association of Universities for Research in Astronomy for the National Science Foundation); at Lowell Observatory; and at McDonald Observatory (operated by the University of Texas at Austin).

stars in the solar neighborhood. In this paper we examine the LF for components which have been spectroscopically identified as WDs and make a similar comparison to the LF of single field WDs.

2 Completeness of the Luyten-Giclas Sample

A number of problems affect the completeness of the CPMB sample and the LFs derived from it. The typical L-G pair consists of a cool MS primary of $m_{pg} \approx +15$ accompanied by a WD that is ~ 2.5 magnitudes fainter. Oswalt and Strunk (1994) concluded that undetected pairs of very close separation and/or large magnitude difference have only a small influence on the CPMB sample completeness. However, one would expect the completeness of the sample to decline rapidly for primaries fainter than $m_{pg} = 21.2 - 2.5 = 18.7$, where companions begin to fall below the plate limit. Also, cool WDs, subdwarfs and cool MS stars are hopelessly intermingled in reduced proper motion diagrams.

Implicit in our determination of the CPMB sample completeness is the assumption that they are true physical pairs. At present we have obtained BVRI data for components of ~ 200 CPMBs. Photometric parallaxes have been determined from our V-I color indices by fitting low-order polynomials to the Monet et al. (1992) MS, subdwarf and WD sequences. These relations are indicated by the solid curves in Figure 1. Each CPMB component was fixed (but not plotted) on the calibration line most consistent with its spectral class and V-I color. Its companion (primary or secondary) was then plotted relative to this locus using its observed difference in V and V-I.

The observed data agree well with the Monet et al. fits— especially considering that this procedure imposes on each plotted point the scatter inherent to both components. We conclude that >90% of CPMBs are indeed physical pairs and that a significant fraction of CPMB components are members of the old disk or halo population. In addition, at least a dozen primaries (those with $M_v < +7.0$) are likely to have massive WD companions.

We have assessed the completeness of the L-G sample as follows. Assuming a constant number density of CPMBs in the solar neighborhood, the cumulative number (ΣN) of CPMBs in the sample that exceeds a particular limiting proper motion μ should be proportional to μ^{-3}. In this way, Oswalt et al. (1991) showed that the sample is essentially complete for $\mu > 1.0''/y$. Figure 2 shows that cumulative star counts below this limit increase log-linearly with decreasing μ, though at a less than complete rate, to $\mu \approx 0.2''/y$, where a marked change in slope occurs. The sample completeness is also a function of apparent magnitude. Smith and Oswalt (1994) showed that cumulative counts by apparent magnitude increase log-linearly until $m_{pg} \sim +18$, where faint companions first fail to be detected. Within both these limits ($\mu \geq 0.2''/y$, $m_{pg} \leq +18$) the CPMB sample appears to be at most $\sim 10\%$ complete. However the effective search volume of this subset of the L-G sample is at least 50 times larger than those normally used to derive stellar LFs, hence truly rare WDs of low luminosity are more likely to be found in it. Fortunately, the completeness of the CPMB sample is a

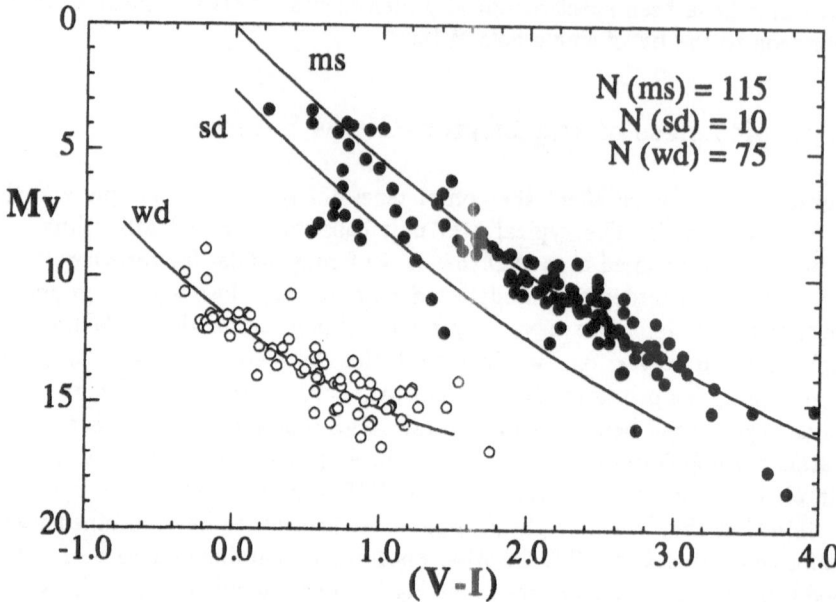

Fig. 1. H-R diagram for CPMBs with V,I photometric data. Solid lines indicate the MS, subdwarfs and WD cooling track derived from trigonometric parallax work by Monet et al. (1992). Filled and open symbols represent our V,I photometry of the nondegenerate and WD components, respectively.

well-behaved function of both proper motion and apparent magnitude, providing a straightforward means of correcting the raw LFs.

3 The White Dwarf Luminosity Function

We have computed LFs for CPMB components using the $1/V_m$ method of Schmidt (1975). Figure 3 compares our raw LF for WD components of CPMBs (filled circles) having $m_{pg} \leq +18$ and $\mu \geq 0.2''/y$ and the LFs for single field WDs determined by Liebert et al. (1988), Fleming et al. (1986) and Evans (1992). Though our raw LF has a similar shape, it is an order of magnitude below the LFs determined for single field WDs. The raw LF also exhibits some potentially important fine structure. The bin at $M_v = +12$ is ~2σ above the trend set by the rest of the data. Also, there is no evidence for a sharp downturn in the WD LF in the final two non-empty bins— more in accord with the Evans (1992) results than those of Liebert, Dahn and Monet (1988). If real, this implies an older age for the local Galactic disk than that suggested by Winget et al. (1987). However, for our sample $\langle V/V_m \rangle \neq 0.5$, confirming that it is incomplete.

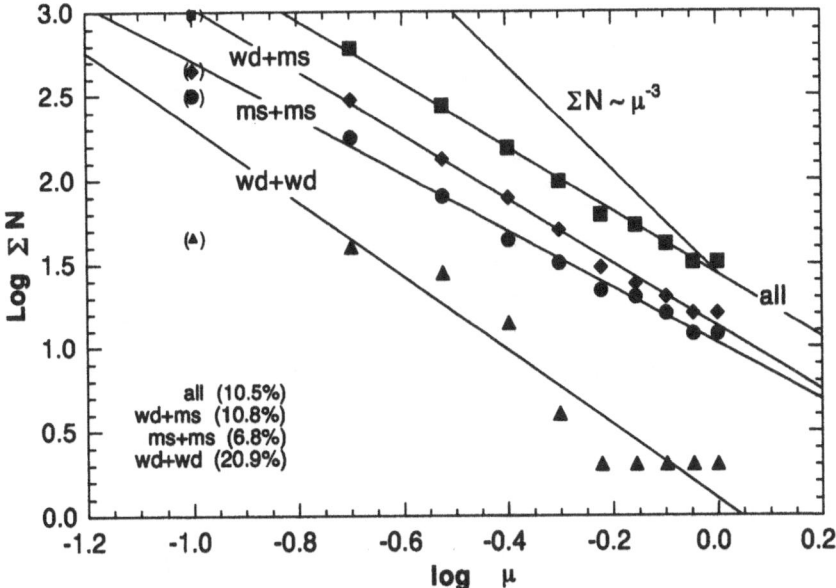

Fig. 2. Cumulative counts (log ΣN) vs. proper motion (log μ) for the CPMB sample. The regression lines were used to compute the listed completeness of each spectroscopically identified subgroup as described in the text. Zero points were set by assuming each sample is complete at log $\mu = 0$. Bins for $\mu < 0.2''/y$ were omitted from fits. Solid line indicates expected slope for a uniform space density.

Are the differences between the single and wide binary WD LFs artifacts of the incompleteness of our sample? We have attempted to address this question by constructing plots of *differential counts* vs. log μ for each magnitude bin in m_{pg}. Following the procedure outlined in Smith and Oswalt (1994) we determined completeness correction factors for each star in the LF of CPMB WDs. The inverse of each star's completeness factor was used as its weight in the $1/V_m$ sums. In effect this counts stars in each (m_{pg}, μ) bin that are missing due to the sample's incompleteness.

The resulting *completed* LF is plotted in Figure 3 with open circles. As before, our LF compares best with the LF for single WDs published by Evans (1992), in that the LF is still increasing even in the last two non-zero bins. Our LF also suggests a space density of WD binary components that is roughly three times that of single field WDs implied by the Evans LF. Using the same technique outlined here, Smith and Oswalt (1994) found very good agreement between the LFs of MS CPMB components and single MS stars. We are led to conclude that WDs are indeed found more frequently in binaries than as single stars.

The *completed* bin at $M_v = +12$ remains $\sim 2\sigma$ above the trend set by the

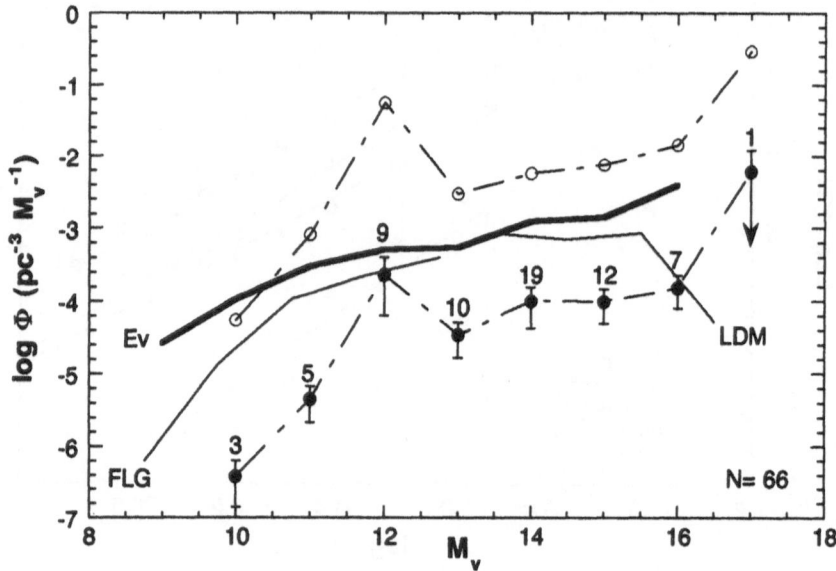

Fig. 3. Comparison of our *raw* LF (filled circles) and *completed* LF (open circles) for WD components of CPMBs to the LFs for single field WDs determined by Liebert et al. (LDM; 1988), Fleming et al. (FLG; 1986) and Evans (Ev; 1992).

other data. To explain it, preliminary calculations by Wood (1994) require a twenty-fold increase in the wide binary formation rate between 0.5-1.0 Gigayears ago, with no corresponding increase in the single star formation rate. Thus, at present we regard this bin as a statistical outlier. Ultimately, the LFs derived from our photometric program will include >1000 CPMB components and the nature of this bin will be more apparent.

4 Work in Progress

The *completed LF technique* introduced here is especially useful for large samples which have a quantifiable and well-behaved level of completeness. It permits the use of a very large search volume that is crucial to defining the faint end of the WD LF, where objects are exceedingly rare.

Our spectroscopic classifications and photometric colors indicate that about 15% of the L-G nondegenerate CPMB components are metal-poor (see Smith and Oswalt 1994). Using these stars as indicators of their companion WDs' population membership, we are now working to define the LF of WDs among the old disk and halo population.

Our current sample of CPMBs was drawn from the Luyten Double Star (LDS) catalog, which lists 6124 pairs (see Luyten 1987 and references therein). Using a machine-readable version of the LDS catalog provided by Halbwachs (1986) and procedures similar to those outlined above, we have determined that the LDS is > 40% complete to $m_{pg} \leq +18$ and $\mu \geq 0.2''/y$, implying that Luyten's crude proper motion and color criteria missed many WDs among the LDS pairs. Supported by 40-60 nights of dedicated time per year on the SARA 0.9-m automated telescope at Kitt Peak (see Oswalt et al. 1992) in addition to our ongoing requests for larger telescope time at KPNO and CTIO, we will add at least 500 previously unobserved LDS pairs to our current photometric program. Our goal is a definitive LF for WDs in wide binaries, extending at least two magnitudes beyond the turndown in the WD LF suggested by previous studies.

5 Acknowledgments

We are indebted to W.J. Luyten, who spent decades identifying the CPMBs. The authors gratefully acknowledge support from NSF grant AST-9016284 (TDO), NASA Graduate Student Research Fellowship NGT-51086 (JAS), and an Ernest F. Fullam Award from the Dudley Observatory, as well as partial travel support from the 9th European Workshop on White Dwarfs and the American Astronomical Society.

References

Evans, D.W. 1992, MNRAS 255, 521.

Fleming, T.A., Liebert, J., Green, R. 1986, ApJ 308, 176.

Giclas, H.L., Burnham, R., Thomas, N.G. 1971, Lowell Proper Motion Survey, Northern Hemisphere Catalog (Flagstaff: Lowell).

Giclas, H., Burnham, R., Thomas, G. 1978, Lowell Proper Motion Survey, Southern Hemisphere Catalog. Lowell Obs. Bull. 164, Vol. VIII, p.89.

Halbwachs, J.-L. 1986, A&AS, 66, 131.

Jones, E.M. 1972, ApJ 173, 671.

Leggett, S.K., Smith, J.A., Oswalt, T.D., Hintzen, P.M., Liebert, J., Sion, E.M. 1993, in White Dwarfs: Advances in Observation & Theory, ed. Barstow, (Kluwer), p.427.

Liebert, J., Dahn, C., Monet, D. 1988, ApJ 332, 891.

Luyten, W.J. 1969, Proper Motion Survey with the 48" Schmidt Telescope XXI. (U. Minn.).

Luyten, W.J. 1970, in IAU Coll.No.7, Proper Motions, ed. W.J. Luyten, (U. Minn.).

Luyten, W.J. 1974, Proper Motion Survey with the 48" Schmidt Telescope XXXVIII. (U. Minn.).

Luyten, W.J. 1979, Proper Motion Survey with the 48" Schmidt Telescope LII. (U. Minn.).

Luyten, W.J. 1987, Proper Motion Survey with the 48" Schmidt Telescope LXXI. (U. Minn.).

Monet, D.G., Dahn, C.C., Vrba, F.J., Harris, H.C., Pier, J.R., Luginbuhl, C., Ables, H. 1992, AJ 103, 638.

Oswalt, T.D., Rafert, J.B., Wood, M., Castelaz, M., Collins, L. Henson, G., Powell, H.,
    Caillault, J-P., Shaw, S., Leake, M., Marks, D., Rumstay, K. 1992, in ASP Conf.
    Ser., ed. Adelman & Dukes, p111.
Oswalt, T.D., Sion, E.M., Hintzen, P.M., Liebert, J. 1991, in White Dwarfs, ed. G.
    Vauclair, E. Sion (Kluwer), p.379.
Oswalt, T.D., Strunk, D. 1994, A Catalog of White Dwarfs in Wide Binaries, BAAS
    26, 901.
Schmidt, M. 1975, ApJ 202, 22.
Smith, J.A., Oswalt T.D. 1994, in The Lower Main Sequence− and Beyond, ed. C.
    Tinney, Springer-Verlag (in press).
Winget, D.E, Hansen, C.J., Liebert, J., Van Horn, H.M., Fontaine, G., Nather, R.E.,
    Kepler, S.O., Lamb, D.Q. 1987, ApJL 315, L77.
Wood, M. 1994, private communication.

A New Determination of
the Luminosity Function of Hot White Dwarfs

Alessandra G. Gemmo[1,2],
Stefano Cristiani[2], *Fabio La Franca*[2] and *Paola Andreani*[2]

[1] European Southern Observatory, Karl Schwarzschild Str. 2, D–85748 Garching, Germany
[2] Dipartimento di Astronomia, Vicolo dell'Osservatorio 5, I–35122 Padova, Italy

1 Introduction

The current state–of–the–art about the Luminosity Function of White Dwarfs (hereafter WD LF) has been achieved through the work of Fleming et al. (1986), based on a statistically complete sample of 353 DAs identified by the Palomar–Green (PG) survey. The resulting LF –down to $M_V \leq 12.75$– has been extended to lower luminosities by studies of Liebert, Dahn and Monet (1988, hereafter LDM) concerning proper–motion selected cool WDs.

Although it is obvious that the observed WD LF has to be improved at the faint end, all new samples of WDs published after LDM contain almost exclusively hot WDs. This is mostly a consequence of the large number of surveys of blue objects which have been completed in recent years (see e.g. the AAT QSO survey by Boyle et al. 1990, the LBQ survey by Hewett et al. 1993, and the EC Blue Object survey by Stobie et al. 1992). Most new samples of WDs are complete down to magnitude limits much fainter than the B=16.5 achieved by the PG survey. Their importance goes beyond the simple fact that they go fainter and are better defined (as a natural consequence of the improvement in instrument and detector technology since the completion of the PG survey): this bears mainly on the large number of previously unknown WDs they discover and which potentially include new and/or interesting types of objects (e.g. PG 1159 stars, magnetic WDs, etc). Moreover, one should not forget that some studies crucially depend on the availability of deep and statistically complete samples of intrinsically bright WDs. The scaleheight of the WD population can only be reliably determined by observing hot WDs at faint magnitudes, that is at z distances high above the galactic plane. Also, the hottest WDs are particularly interesting because the dominant role in their evolution is played by heat losses due to neutrino emission. They provide an opportunity to place constraints on neutrino properties, such as the magnetic moment, which are very difficult to measure in laboratory experiments (Blinnikov and Dunina–Barkovskaya, 1994). In this work, although preliminary, we present a new determination of the WD LF, obtained from 112 hot WDs identified in 8 fields on the southern sky by the Homogeneous Bright QSO Survey (see e.g. Cristiani et al. 1993). With respect to the other blue surveys that have followed the PG, the significance of the HBQS in the context of WD research derives from the particular care taken in order to secure complete samples not only of QSOs –which were the primary goal of the survey– but also of WDs.

For example, although it is well known that the number of QSOs in the magnitude bin 14<B<15 is extremely low, making it very unpractical to look for them in only 500 deg^2 of sky –the total area covered so far by the HBQS– blue candidates have been selected and observed in this magnitude range for almost all survey fields. Furthermore, in the phase of the follow–up spectroscopy, low S/N spectra would have been sufficient for the purposes of the QSO studies, as long as an indisputable determination of the redshift from the emission lines was possible. However, whenever it was achievable with reasonable exposure times, we always tried to obtain high S/N spectra so that model atmospheres could be fitted in order to obtain optimal values for the effective temperature T_{eff} and surface gravity g.

2 The HBQS White Dwarf Sample

2.1 Selection

The WDs used in the following analysis were originally selected as part of a sample of UVX stellar images obtained from COSMOS machine measurements of ESO and UK Schmidt photographic plates. These UVX objects were subsequently observed spectroscopically at low–resolution (10–12 Å, spectral range 3500–8500 Å) using the Boller & Chivens spectrograph and EFOSC 2 (ESO Faint Object Spectrograph and Camera), respectively at the 1.52m and 2.2m telescopes in La Silla. Full details of this project, which was carried out as an ESO Key Programme, are given in Cristiani et al. 1993, and so only a brief description of the survey will be given here.

Machine measurements of pairs of U,B(or J),V,R and I Schmidt plates centered on high galactic latitude $|b|$>50° ESO/SERC fields, were used to produce positions and magnitudes for several tens of thousands stellar images in each 5°×5° field. Machine magnitudes were calibrated using standard photoelectric or CCD stellar sequences on each field. On average the accuracy of the B magnitudes and colors is 0.1 mag at B=18.5, the typical magnitude limit of the survey. The magnitude and color criteria used to define the UVX sample varied from field to field but were typically 14<B<18.5 and U−B and B−V satisfying the so–called *Braccesi less restricted criterium*, by Braccesi et al. 1980. For the purpose of this work, we analysed 145 WDs observed in 8 fields (SA 94, SGP and SERC no. 290, 295, 296, 297, 351, 534) of the HBQS. The vast majority of these WDs were previously unknown. Most of them (128) are DAs, easily identified by the presence of broad Balmer lines. There are then 15 DBs, with the typical He I broad absorbtion line at 4471 Å, and 2 DC WDs, with featureless spectra. After rejecting all WDs beyond the survey completeness limit (see Section 3) and/or showing an unresolved binary spectrum, we were left with 112 WDs. We decided not to include close binary WDs in the determination of the WD LF because they may have previously interacted with the companions and, as a consequence, their overall characteristics and LF may differ from those of single stars.

2.2 Determination of M_V

Three different empirical techniques have been adopted to compute the absolute magnitudes of the WDs, with the purpose of considering only the median value in a preliminary determination of the LF. The techniques were the following:

1. Having the B−V colors for the sample, we applied the empirical relation derived by Sion and Liebert (1977):

$$M_V = 11.246(B - V + 1)^{0.60} - 0.045$$

   to obtain an absolute M_V magnitude.
2. Another empirical relation which ties the absolute magnitude to the colors of WDs is:

$$M_V = 7.56(b - y) + 11.50$$

   (see e.g. Graham, 1972) where b and y are Strömgren magnitudes. In our case, no Strömgren photometry was available, so b and y have been calculated by convolving the WD spectra with the passbands of the Strömgren b and y filters.
3. M_V derived from T_{eff}: we obtained T_{eff} by fitting model atmospheres to the WD spectra. The model atmospheres published by Kurucz (1979) and Wesemael et al. (1980) were used for $T_{eff} < 20,000$ K and $T_{eff} > 20,000$ K respectively. Kurucz's atmospheres, however, are more suited for low gravity stars and the published grid of Wesemael et al. models is too loose to permit very accurate determinations of T_{eff}. Making assumptions for the values of the logarithm of the surface gravity $\log g$ and the radius R of the WD (in this work they have been assumed equal to 8 and 0.012 $R_\odot$ respectively), it is possible to derive M_{bol} from T_{eff}. Once M_{bol} is known, M_V is computed by subtracting the value of the bolometric correction (BC). In this case we took the BC values for H−rich atmospheres (DAs) from Greenstein (1976) and for DB and DC WDs we used the BC corresponding to a black body at the same T_{eff}.

In most cases, and especially when the WD spectra were well calibrated and had a very good S/N, the three techniques gave results that agreed within 0.2–0.3 magnitudes. Finally, the typical accuracy of M_V for this sample of WDs is estimated to be 0.2 mag.

3 The Luminosity Function

As previously discussed by e.g. Green (1980), the crucial quantity in the definition of a complete sample for a statistical study is the limiting magnitude. The V/V_m method (Schmidt, 1968) is a powerful test of the completeness of a sample. We applied the V/V_m test to our sample of WDs and corrected for the effect of the nonuniform space distribution by assuming it to be an exponential disk with a scaleheight $z_0 = 250$ pc. We used this value of z_0 for comparison purposes, being the same used by LDM. The 8 HBQS fields turned out to have an average completeness limit $B_{lim}=17.9$. The major source of incompleteness in the HBQS white dwarf sample comes from the loss of objects with redder U−B colors than the UVX criteria set for inclusion in the spectroscopic survey. Such objects –the cooler WDs– need to be selected on the basis of their proper motion.

The 112 single WDs entering the completeness limit were used to derive the local LF for hot WDs by means of the method discussed by Schmidt (1975). The results are presented in Fig.1, together with the ones published by LDM; the data are binned into half-magnitude bins. The error bars are simply Poissonian, and do not include the effects of errors in the photometry or absolute magnitude calibration.

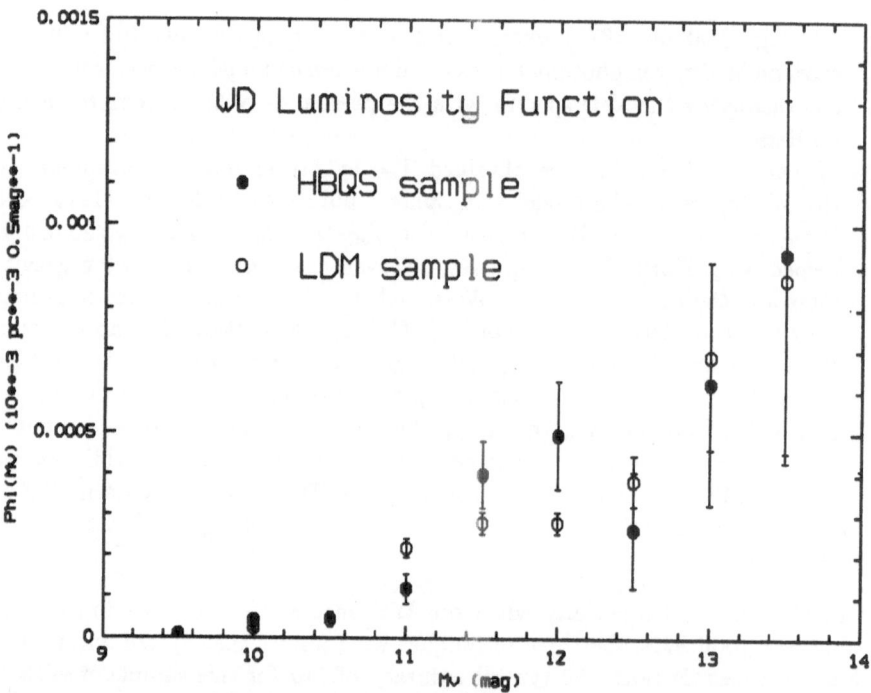

Fig. 1. The white dwarf LF derived from the HBQS and the LDM samples.

We note that the agreement between the LFs derived from the HBQS and LDM samples is good at bright and faint absolute magnitudes. A relatively small discrepancy exists in the range $11<M_V<12$. The higher space densities derived by the HBQS in this range can partially be attributed to the fact that the HBQS is likely to be less incomplete at faint absolute magnitudes because of the redder U−B limit used during the selection of candidate objects (the *Braccesi less restricted criterium* for the HBQS compared to U−B<−0.44 for the PG survey). That our selection criterion is able to pick up redder objects with respect to other surveys is also suggested by the relatively high number of WDs with a cool companion that we find (at least 20%, confirmed by infrared data, Gemmo 1995). The presence of a very red, late type companion changes the overall appearance of a WD spectrum, by enhancing the continuum at longer wavelengths, in such a way that many of these binaries are usually missed by very blue surveys.

4 Conclusions

We have obtained a new determination of the WD LF from a subsample of the HBQ survey (8 fields, that is 200 deg^2). Our WD counts appear to be more complete than those used to derive the LF published by Liebert, Dahn and Monet (1988). Further, more detailed comparisons of previous data with our WD LF will be possible when the more than 300 WDs discovered in the full area of the HBQS (500 deg^2) will be included in the calculation. We also plan to improve the accuracy in the determination of M_V by fitting our WD spectra with more suitable and up-to-date model atmospheres.

References

Blinnikov, S.I., Dunina–Barkovskaya, N.V.: 1994, *Monthly Notices Roy. Astron. Soc.* **266**, 289.

Boyle, B.J., Fong, R., Shanks, T., Peterson, B.A.: 1990, *Monthly Notices Roy. Astron. Soc.* **243**, 1.

Braccesi, A., Zitelli, V., Bonoli, F., Formiggini, L.: 1980, *Astron. Astrophys.* **85**, 80.

Cristiani, S., La Franca, F., Miller, L., Goldschmidt, C., Andreani, P., Gemmo, A., Vio, R., Barbieri, C., Bodini, L., Lazzarin, M., MacGillivray, H., Clowes, R., Gouiffes, C., Iovino, A., Savage, A.: 1993, in *First Light in the Universe: Stars or QSOs?*, ed. B. Rocca–Volmerange et al., 8th IAP Astrophysics Meeting, 353.

Fleming, T.A., Liebert, J., Green, R.F.: 1986, *Astrophys. J.* **308**, 176.

Hewett, P.C., Foltz, C.B., Chaffee, F.H.: 1993, *Astrophys. J.* **406**, L43.

Gemmo, A.G.: 1995, in preparation.

Graham, J.A.: 1972, *Astron. J.* **77**, 144.

Green, R.F.: 1980, *Astrophys. J.* **238**, 685.

Greenstein, J.: 1976, *Astron. J.* **81**, 323.

Kurucz, R.L.: 1979, *Astrophys. J. Suppl* **40**, 1.

Liebert, J., Dahn, C.C., Monet, D.G.: 1988, *Astrophys. J.* **332**, 891 (LDM).

Schmidt, M.: 1968, *Astrophys. J.* **151**, 393.

Schmidt, M.: 1975, *Astrophys. J.* **202**, 22.

Sion, E.M., Liebert, J.: 1977, *Astrophys. J.* **213**, 468.

Stobie, R.S., Chen, A., O'Donoghue, D., Kilkenny, D.: 1992, in *Variable Stars and Galaxies*, ed. B. Warner, Astron. Soc. Pac. Conf. Series.

Wesemael, F., Auer, L.H., van Horn, H.M., Savedoff, M.P.: 1980, *Astrophys. J. Suppl.* **43**, 159.

The Luminosity Function of Dim White Dwarfs

M. Hernanz[1], E. García-Berro[2], J. Isern[1], R. Mochkovitch[3]

[1] Centre d'Estudis Avançats (CSIC), Camí de Santa Bàrbara, 17300 Blanes, Spain
[2] Dept. de Física Aplicada, Universidad Politécnica de Cataluña, Jordi Girona Salgado s/n, Módul B-4, Campus Nord, 08034 Barcelona, Spain
[3] Institut d'Astrophysique de Paris (CNRS), 98bis bvd. Arago, 75014 Paris, France

1 Introduction

The luminosity function of dim white dwarfs is a powerful tool to determine some aspects of galactic evolution, such as the age of the disk or the star formation rate. Such analysis requires a good knowledge of both the observational and the theoretical luminosity functions, in order to extract the corresponding information from the comparison between them.

The most reliable observational luminosity function for white dwarfs comes from Liebert et al. (1988). One of its most important features is the shortfall that it shows at a luminosity in the range $-4.7 \leq \log(L/L_\odot) \leq -4.3$. The uncertainty in the location of the cutoff is related to the poor knowledge of the bolometric corrections of very dim white dwarfs leading to an uncertain determination of the age of the disk by about 2 Gyr.

Concerning the theoretical luminosity function, its computation requires the evaluation of cooling sequences and, therefore, the inclusion of a detailed treatment of crystallization. Two physical processes essentially affect the crystallization of white dwarfs: the release of latent heat and the chemical fractionation in the mixture. The last one is related mainly to the shape of the phase diagram of the main components (carbon–oxygen) but it also depends on the presence of some minor chemical species, such as ^{22}Ne, which are also present in white dwarf interiors. All these processes provide extra energy sources which delay the cooling process of white dwarfs. For instance, carbon–oxygen separation delays the cooling of a 0.6 M$_\odot$ white dwarf by more than 2 Gyr, whereas ^{22}Ne deposition introduces and extra delay of about 3 Gyr. But to evaluate how these delays affect the luminosity function, the cooling sequences have to be connected with a model of galactic evolution. This requires the inclusion of the full mass–spectrum of white dwarfs, the star formation rate and the evolution of the minor chemical species abundances. With all these ingredients, a theoretical luminosity function for dim white dwarfs can be built which accurately takes into account the effect of solidication and from which information about galactic history can be retrieved.

2 The Luminosity Function

White dwarfs with masses in the range between $0.4\,M_\odot$ and $1.2\,M_\odot$ are composed mainly of carbon and oxygen, but some minor chemical species reflecting the metallicity of the parent star are also present. The most important one is ^{22}Ne, which is a product of the helium burnig of the ^{14}N left by the CNO cycle.

The luminosity function has been computed from a generalization of the method developed by Noh and Scalo (1990). The number of white dwarfs per unit bolometric magnitude, per cubic parsec, with luminosity L, $n(L)$, is given by

$$n(L) = \int_{M_{inf}}^{M_{sup}} \tau_{cool}(L, M, Z(t_b)) \psi(t_{disk} - t_{cool}(L, M, Z(t_b)) - t_{ms}(M)) \phi(M) dM$$

where τ_{cool} is the characteristic cooling time, defined as $\tau_{cool} = dt_{cool}/d\log M_{bol}$, t_{cool} the cooling time, t_b the birth time of the progenitor star, Ψ the star formation rate per unit volume, $\Phi(M)$ the initial mass function. M_{sup} and M_{inf} denote respectively the maximum and the minimum mass of of the white dwarf progenitors contributing at luminosity L. M_{inf} is obtained from

$$t_{ms}(M_{inf}) + t_{cool}(L, M_{inf}, Z_0) = t_{disk}$$

where t_{disk} is the age of the disk, t_{ms} the main sequence lifetime and Z_0 the initial metallicity of the interstellar medium, usually taken as zero. Other details about the parameters related to the model of galactic evolution can be found elsewhere (Hernanz et al., 1994).

The cooling sequences are obtained directly from the computation of the binding energy of the white dwarf (see Díaz–Pinto et al., 1993 and Ségretain et al., 1994 for the details), including a proper treatment of carbon–oxygen crystallization, given by the corresponding phase diagram (Ségretain and Chabrier, 1993). An internal temperature–luminosity relationship obtained from detailed envelope models is adopted (Wood and Winget, 1989).

Figure 1 shows the luminosity function for carbon–oxygen white dwarfs obtained when the effect of chemical differentiation is neglected. The initial carbon–oxygen profile adopted is that of Mazzitelli and D'Antona (1986). The corresponding ages of the disk, 8.5 and 10.3 Gyr, are those that better fit the position of the cutoff, for the two approximations to the bolometric corrections of dim white dwarfs considered (Liebert et al., 1988). The effect of carbon–oxygen separation is clearly seen when these results are compared with those shown in figure 2, where the corresponding ages are 9.5 and 12 Gyr. The ages of the galactic disk obtained are roughly 1.5 Gyr older. The amplitude of the bump at $\log(L/L_\odot) \simeq -4$ is considerably reduced when selection effects due to the scale height of old white dwarfs are included (García-Berro et al., 1988a and b).

The importance of ^{22}Ne crystallization on white dwarf cooling was first pointed out by Isern et al. (1991), using preliminary phase diagrams for the C/^{22}Ne, O/^{22}Ne and N/^{22}Ne binary mixtures. Because of the lack of phase

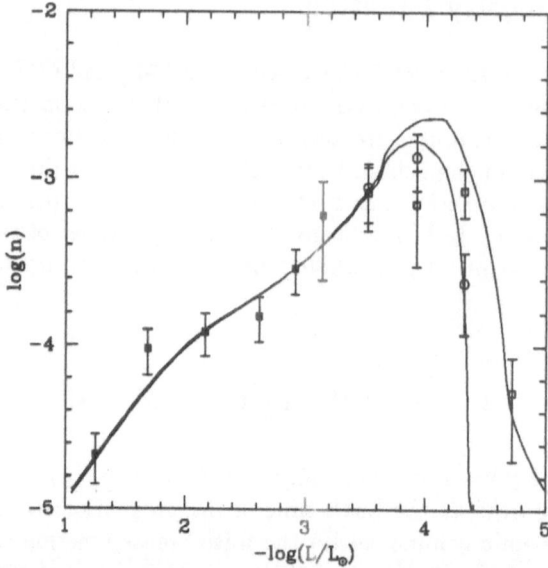

Fig. 1. Theoretical luminosity functions for C/O white dwarfs with no fractionation, for ages of the disk t_{disk} = 8.5 (left line) and 10.3 Gyr (right line). Solid squares represent the bright end of the luminosity function. For the low–luminosity objects, two approximations to the bolometric correction are taken: either model BC for cool DA's and blackbody BC for cool non-DA's (open circles) or zero BC for all cool white dwarfs (open squares).

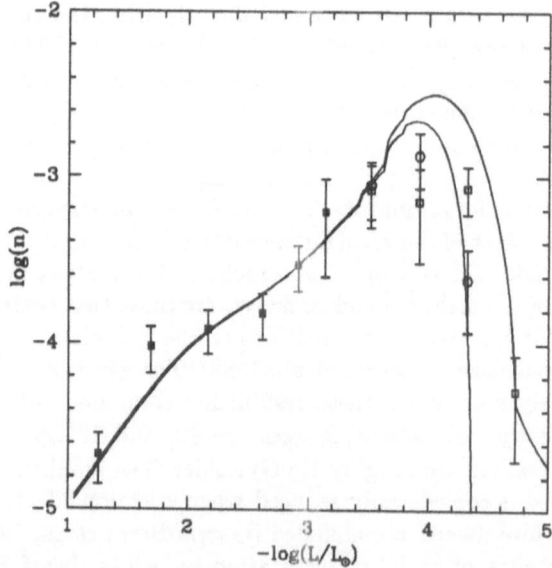

Fig. 2. Same as figure 1 but with the effect of C/O separation included

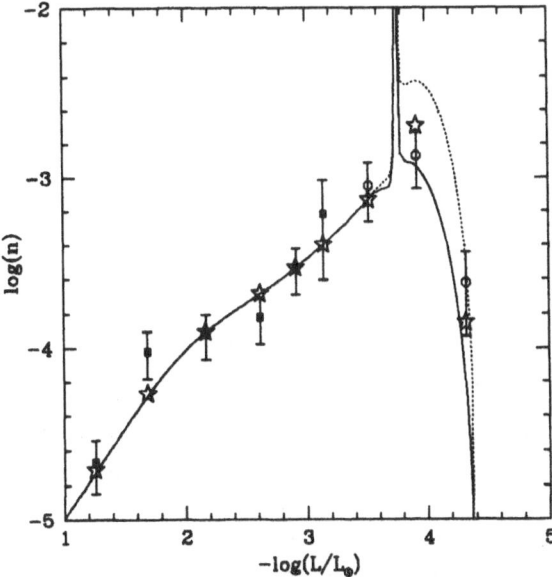

Fig. 3. Theoretical luminosity function when both ^{22}Ne crystallization and then C/O crystallization are taken into account succesively. The effect of scale height inflation is either included (solid line) or ignored (dotted line), for t_{disk} =10.5 Gyr. The observational points for cool white dwarfs are obtained with blackbody BC's for non DA's and model BC's for cool DA's. The open stars are the values of the binned theoretical luminosity function.

diagrams for three–component plasmas, the C/O/^{22}Ne system has been approximated by an effective two-component N/^{22}Ne mixture. This has as a consequence that all ^{22}Ne is collected in the core, with the azeotropic composition. The corresponding release of gravitational energy is of the same order of that associated with carbon–oxygen separation, leading to a dramatic increase of the cooling times down to the cutoff luminosity (of the order of 3 Gyr for a 0.6 M$_\odot$ white dwarf). However, one has to keep in mind that the exact behaviour of the ternary mixture C/O/Ne is still not known and that perhaps the simultaneous enrichment in oxygen presumably occuring during solidification can balance the effect of neon sedimentation and reduce its effect on the cooling history. The luminosity function which takes into account first ^{22}Ne crystallization and then carbon–oxygen separation is shown in figure 3. As can be deduced from a detailed comparison with the previous figures, the position of the cutoff is not altered by the presence of ^{22}Ne. The reason for this is that stars old enough to reach the cutoff luminosity have progenitors that were born when the Galaxy formed, i.e. with zero metallicity, and therefore without ^{22}Ne. The presence of a spike in the luminosity function at $\log(L/L_\odot) \simeq -3.8$ could be used as an observational signature of the process of ^{22}Ne deposition, but with some caution. First of all, the luminosity functions have been computed assuming that the cooling times are those of a 0.6 M$_\odot$ white dwarf with different ^{22}Ne abundances. When the effect of the mass distribution of white dwarfs is fully taken into account, the spike spreads over a larger range of luminosities. Second, as already mentioned, old white dwarfs have larger scale heights over the galactic plane than young ones. This also leads to a further reduction of the bump. Moreover, a correct comparison between theory and observations requires the binning of the theoretical results in the same way as the observations, i.e. in half or one visual magnitude intervals. The results of this procedure are shown as asterisks in figure 3.

3 Conclusions

Theoretical luminosity functions for white dwarfs have been computed for pure C/O white dwarfs and also including the effect of minor species such as ^{22}Ne. The comparison with the observational luminosity function allows to determine the age of the galactic disk. The time delay introduced by C/O fractionation at crystallization directly translates into ages of the disk 1.5 to 2 Gyr older than those obtained from the standard models (Wood, 1992).

The crystallization of minor species is found to modify drastically the cooling times of the star, but this effect does not affect the age of the disk obtained from the luminosity function. However, the important release of gravitational energy associated with the sedimentation of minor species produces a sharp peak at around $\log(L/L_\odot) \sim -3.8$. Though not detectable with present-day observations, this peak provides a future observational test of the mentioned effect. Moreover, if the initial metallicity of the galaxy is nonzero, the presence of minor chemical species modifies substantially the position of the cutoff of the luminosity function. The ages of the disk would be increased by around 3 Gyr for an initial metallicity $Z_0=1\%$.

References

Díaz–Pinto, A., García–Berro, E., Hernanz, M., Isern J., Mochkovitch, R., 1994, *A&A*, **282**, 86

García–Berro E., Hernanz M., Isern J., Mochkovitch R., 1988a, *Nature*, **333**, 642

García–Berro E., Hernanz M., Mochkovitch R., Isern J., 1988b, *A&A*, **193**, 141

Hernanz, M., García–Berro, E., Isern, J., Mochkovitch, R., Ségretain, L., Chabrier, G., 1994, *Ap.J*, **434**, 652

Isern, J., Mochkovitch, R., García–Berro, E., Hernanz, M., 1991, *A&A*, **241**, L29

Liebert, J., Dahn, C.C., Monet, D.G., 1988, *Ap.J*, **332**, 891

Mazzitelli, I., D'Antona, F., 1986, *Ap.J*, **308**, 706

Noh, H.R., Scalo, J., 1990, *Ap. J*, **352**, 605

Ségretain, L., Chabrier, G., 1993, *A&A*, **271**, L13

Ségretain, L., Chabrier, G., Hernanz, M., García–Berro, E., Isern, J., Mochkovitch, R., 1994, *Ap.J*, **434**, 641

Wood, M.A., Winget, D.E., 1989, *in IAU Colloq. 114, White Dwarfs* (Berlin: Springer–Verlag), 282

Wood, M.A., 1992, *Ap. J*, **386**, 536

Theoretical White Dwarf Luminosity Functions: DA Models

Matt A. Wood

Department of Physics and Space Sciences
Florida Institute of Technology
Melbourne, FL 32901–6988 USA

Abstract: New DA white dwarf models with thick H and He surface layers are presented. Uncertainties in cool WD ages resulting from the lack of opacities in the weakly degenerate regime are estimated at ~10%. Luminosity functions computed using these DA sequences suggest that the age of the local Galactic disk is in the range 6.5 to 11 Gyr. Phase separation of crystallizing C/O-core models has the potential to add up to ~20% to the ages of the coolest WDs, but the suggested process by which the matter is redistributed in the core is not sufficiently compelling.

1 Introduction

The observed white dwarf luminosity function (WDLF; see Liebert, Dahn, & Monet 1988) contains information on the age and star-formation history of the local Galactic disk which is independent of other cosmochronological techniques (Winget et al. 1987, Wood 1992). In order to extract this information, however, theoretical WD model sequences must be computed which are representative of the stars which are observed; that is, the models must include realistic composition profiles from core to surface and the best constitutive physics available. Once the sequences are in hand, they must then be included in a reasonable model for Galactic evolution to compute theoretical LFs which can then be compared with the observed sample.

Early non-adiabatic pulsation calculations (Winget & Fontaine 1982) and the observed changing ratio of DAs to non-DAs as a function of effective temperature (Fontaine & Wesemael 1987) suggested that the DAV H-layer masses were uniformly "thin," with a fractional mass of $q_H \equiv M_H/M_\star \lesssim 10^{-8}$. Because a star with a hydrogen layer this thin would mix convectively with the underlying helium layer at $T_{eff} \sim 10^4$ K and thus result in a DB WD for the remainder of that star's lifetime, the LFs presented in Wood (1992) were computed using DB model sequences with (thin) fractional He-layer masses of $q_{He} = 10^{-4}$.

Recent results from asteroseismology of DAV WDs, however, wherein the observed pulsation periods of DAVs are compared with *adiabatic* pulsation spectra (Fontaine et al. 1992, Clemens 1994), suggest convincingly that the surface H layers have masses that are in fact of order 10^{-4} $M_\star$ (i.e., "thick"), consistent with predictions from most evolutionary model calculations (see, for example, D'Antona & Mazzitelli 1990, and references therein). Therefore, because we believe that the DAVs are representative of the class of DA WDs as a whole, and because the DAs vastly outnumber non-DAs, it is useful (and past time) to compute LFs for WDs with thick H and He layer masses.

2 DA Model Sequences

I computed the sequences presented here using the White Dwarf Evolution Code (Lamb & Van Horn 1975, Wood 1990) updated as described in Wood (1994). The sequences include C-core, O-core, and C/O-core models with masses 0.4, 0.5, 0.6, 0.7, 0.8, and 1.0 $M_\odot$. The C/O-core models have interior O profiles similar to that in Figure 4 of D'Antona & Mazzitelli (1990):

$$X_O = \begin{cases} 0.75, & 0.0 \leq q \leq 0.5, \\ 0.75 - 1.875(q - 0.5) & 0.5 < q \leq 0.9, \\ 0. & 0.9 < q \leq 1.0, \end{cases} \tag{1}$$

where $q \equiv M_r/M_\star$ and $X_C = 1 - X_O$ for all but the outermost $10^{-2}\ M_\star$. The models all have fractional H- and He-layer masses of 10^{-4} and 10^{-2}, respectively, and the metallicity used is $Z = 0.000$. The suggested process of phase separation is not included in these models (but see §4 below). Table 1 lists an abbreviated summary of the 0.6 $M_\odot$ C/O-core model sequence[1]. Because the evolution code's current envelope equation of state (EOS) tables do not extend to the crystallization regime, the sequences terminate when the crystallization front reaches the core/envelope boundary. Because this occurs at a luminosity just fainter than the observed turndown luminosity of $\log(L/L_\odot) \approx -4.4$, it is necessary to extrapolate the luminosity versus age curves. For this work, I have extrapolated the curves by fitting a third-order polynomial to the last $\sim$5 converged models, all characterized by $q_{xtal} \geq 0.95$.

One continuing problem with the WD models is the lack of opacities in the low-temperature ($10^4 \lesssim T \lesssim 10^6$) weakly-degenerate regime ($T \approx T_{Fermi}$) (see Wood 1994). While this void should be filled by the time these proceedings are in print (F. Rogers, personal communication), it is important to know the maximum uncertainty caused by the current extrapolations into this regime, particularly since for some temperatures the opacities have maxima here. To determine this uncertainty, I ran two otherwise identical C-core DA sequences where one sequence used the standard linear-in-the-logarithm extrapolations, and the other simply used the radiative opacity at the edge of the table (i.e., a flat extrapolation). The differences in the model ages at a given luminosity are always 5% or less above $\log(L/L_\odot) = -4.0$, but can be as large as 30% for fainter models. Because this is an extreme test, I estimate that the true uncertainty in the WD cooling ages resulting from the lack of opacities in the weakly degenerate regime is $\lesssim$10%.

---

[1] Full summary listings of all sequences used in this work are available via anonymous ftp at the site kepler.pss.fit.edu in the directory /pub/wd.

Table 1. Evolutionary summary of a 0.6 $M_\odot$, C/O-core, DA model sequence[a]

$\log(L/L_\odot)$	Age	T_{eff}	$\log T_c$	$\log(\rho_c)$	$\log(R)$	$\log(L_\nu/L_\odot)$	q_{xtal}
2.0	5.673 (4)	106380	8.023	6.354	9.315	1.648	0.00
1.5	2.297 (5)	90351	7.993	6.401	9.207	1.548	0.00
1.0	6.630 (5)	74466	7.928	6.456	9.125	1.231	0.00
0.5	1.432 (6)	59396	7.866	6.491	9.072	0.796	0.00
0.0	2.774 (6)	46445	7.813	6.512	9.035	0.330	0.00
−0.5	5.779 (6)	35899	7.743	6.527	9.009	−0.165	0.00
−1.0	1.541 (7)	27548	7.651	6.539	8.989	−0.986	0.00
−1.5	5.858 (7)	21018	7.475	6.549	8.974	−2.465	0.00
−2.0	1.677 (8)	15961	7.256	6.556	8.963	−4.436	0.00
−2.5	3.725 (8)	12081	7.044	6.560	8.955	−6.495	0.00
−3.0	7.650 (8)	9125	6.843	6.562	8.949	−9.999	0.00
−3.5	1.665 (9)	6888	6.641	6.563	8.943	−	0.13
−4.0	4.515 (9)	5211	6.332	6.565	8.935	−	0.75
−4.2	6.010 (9)	4653	6.166	6.565	8.933	−	0.90
−4.4	7.298 (9)	4151	6.012	6.565	8.933	−	0.98
−4.6	8.372 (9)	3702	−	−	−	−	−
−4.8	9.356 (9)	3299	−	−	−	−	−
−5.0	1.029(10)	2940	−	−	−	−	−
−5.5	1.255(10)	2205	−	−	−	−	−
−6.0	1.480(10)	1653	−	−	−	−	−

a) Luminosities for models with $\log(L/L_\odot) = -4.6$ and fainter are extrapolated using a third-order polynomial fit to the final 5 models of the luminosity vs. age relation.

3 DA White Dwarf Luminosity Functions

The parameters I use for computing the LFs are as follows: a Salpeter IMF, the Clayton (1988) infall star-formation rate, the Iben & Laughlin (1989) pre-WD evolutionary timescales, the initial-to-final mass relation from Iben and Renzini (1984) with $\eta = 3.0$ and no disk inflation with time. With these parameter choices the mean mass of the theoretical sample is 0.61 $M_\odot$. Figure 1 shows the LFs computed using the C- and O-core models. The C/O-core results are similar to the O-core results and are not shown. The input ages to the LFs range from 6 to 12 Gyr in both cases, and it is evident that there is a roughly 2 Gyr offset between the two curves. The curves which pass through the lowest-luminosity box suggest that the ages of the local WDs are in the range 6.5 to 8.5 Gyr if the cores are O-rich, and 8.5 to 11 Gyr if they are C-rich. The C/O-core results suggest ages in the range 6.5 to 9 Gyr. These ages are roughly 1.5 Gyr *less* than the DB model results from Wood (1992), resulting from the 100× thicker He layer and to a lesser extent the improvement of the opacity routines resulting from the inclusion of the low-temperature Lenzuni, Chernoff, & Salpeter (1991) radiative opacities.

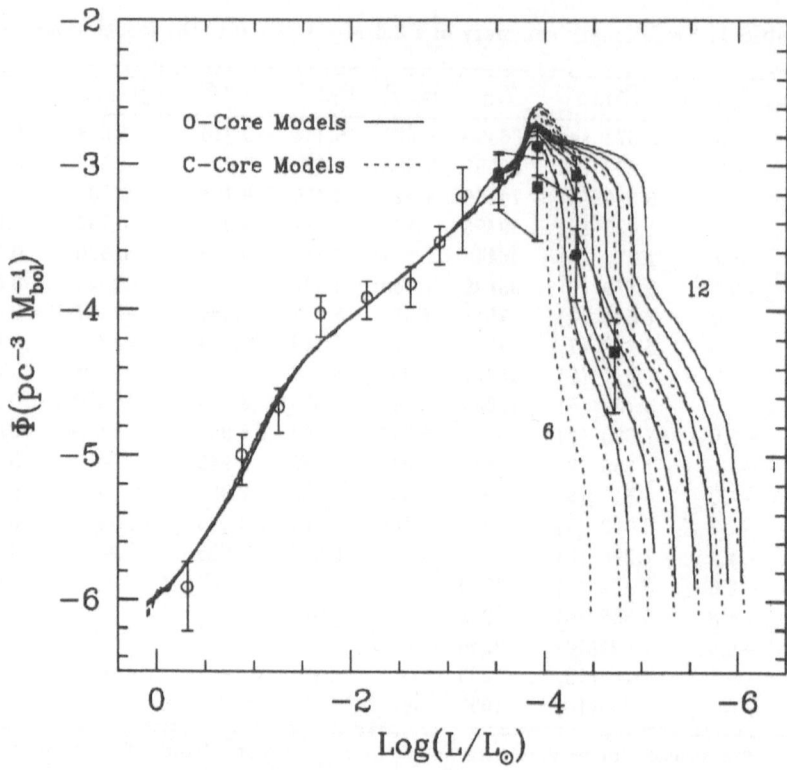

Fig. 1. Luminosity functions computed using the C- and O-core DA model sequences, plotted along with the observed LF of Liebert et al. (1988).

4 The Energetics of Phase Separation

Based on the crystallization phase diagram for binary ionic mixtures, several authors in recent years have suggested that WD interiors undergo phase separation during crystallization (see Segretain et al. 1994 and references therein), and that there is a subsequent release of gravitational binding energy which can lengthen by up to several gigayears the time required to reach the LF turndown luminosity. While these new EOS calculations have not been published in a form suitable for inclusion in evolution codes, and while the suggested mechanism for the redistribution of matter in the core — convective instability (Mochkovitch 1983) — is still less than compelling, it is instructive to explore the energetics which are potentially involved in the crystallization and phase separation process using the currently available tools. I computed 2 otherwise identical C/O core model sequences, one with a homogeneous 50:50 C/O mixture throughout the core, the other with a global C/O ratio of 50:50, but with a "post-crystallization" profile as given Figure 4 of Segretain et al. Averaging over the crystallized models, I find that the photon luminosity L is approximately 1/2 the change in gravitational binding energy $\Delta\Omega$, as expected from the virial theorem. Comparing the difference in the mean binding energies of the two sequences with total $\Delta\Omega$ over the course of crystallization for the homogeneous sequence suggests that phase separation *could* lengthen the cooling time of WDs by up to $\sim 20\%$, *if* it can be demonstrated that redistribution of matter can occur. In my opinion, this is the primary uncertainty with regard to the suggested process of phase separation.

References

Clayton, D.D. 1988, MNRAS, 234, 1

Clemens, J.C. 1994, PhD thesis, The University of Texas at Austin

D'Antona, F., & Mazzitelli, I. 1989, ApJ, 347, 934

D'Antona, F., and Mazzitelli, I. 1990, ARA&A, 28, 139

Fontaine, G., Brassard, P., Bergeron, P., & Wesemael, F. 1992, ApJL, 399, 91

Fontaine, G., and Wesemael, F. 1987, in IAUColl 95, The Second Conference on Faint Blue Stars, ed. A.G.D. Philip, D.S. Hayes, and J. Liebert (Schenectady: Davis Press), p. 319

Iben, I., Jr., and Laughlin, G. 1989, ApJ, 341, 312

Iben, I., Jr., and Renzini, A. 1983, ARA&A, 21, 271

Lamb, D.Q., and Van Horn, H.M. 1975, ApJ, 200, 306

Lenzuni, P., Chernoff, D.F., and Salpeter, E.E. 1991, ApJS, 76, 759

Liebert, J., Dahn, C.C., and Monet, D.G. 1988, ApJ, 332, 891

Mochkovitch, R. 1983, A&A, 122, 212

Segretain, L., Chabrier, G., Hernanz, M., García-Berro, E., Isern, J., & Mochkovitch, R. 1994, ApJ, 434, 641

Winget, D.E., and Fontaine, G. 1982, in Pulsations in Classical and Cataclysmic Variable Stars, ed. J.P. Cox and C.J. Hansen (Boulder: University of Colorado Press), p. 46

Winget, D.E., Hansen, C.J., Liebert, J., Van Horn, H.M., Fontaine, G., Nather, R.E., Kepler, S.O., and Lamb, D.Q. 1987, ApJ, 315, L77

Wood, M.A. 1990, PhD thesis, The University of Texas at Austin

Wood, M.A. 1992, ApJ, 286, 539

Wood, M. A. 1994, in IAU Coll. 147: The Equation of State in Astrophysics, eds. G. Chabrier and E. Schatzman (Cambridge U. Press: Cambridge) p. 612

Cool White Dwarfs in a Southern Proper Motions Survey

Maria Teresa Ruiz

Departamento de Astronomia, Universidad de Chile

In an effort to identify low luminosity stars in the southern hemisphere, a search program was started for proper motion stars in red plates (IIIa F) taken with the ESO Schmidt Camera at La Silla. The magnitude limit of these plates is $m_R \sim 20.5$.

The spectroscopic follow-up of stars in the first 3 ESO areas surveyed (207, 439, and 440) having $\mu \geq 0.25$ "yr^{-1} , revealed the existence of 8 cool WDs not previously identified. Table 1 gives a summary of the characteristics of these stars. Except for ESO439-26, for which we have a trigonometric parallax determination, absolute magnitudes have only been estimated. This survey is being

TABLE 1

Star Number	V	ST	M_V ± .5 mag	d pc	μ " yr^{-1}	v_T km s^{-1}
207-124	17.00	DC7	14.5	32 ± 7	.58	88 ± 19
207-21	19.60	DC9	16.0	52 ± 11	.35	86 ± 19
207-149	18.82	DA8	14	92 ± 20	.27	118 ± 25
439-162	18.77	DQ7	14.9	59 ± 12	.38	94 ± 18
439-163	19.84	DC9	16.0	59 ± 12	.38	94 ± 18
439-26	20.64	DC9	$17.6 \pm .1$	41 ± 1.6	.34	43 ± 2.5
439-80	15.02	DA6	13.5	20 ± 5	.34	35 ± 6
440-146	17.30	DA9	14.0	46 ± 11	.28	85 ± 15

extended to include a total of 15 areas. Our goal is to have a well defined, complete sample of Cool White Dwarfs that would allow a better determination of the faint end of the Luminosity Function for degenerates.

This work received partial support from FONDECYT grant 880-92.

Part II
Evolution and Interior Structure

The Born Again AGB Phenomenon

Icko Iben, Jr.[1] *and Jim MacDonald*[2]

[1] University of Illinois at Urbana-Champaign, Astronomy Department
   1002 West Green St., Urbana, IL 61801, USA
[2] University of Delaware, Department of Physics and Astronomy
   Newark, DE 19716, USA

1 The Overall Picture

After completing the core helium-burning phase, single stars less massive than $\sim 10.5\ M_\odot$ develop electron-degenerate cores composed predominantly of carbon and oxygen (if initial mass is less than $\sim 8\ M_\odot$) or of oxygen and neon (if initial mass is in the range $\sim 8\text{-}10.5\ M_\odot$). The star is called an asymptotic giant branch (AGB) star because, for small stellar mass, it evolves toward the shell hydrogen-burning giant branch in the Hertzsprung-Russell diagram as it brightens. During the AGB phase, a star alternately burns hydrogen and helium in shells. Hydrogen burns quiescently, but a thermonuclear runaway precedes each quiescent helium-burning stage.

During a helium shell flash, the flux of energy released becomes so high that a convective shell forms and extends from the base of the helium-burning region almost to the hydrogen-helium discontinuity. As the flash dies down, convection in a shell disappears, leaving behind a region rich in carbon ($\sim 20\%$ in mass) and s-process elements. With each thermal pulse, the peak helium-burning luminosity increases, and, as the flash "powers down", the increase in the outward energy flux at the hydrogen-helium discontinuity and beyond causes the base of the convective envelope to extend inward in mass until it reaches into the carbon-rich region. A dredge up phase ensues, during which fresh carbon and s-process elements are brought to the surface.

At some point in its life, an AGB star develops a pulsational instability in its envelope and becomes a Mira or long period variable, with a pulsation period from a few hundred days to several thousand days. Empirically, it is found that such stars lose mass at a very high rate and that the rate of mass loss is, on average, larger for larger periods. As period increases from ~ 200 days to ~ 500 days, the average mass-loss rate increases from $\sim 10^{-7}\ M_\odot$ yr^{-1} to $\sim 10^{-5}\ M_\odot$ yr^{-1}. In the case of the most massive AGB stars, the mass-loss rate approaches $\sim 10^{-4}\ M_\odot$ yr^{-1} and the star develops an expanding wind envelope that radiates as an OH/IR radio source which often hides the interior AGB star from view at optical wavelengths. The wind is sometimes referred to as a "superwind". Wind velocities are of the order of several times the escape velocity from a giant

(20-30 km s^{-1}). The mechanism for mass loss is thought to involve (1) a sharp increase in the density in the atmosphere due to pulsation-induced shocks and (2) radiation pressure on grains.

Once the mass of the hydrogen-rich envelope of the AGB star drops below a critical value, which is about 10% of the mass of hydrogen burned between helium shell flashes, the envelope begins to contract (in order to maintain sufficiently high densities and temperatures for hydrogen to continue to burn) and a detachment between the remnant of the AGB star and the shell of ejected matter occurs. Thanks to high mass-loss rates and low wind velocities, the shell of ejected matter is at relatively high densities. As the stellar remnant contracts at high luminosity, its surface temperature increases until the photons it emits are hard enough to ionize the hydrogen in the ejected shell. At the same time, the remnant continues to lose mass via a stellar wind, now driven by radiation pressure on ionized hydrogen and metals, and wind velocities are of the order of 1000-3000 km s^{-1}. Mass-loss rates are of the order of 10^{-9}-10^{-7} $M_{\odot}$ yr^{-1} (e.g., Perinotto 1993). If the wind is thought of as a fluid, steady state models can be constructed (e.g., Kahn 1983, 1989) in which the wind particles form a shock-heated bubble which presses on the base of the shell formed by the superwind and compresses it, thus helping shape the object now called a planetary nebula.

The stellar remnant continues to contract as the mass of its hydrogen-rich envelope decreases in response to both wind mass loss and hydrogen burning. As the remnant contracts, its surface temperature increases, and the chemistry of the nebula changes in response to a bath of harder and harder photons. Eventually, when the envelope mass decreases below a critical value (of the order of 1-2 $\times 10^{-4}$ for a remnant of mass ~ 0.6 $M_{\odot}$), the rate at which hydrogen burns drops abruptly. This occurs after a time (following the first departure from the AGB) of the order of 10^4 yr $(0.6$ $M_{\odot}/M_{\rm rem})^{10}$, where $M_{\rm rem}$ is the mass of the remnant. The envelope of the star contracts and the bolometric luminosity of the star drops by a factor of ten in about a thousand years. The nebula, in response, declines by a factor of ten in bolometric luminosity.

Under certain conditions, the helium-rich and helium-carbon layers which are below the hydrogen-helium discontinuity are heated (during the contraction of the envelope layers which immediately follows the abrupt drop in the hydrogen-burning rate) to a sufficiently high temperature to initiate a final helium shell flash. The convective shell which is created because of the high fluxes generated by helium burning grows until its outer edge reaches the hydrogen-helium discontinuity and, because there is no longer an entropy barrier (Iben 1976) to surmount, it extends further into the hydrogen-rich envelope. This phenomenon was first predicted by Fujimoto (1977). In the first work to describe the phenomenon quantitatively (Iben, Kaler, Truran, & Renzini 1983), it was assumed that the hydrogen ingested by the convective shell is converted immediately into heavy elements, with no release of nuclear energy. The model, relying solely on helium burning, expanded and dimmed for a very short time and then expanded and brightened as it evolved back to the region of the AGB. The chemistry expected at the surface is similar to that at the surface of an R CrB star and at the surface of FG Sge, and the evolution in the H-R diagram is qualitatively similar to that of FG Sge over the past 100 yr. These developments are described in detail elsewhere (Iben 1995, and references therein).

In reality, of course, as hydrogen convectively diffuses into the convective helium-burning shell, it begins to burn when it reaches regions at high enough temperature. The entropy generated by the release of nuclear energy by proton capture and beta decay forces the development of two detached convective zones, one powered by helium burning and the other powered by hydrogen burning (e.g., Hollowell, Iben, & Fujimoto 1990). The chemistry of the convective zone driven by hydrogen burning is rather exotic, as the zone consists of matter which has been processed first by complete hydrogen burning, next by partial helium burning, and then once again by hydrogen burning. A calculation of these burning phases has been done, and the results are described in §3.

2 Details from the Theory

The real world is much more complicated than indicated in §1, and this is true also of the theoretical world. There are five distinct ways in which the stellar remnant of a single AGB star can evolve after the superwind phase (Iben 1984). The particular way depends on the precise phase in the thermal pulse cycle when departure from the AGB occurs. If departure occurs during the hydrogen-burning phase and the mass δM_H of hydrogen which has been processed since the immediately preceding pulse satisfies $0 < \delta M_H < 0.75 \, \Delta M_H$, where ΔM_H is the total amount of hydrogen processed between thermal pulses, the remnant does not reignite helium and does not become a born again AGB star (case 1). (The numbering of cases here is the same as in Iben 1995)

If departure from the AGB occurs during the quiescent helium-burning phase (case 2), the remnant continues to burn helium as it evolves to higher surface temperatures on roughly the same timescale as a case 1 remnant of the same mass burns hydrogen. The planetary nebula formed when the matter which was ejected during the AGB phase is excited by photons which owe their existence to helium burning is, in practice, indistinguishable from the planetary nebula formed in case 1. If departure from the AGB occurs during a helium shell flash or during the following power-down phase (case 3), the behavior of the remnant is nearly the same as that of a case 2 remnant of the same mass, especially during the quiescent helium-burning phase.

In both cases 2 and 3, the mass of the helium "buffer" layer is at a minimum. The buffer layer separates the carbon-rich ($\sim 20 \%$ abundance by mass, independent of population type) layer below it from the hydrogen-rich envelope. Iben & MacDonald (1986) took diffusion into account and found that, after nuclear burning appears to have been completed, chemical diffusion carries hydrogen inward and carbon outward through the buffer layer until the product of the hydrogen and carbon abundances becomes large in a region of high temperature and density. A shell hydrogen-burning flash is ignited and the star returns to the AGB burning hydrogen. Iben & Tutukov (1986) found a similar result.

In retrospect, it does not appear that the inclusion of diffusion is necessary to produce this "self-induced" nova phenomenon. The literature on binary star evolution contains a number of examples of the occurrence of multiple hydrogen shell flashes when diffusion is not included (e.g., Kippenhahn, Kohl, & Weigert 1967, Iben, Fujimoto, Sugimoto, & Mayaji 1986). The flashes are a consequence of the changing thermal and density structure of the hydrogen-exhausted core (Iben & Tutukov 1995).

If departure from the AGB occurs during the hydrogen-burning phase and $0.75\Delta M_H \stackrel{\sim}{<} \delta M_H \stackrel{\sim}{<} 0.85\Delta M_H$ (case 5), a final helium shell flash occurs after hydrogen burning has ceased to be a major source of surface luminosity and the remnant becomes a born again star with an interesting composition, as described in §1 & §3. If $0.85\Delta M_H \stackrel{\sim}{<} \delta M_H \stackrel{\sim}{<} 1.0\Delta M_H$, a final helium shell flash occurs while hydrogen is burning strongly (case 4), but the outer edge of the convective shell produced by helium burning does not pass through the existing entropy barrier at the hydrogen-helium discontinuity and a layer containing hydrogen and products of partial helium burning is not formed. The remnant expands to become a born again AGB star for a short while, but this time without an especially exotic surface composition. Since the luminosity provided by helium burning can help maintain the remnant at large radius, a very short-lived superwind may operate, even when the mass of the hydrogen-rich layer is much less than $\sim 0.1 \ \Delta M_H$. Subsequent evolution takes the remnant to the blue where photons from its surface can cause the nebula ejected in the first superwind phase to become visible again, provided that nebular densities have not been reduced too far by expansion. Radiatively driven mass loss from the star may remove the hydrogen-rich envelope entirely, and the star may evolve into a DB white dwarf.

To complicate the situation still further, the theory of binary stellar evolution shows that there are many ways in which binaries with different initial configurations can evolve into R CrB and related stars. The number of such stars for which binary paths are responsible appears at first sight to be comparable to the number for which the single star born again scenario is responsible (Iben, Tutukov, & Yungelson 1995). Thus, there are "many ways to skin a cat", and considerably more theoretical and observational work remains to be done before we will be in a position to establish those characteristics which will allow us to distinguish the ancestry of any given observed star.

3 A Theoretical Calculation with Time-Dependent Mixing

A theoretical calculation (Iben & MacDonald 1995) of the final helium shell flash of the case 1 variety begins with a model from Iben et al. (1983) of mass 0.6 $M_\odot$ and metallicity $Z = 0.001$. In the first model, helium has begun to burn, but shell convection has not yet appeared. When, in any region, the adiabatic temperature gradient is less than the radiative gradient, convective mixing is calculated with a diffusion equation (see Iben & MacDonald 1985) using a diffusion coefficient $D = 0.1 \times l_{mix} \times v_{conv}$, where l_{mix} is the mixing length (taken to be 1.5 times the pressure scale height) and v_{conv} is the convective velocity in the mixing length approximation.

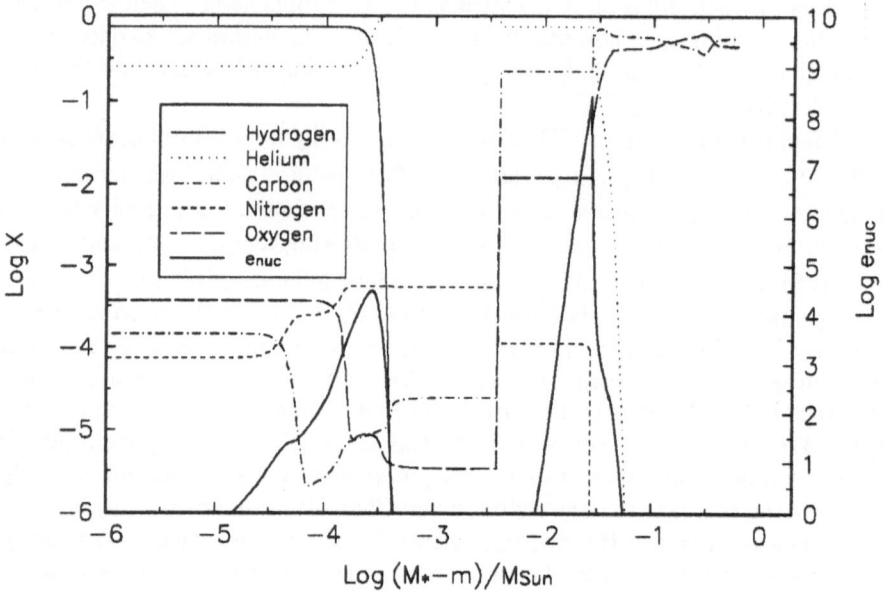

Fig. 1. Composition (abundances by mass) and the energy generation rate for an early model.

In Figure 1 are shown composition profiles in a model after a convective shell driven by helium burning has formed. The region over which convection is occurring can be identified by the flat portion of, say, the carbon abundance profile at $X_C = 0.23$ ($\log X_C = -0.64$). The nuclear energy generation rate is also shown in Figure 1. Note that, even though hydrogen burning contributes negligibly to the surface luminosity, hydrogen has actually not ceased to burn.

A hydrogen shell flash begins after the outer edge of the convective shell driven by helium burning has extended beyond the hydrogen-helium discontinuity and hydrogen has convectively diffused through the outer $\sim 25\%$ of the mass of the convective shell. A second convective shell driven by hydrogen burning mixes $\sim 6 \times 10^{-3}\ M_\odot$ of matter which has experienced partial helium burning with $\sim 3 \times 10^{-4}\ M_\odot$ of hydrogen-rich matter. In the course of the flash, $\sim 20\%$ of the carbon initially in the second convective zone is converted into nitrogen. Oxygen originally in the outer edge of the first convective zone is spread out over the second convective zone without experiencing much burning.

Motion in the H-R diagram is shown in Figure 2. The effects of the helium shell flash become apparent as the model begins to depart from the white dwarf cooling curve at $\log T_e \sim 4.98$. The time for motion to the lowest luminosity position on the track is only a matter of several weeks, and the time to reach the next minimum in surface temperature at $\log T_e \sim 4.58$ is only three weeks.

The remainder of the track requires about seventeen years and, over this time interval, the model brightens by about 2.5 mag bolometrically or about 5 mag visually. The details of the track must not be taken too seriously since the opacity which we use is not designed for matter of the exotic composition encountered.

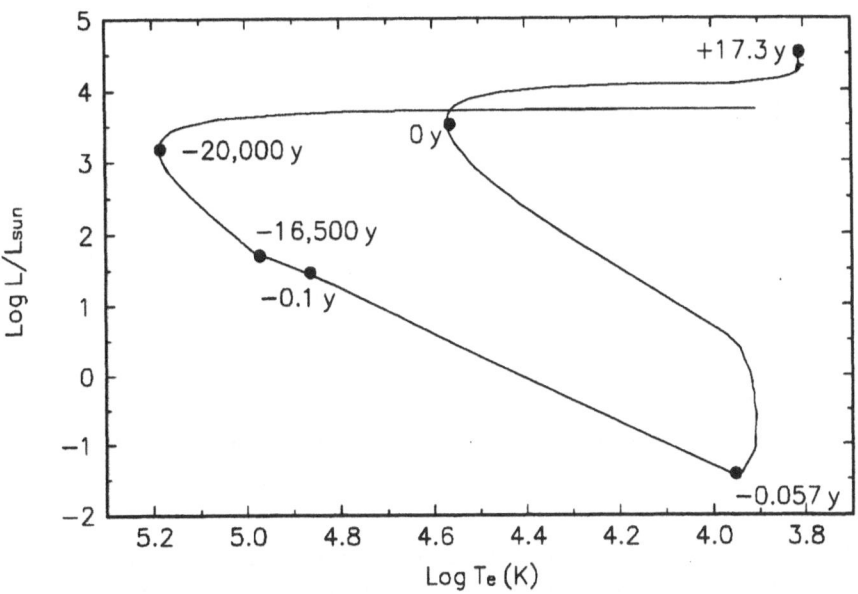

Fig. 2. Model track in the Hertzsprung-Russell diagram. Time is indicated at several points along the track in units of years

Composition profiles and the energy generation rate in the last model computed are shown in Figure 3. By this time, the two convective shells have died out, the helium-burning region is in a power-down phase, and hydrogen is burning quiescently. The model has a convective envelope which extends almost down to the hydrogen-burning shell. From Figure 3, it is evident that the model star has become hydrogen deficient, carbon rich, and nitrogen rich $(X_H \sim 0.03, X_{He} \sim 0.76, X_C \sim 0.15, X_N \sim 0.05,$ and $X_O \sim 0.01)$.

These calculations could be continued until both hydrogen and helium burning have ceased, but it is not clear that much more is to be gained other than an estimate of the upper limit on the lifetime of the red star phase. For one thing, wind mass loss in a real counterpart could drastically reduce the lifetime of the hydrogen-burning phase. For another, wind mass loss during the time between the cessation of hydrogen burning as a major source of surface luminosity and the onset of the born again event could considerably reduce the mass of hydrogen-rich matter which mixes with products of partial helium burning during the final helium shell flash, thus changing the final surface chemistry. In the future, it will be important to include the effect of wind mass loss and to

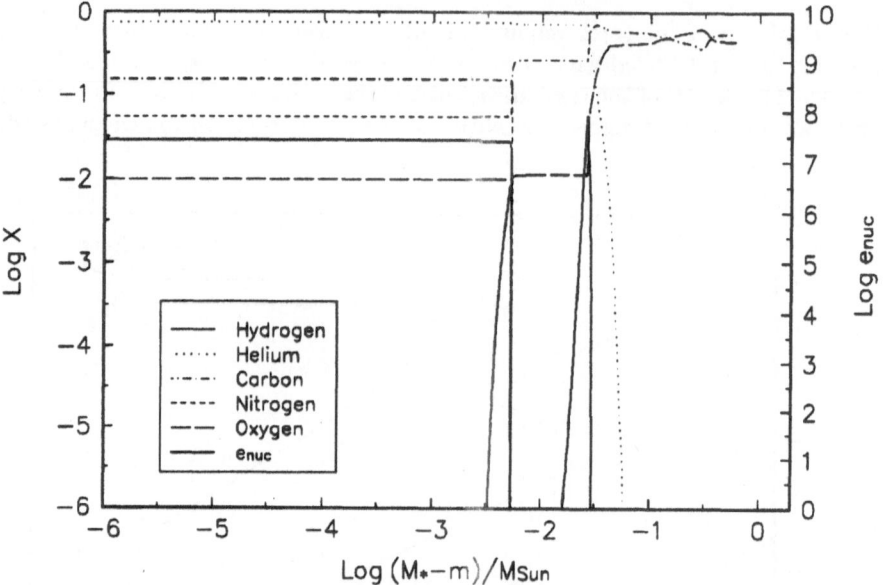

Fig. 3. Composition (abundances by mass) and the energy generation rate for the last model.

explore the case 5 born again phenomenon with models of different mass and initial composition. Most important of all is an exploration of how evolution in the H-R diagram and evolution of the surface composition depend on the precise phase in the AGB thermal pulse cycle that departure from the AGB occurs.

4 Examples from the Observations

There are a number of real stars which show behavior or exhibit characteristics which are qualitatively similar to those of the model just described.

In fact, the identification of knots of hydrogen-deficient, nitrogen-rich matter close to, and apparently emitted by, the central stars of the planetary nebulae Abell 30 (Hazard et al. 1980) and Abell 78 (Jacoby & Ford 1983) served as motivation for the initial calculations of the born again phenomenon (Iben et al. 1983, Iben 1984). The knots have velocities ~ 30 km s^{-1}, typical of red giant winds, and this suggests that the central stars expanded to giant dimensions long after the formation of the original nebular shells.

The composition of the central star of Abell 78 bears some resemblance to the compositions found in the model calculation. Using model atmospheres with wind emission taken into account, Werner & Koesterke (1992) have estimated abundances by mass in the atmosphere of the central star, obtaining $X_{\rm H} \sim 0.1, X_{\rm He} \sim 0.33, X_{\rm C} \sim 0.5, X_{\rm N} \sim 0.02$, and $X_{\rm O} \sim 0.05$. While this composition differs in detail from the theoretically theoretically calculated one, the general pattern of processing the products of partial helium burning by hydrogen burning is quite evidently the same in the two sets of composition. Furthermore, the theoretical calculation has been done for only one mass and one initial composition, and wind mass loss has not been taken into account. Finally, there is large uncertainty in the cross section of the $^{12}C(\alpha, \gamma)^{16}O$ cross section and this cross section determines the ratio of oxygen to carbon produced in the convective shell during partial helium burning.

There is conceivably also a large variation from one born again system to another in the surface abundance pattern of the central star. Hamann & Koesterke (1993) estimate for the central star of NGC 6751 $X_{He} \sim 0.615, X_C \sim 0.27, X_N \sim 0.015$, and $X_O \sim 0.1$. Again, the pattern indicates the same type of nucleosynthesis history, but the details are not the same as found in the one theoretical calculation or in the central star of Abell 78. It should also not be forgotten that there can be large errors in the abundance analyses of real stars.

Evidence for a star actually in the process of undergoing a born again event of the case 5 variety is V605 Aql, the central star of Abell 58. The central star presumably erupted 65 years ago as Nova Aql 1919. Bond et al. (1993) have detected a clumpy nebulosity at a distance of $\sim 10^{16}$ cm from the central star, and Pollacco et al. (1993) have analysed a hydrogen-deficient knot near this same star, finding a velocity of ~ 100 km s^{-1}. Performing the operation $D \sim 100$ km s$^{-1} \times 65$ yr $\times 3 \times 10^7$ s yr$^{-1} \sim 2 \times 10^{16}$ cm, we find a suggestive similarity between the distance a knot travels since the nova eruption and the dimensions of the clumpy nebulosity. The final feature of the central star, which strengthens its identification as a born again star, is the fact that the spectrum of Nova Aql 1919 was similar to that of an R CrB star, namely, hydrogen deficient and carbon rich (see Bond et al. 1993).

Another very interesting and relevant system is FG Sge. This star is in a wide binary surrounded by a faint nebula of diameter ~ 0.4 pc (Herbig & Boyarchuck 1968) and age ~ 6000 yr (Flannery & Herbig 1973). The central star was detected 100 yr ago and it has been brightening steadily since, with a sharp but transitory dimming in the optical beginning over two years ago (Jurcsik 1992). This dimming has been established as the consequence of a puff of mass emitted by the central star (Woodward et al. 1993, Stone, Kraft, & Prosser 1993), much like the puffs emitted by R CrB stars, rather than as a consequence of the rapid contraction of the central star as suggested by Iben & Livio (1992). The steady brightening of FG Sge by about 6 mag over a 100 yr period is probably due primarily to a decrease in the bolometric correction as the star expands at nearly constant bolometric luminosity. The difference between the observed expansion time scale and the 17 yr expansion time scale given by the model calculation (§3) is probably related to the fact that the model parameters are simply not appropriate for FG Sge. Over a six year period ending in 1973, the surface abundance of s-process elements increased by a factor of ~ 25 (Langer, Kraft, & Anderson 1974). This can be explained only in the context of the case 5 born again scenario (Livio & Iben 1992).

Still another relevant star is N66 in the LMC (Peña et al. 1994, 1995). This might be a star which has been caught in the act of experiencing a final helium shell flash while it is still burning hydrogen at a high rate (case 4), with its luminosity decreasing from $\sim 2.5 \times 10^4 \, L_\odot$ to $\sim 1.5 \times 10^4 \, L_\odot$ and its surface temperature decreasing from 120,000 K to 50,000 K over a 6 yr interval. This is in qualitative accord with the initial behavior of born again models of all types, but the high initial luminosity argues for a case 4 rather than a case 5 scenario. The nebular abundances (Peña et al. 1995) are He/H ~ 0.12, $\log(C/H) = 7.45, \log(N/H) = 7.95, \log(O/H) = 8.24$, and $\log(Ne/H) = 7.70$. The helium and nitrogen enhancements are probably the consequence of a second dredge up episode in a precursor of mass $\overset{\sim}{>} 5 \, M_\odot$ (e.g., Iben 1972, Kaler, Iben, & Becker 1978).

Another class of stars which show evidence of prior born again behavior are the PG 1159 stars. Werner, Heber, & Hunger (1991) find (C/He, N/He, O/He) = $(0.5, < 0.002, 0.13)$ for 4 such stars. Werner & Heber (1991) find (C/He, N/He, O/He) = $(0.5, 0.01, 0.01)$ for PG144+005, and Werner & Rauch (1994) find $X_{Ne} \sim 0.02$ in RXJ2117+3412, NGC 246, and K1-16. An explanation of these abundance mixes may require invoking an extensive phase of mass loss after a born again episode of the case 4 or 5 varieties.

In summary, there is a steadily growing body of evidence for the occurrence of the born again phenomenon in nature. The theoretical study of this phenomenon from the point of view of stellar interiors is in its infancy and the topic deserves much more attention by the theoretical astrophysics community than it has thus far received. Observers have provided a wealth of examples which indicate that many different stellar parameters are involved and it is important that theoretical models be constructed for a large range in parameter space.

This work was supported in part by the NSF grant AST91-13662 and the NASA Astrophysics Theory Program grant NAGW-2456. It is a pleasure to thank A. V. Tutukov for many useful discussions, J. Liebert and D. Kent for helping us to understand some of the mysteries of the LaTeXcompiler, Doug Swesty for helping us to manipulate postscript files, and Klaus Werner for gently persuading us to make this contribution.

Most of all, thanks to Volker Weidemann, whose magnificent work over the years inspired the organization of this conference. Happy Birthday, Volker!

References

Bond, H. E., Meakes, M. G., Liebert, J. W., & Renzini, A. 1993, in Planetary Nebulae, ed. R. Weinberger & A. Acker (Dordrecht: Kluwer), 499

Clegg, R. E. S., Devaney, M. N., Doel, A. P., Dunlop, C. N., Major, J. V., Meyers, R. M., & Sharples, R. M. 1993, in Planetary Nebulae, ed. R. Weinberger & A. Acker (Dordrecht: Kluwer), 388

Flannery, B. P., & Herbig, G. H. 1973, ApJ, **183**, 491

Fujimoto, M. Y. 1977, PASJ **29**, 331

Hamann, W. R., & Koesterke, L. 1993, in Planetary Nebulae, ed. R. Weinberger & A. Acker(Dordrecht: Kluwer), 87

Hazard, C., Terlevich, B., Morton, D. C., Sargent, W. L. W., & Ferland, G. 1980, Nature, **285**, 463

Herbig, G. H., & Boyarchuck, A. A. 1968, ApJ, **153**, 397

Hollowell, D., Iben, I. Jr., & Fujimoto, M. Y. 1990, ApJ, **351**, 245

Iben, I. Jr. 1972, ApJ, **178**, 433

—— 1984, ApJ, **277**, 333

—— 1996, ApJ, **208**, 165

—— 1995, Phys. Rep., to appear in January

Iben, I. Jr., Fujimoto, M. Y., Sugimoto, D., & Miyaji, S. 1986, ApJ, **304**, 217

Iben, I. Jr., Kaler, J. B., Truran, J. W., & Renzini, A. 1983, ApJ, **264**, 605

Iben, I. Jr., & Livio, M. 1992, ApJL, **406**, L15

Iben, I. Jr., & MacDonald, J. 1985, ApJ, **296**, 540

—— 1986, ApJ, **301**, 164

—— 1995, in progress

Iben, I. Jr., & Tutukov, A. V. 1986, ApJ, **311**, 742

—— 1995, ApJ, submitted

Iben, I. Jr., Tutukov, A. V., & Yungelson, L. R. 1995, in progress

Jacoby, G., & Ford, H. 1983, ApJ, **266**, 298

Jurcsik, J. 1992, Inf. Bull. Var. Stars, 3775

—— 1993, Acta Astron., in press

Kahn, F. D. 1983, in Planetary Nebulae, ed. D. R. Flower (Dordrecht:Reidel), 305

Kaler, J. B., Iben, I. Jr., & Becker, S. A. 1978, ApJ, **224** L63

—— 1989, in Planetary Nebulae, ed. S. Torres-Peimbert (Dordrecht: Kluwer), 411

Kippenhahn, R., Kohl, K., & Weigert, A. Zf. Ap., **66**, 58

Langer, G. E., Kraft, R. P., & Anderson, K. S. 1974, ApJ, **189**, 509

Peña, M., Torres-Peimbert, S., Peimbert, M., Ruiz, M. T., & Maza, J. 1994, ApJ, **428**, L9

Peña, M., Peimbert, M., Torres-Peimbert, S., Ruiz, M. T., & Maza, J. 1995, ApJ, in press

Pollacco, D. L., Hill, P. W., & Clegg, R. E. S. 1993, in Planetary Nebulae, ed. R. Weinberger & A. Acker (Dordrecht: Kluwer), 387

Rauch, T., Köppen, J., & Werner, K. 1994, A&A, **286**, 543

Schwarzschild, M., Härm, R. 1965, ApJ, **142**, 158

Stone, R. P. S., Kraft, R. P., & Prosser, C. F. 1993, PASP, **105**, 755

Werner, K., & Koesterke, L. 1992, in Atmospheres of Early-type Stars, ed. U. Heber & C. S. Jeffery, (Berlin: Springer), 305

Werner, K., & Heber, U. 1991, A&A, 247, 476

Werner, K., Heber, U., & Hunger, K. 1991, A&A 1991, **244**, 437

Werner, K., & Rauch, T. 1994, A&A, **284**, L5

Woodward, C. E., Lawrence, G. F., Gehrz, R. D., Jones, T. J., Kobulnicky, H. A., Cole, J., Hodge, T., & Thronson, H. A. 1993, ApJL, **408**, L37

What is Wrong With Stellar Modeling

Italo Mazzitelli

Istituto di Astrofisica Spaziale C.N.R.,
c.p.67 I-00044 Frascati, Italy,
E-mail: aton@hyperion.ias.cnr.fra.it

Abstract: Numerical stellar modeling dates back since at least 45 years, when the first generation of digital computers became available. By now, we have a reasonably good *qualitative* understanding of what is going on in the vast majority of the cases, and we also *quantitatively* know many details on simple stellar structures (even if the solar neutrino problem is always a warning to be frequently reminded). In the following, a short review of the basic problems with stellar modeling will be provided, in the attempt of convincing the reader that, switching from *local* stellar models to *non–local, turbulent* ones, will be a major, generational leap, which will greatly extend the boundaries of our *quantitative* understanding of stellar evolution

1 The Basic Approximation: Spherical Symmetry

The present generation of stellar models takes into account only *spherically symmetric* configurations. Attempts at building two–(or three–)dimensional stellar structures are performed only in a few peculiar cases, especially when sudden development of turbulent chemical mixing during explosive phases can strongly influence the evolution itself. The first results suggest that full three–dimensional models would be required, since the growth in time of turbulence is orders of magnitude faster in this last case than with two–dimensional simulations.

Three–dimensional computations are presently severely inhibited by numerical problems and, mainly, by the huge amount of computer time required also for very short evolutionary sequences, but even if we limit our attention to two–dimensional structures, to account for the effect of rotation only (with all its drawbacks on chemical mixing), the problem still remains terribly time–consuming. Approximated treatment of rotation in the so–called one–point–five dimensional Kippenhahn & Thomas (1970) approximation, has however shown that, at least for the low rotational velocities representative of the majority of the cases, the structural and evolutionary effects due to stellar rotation are of little matter (Pinsonneault et al. 1989).

Assuming spherical symmetry is then certainly wrong in principle, but not so much in practice as to seriously worry stellar modelists. At present, there are in fact other pieces of input physics upon which larger uncertainties are still weighing. Let us discuss the more important of them.

2 The Equations of Stellar Structure

The first two equations governing a self–gravitating equilibrium sphere come out immediately from Poisson's equation, that is:

$$\frac{dP}{dr} = -\frac{GM_r\rho}{r^2} \tag{1}$$

(hydrostatic equilibrium as a whole) where P is the total pressure, r is the radius, G is the gravitational constant, M_r is the mass inside the radius r and ρ is obviously the density, and:

$$\frac{dM}{dr} = 4\pi r^2 \rho \tag{2}$$

(mass continuity). In the radiative equilibrium case, we can also write:

$$\frac{dT}{dr} = \frac{3}{4ac} \frac{\bar{\kappa}\rho}{T^3} \frac{L}{4\pi r^2} \tag{3}$$

being κ the radiative opacity coefficient, and L the total luminosity, for which a fourth, obvious differential equation can be written, that is:

$$\frac{dL}{dr} = 4\pi r^2 \epsilon \tag{4}$$

where ϵ is the energy generated per unit mass and time. Since we have four equations in seven unknowns (plus one independent variable, which is usually assumed to be the mass), we still need three more equations, which are:

$$P = P(\rho, T, [\mu]), \text{ but also } C_p = C_p(\rho, T, [\mu]) \text{ etc} \tag{5}$$

where $[\mu]$ stands for general dependence on the chemical composition, C_p is the specific heat at constant pressure so that, more in general, we are dealing with *thermodynamics* broadly speaking, rather than with a bare *equation of state*.

Then we have the equation for the radiative (and also *conductive*, given the correct formulation for κ) opacity:

$$\bar{\kappa} = \bar{\kappa}(\rho, T, [\mu]) \tag{6}$$

and for the energy generation (nuclear, gravothermal, neutrinos) rate

$$\epsilon = \epsilon(\rho, T, [\mu], \dot{\rho}, \dot{T}) \tag{7}$$

and the game is done. It is worth noting that, up to this point, we have discussed only the *completely radiative* case, that is: either a fully radiative structure, or a region very far from any convective boundary, where any motion of matter induced by convection and overshooting has definitely ceased.

3 Thermodynamics, Opacity and Energy Generation

In the above scheme, the differential equations are straightforward, and the physical uncertainties weigh upon the last three equations only. Let us examine them in some detail.

3.1 Thermodynamics

At least for stars ending their lives as white dwarfs (WDs), that is far from the quantum–ionic phases one can meet just before the supernova explosion (or in the extreme cooling of WDs, when $Log L/L_\odot \leq -5$), thermodynamic computations are being carried on in a variety of physical descriptions (minimization of Free Energy, Monte–Carlo simulations and so on). Even if extensive tabulations of thermodynamical quantities, covering many chemical compositions and reaching up to the highly non-ideal gas regime are not yet available, there are no reasons —in principle— why they should not become available in a very few years (for a wide discussion on the subject, see Van Horn 1994).

Some problems are instead still present with the fractional crystallization in the presence of eutectic mixtures (like $He/C/O$ mixtures in WDs, see Segretain and Chabrier 1993). Work is in progress also on this front but, perhaps, definitive results will have to wait for quantum Monte–Carlo simulations, still to come. However, this problem affects only the cooling of WDs and some pre-supernova (Type I) cases, and not the more general features of stellar evolution.

3.2 Radiative Opacities

In this field, large progresses have been made only recently, with the OPAL project (Rogers and Iglesias 1992) for the evaluation of the radiative opacities (always in the ideal–gas regime) for $T \geq 6000K$, and with the low–T opacities including molecules by Kurucz 1991 and Alexander 1994. At least for metal poor mixtures, these two last seem by now coincide or quite so.

The present situation is then that, at least in ideal–gas conditions, and for relatively low metal abundances, physically reliable tables of radiative opacities are by now available, both at large and low–T. Some problems are instead still present at low–T, with Pop. I metal abundances (D'Antona and Mazzitelli 1994). Completely different is the case in non–ideal gas regime, where we have only a few tables for pure–H or pure–He compositions, absolutely unsatisfactory for mixed WDs, for the low main sequence (MS) and for the subdwarfs. Progress in this field will be probably slower than for the thermodynamics, and this can still have an influence on a number of interesting cases in stellar evolution.

3.3 Nuclear Cross Sections

So far as the nuclear energy generation coefficient alone is concerned, the present situation is reasonably satisfactory. This is partly due also to the functional dependence of the cross sections $\sigma \propto T^\beta$, where $\beta \gg 1$, so that a very small feedback adjustment in T is required to adjust quite a lot the energy generation coefficient.

The above discussion is obviously valid only for the reactions up to the $^{12}C+\alpha$ and, in any case, does not account for some peculiar problems (high-energy solar neutrinos). Moreover, it is not warranted that also the *chemical evolution* is correctly evaluated since, in this case, an error in the cross sections linearly propagates to the burning rates. However, the general feeling is that the present generation of theoretical models of low and intermediate mass stars does not largely suffer from uncertainties in this field.

3.4 Neutrinos and Electron Conduction

These two items have been put together, since it is thanks to the efforts of the Japanese astrophysical school (Itoh et al. 1984, 1992) that by now we dispose of quite excellent physical inputs for both of them. A radical revision of either neutrino rates or electron conduction coefficients seems to be unlikely in the near (or far) future.

4 Turbulent Convection Broadly Speaking

Convection represents today the main uncertainty in stellar modeling, and is the frontier to be painfully and slowly pushed ahead. In recent years, some initial progresses have been made also in this field, and more progresses are expected in the coming years.

The main problem is that any characteristic number, either the Reynolds one (Re) or the Prandtl one (Pr), will show that convection in stars is bound to be highly turbulent, not mainly because of intrinsically low dynamical viscosity, but due to the *huge* size of any convective region, when compared to the laboratory standards. This situation allows the development of an enormous range of convective scales (usually referred to as "eddies").

Turbulent convection should be treated according to the Navier-Stokes equations (NSe), both from the point of view of the heat transfer (overadiabatic convection), and of the chemical mixing at the convective boundaries (overshooting). Unfortunately, some characteristics of NSe have up today prevented their application in general stellar evolutionary codes (even if some attempts have been performed in some peculiar cases). The main problems with NSe are:

— non locality. They contain terms of the type $(\partial/\partial r)$ which, for a stellar structure code, are at least a complication;

— they cannot be linearized, so that one has to invent some physically acceptable approximation to close them, if the goal is to get an analytical solution;

— even so, they lead in the best of the cases to *many* further, highly nonlinear differential equations. The Runge–Kutta scheme is certainly bound to fail in numerically integrating them; will the other numerical integration procedures for stellar evolution (mainly the Newton–Raphson technique) work?

Due to the above reasons, and also to the fact that only in recent years computing stellar evolution has become reasonably fast and cheap, thanks to a number of technological advancements in the hardware for floating point applications, we have up today resorted to *local* approximations, usually taken from engineering–type turbulent flows, giving rise to expressions of the kind:

$$\left(\frac{dT}{dr}\right)_{conv} = f(\rho, T, [\mu], L, \overline{\kappa}) \tag{8}$$

The drawback of this procedure (apart from the physical reliability of the model behind the approximation) is that, making convection a *local* mechanism, i.e.: decoupling the computation of the stellar structure from turbulence, causes the lost of any intrinsic convective scale length, such as the scale for the eddies spectral distribution, and the e-folding distance for the turbulent velocity field outside the formally convective boundaries. If we want to compute stellar models, we have then to reintroduce in the models, on the basis of some physical reasoning, al least the two above scale lengths, which we usually refer to as: the *mixing length* and the *overshooting distance*.

As a simple comment, it is amazing to discover how many stellar modelists seem to be convinced that these two scale lengths lie in the kingdom of intrinsic *incalculability*. This is an incorrect point of view, since the lack of their knowledge is simply the result of the *local* approximation; models computed starting from first principles (i.e.: with the NSe coupled with the stellar structure ones) will spontaneously provide both scale lengths. In the following, the problems connected with the evaluation of the overadiabatic gradient will be addressed, and also a short discussion on overshooting will be included.

5 The Mixing Length Theory

The mixing length theory (MLT) for the evaluation of the overadiabatic gradient is probably the most successful item of input physics ever adopted in stellar modeling, since it has been in use for about 40 years, without any major modification. It was born in Kiel, at the time of Unsöld, and Erika Böhm–Vitense (1958), for the first time gave a viable recipe that allowed stellar modelists to evaluate "reliable" values of stellar radii or T_{eff}.

The theory is well known; it approximates the whole turbulent energy spectrum $E(k)$, where the geometrical size of an eddy is $l \propto k^{-1}$, with $E(k) \sim \delta(k - k_0)$, that is: $l = \Lambda\pi/k_0$, where Λ is the so–called **mixing length**, to be fixed according to physical considerations.

In a star, there are at least two "sensible" local scale lengths, which in principle could be chosen as representative of the convective mixing length, that is:

— the radial depth z inside a convective region (that is: a scale length directly related to convection itself) or:

— the pressure scale height H_p (that is: a scale length not related to convection, but to the hydrostatic equilibrium).

First attempts of use of the MLT (Böhm 1963) were performed with $\Lambda = z$ since, after all, at any given point inside a turbulent region the size of the largest convective eddies is limited by the distance to the top of convection. However, it soon became clear that, with this choice, it is absolutely impossible to fit the observed solar radius. The switch to the parameterization $\Lambda = \alpha H_p$ occurred very soon, where the free parameter α is tuned such to fit the observed solar radius, and then applied to any other stellar structure. The present situation is that, depending on several items of input physics, the various evolutionary codes work with $1.5 \leq \alpha \leq 2.2$.

Apart from some principle inconsistency (the MLT has been developed in the Boussinesq approximation, that is: it requires that pressure does not change along the eddy; sometimes one finds Λ larger than the thickness of the convective region, etc.), this *parametric* use of the MLT has led to the (incorrect) habit of trying to reset, by means of the free parameter α, almost *all* the uncertainties present in stellar modeling, mainly thermodynamics and opacities. During the years, papers appeared in the literature in which the solar tuning of the MLT, performed *with three or more decimal digits*, was blindly applied to Pop. II stars, red giants etc, where the uncertainties in the input physics are completely different from those present in the sun and in which, then, the solar tuning of the MLT was completely meaningless (the Author was not among the sinners, and yet decided avoiding quoting references).

In more recent years, this obvious abuse of the MLT was finally recognized, even if only implicitly. Some observed features could not be fitted by the solar tuning of the MLT (low–mass stars, subdwarfs, red supergiants etc.), and ad hoc values of α for various classes of stars and evolutionary phases were used. However, this procedure led to the lost of *predictive power* of the MLT, since the tuning was performed "a posteriori", to fit the observations.

6 The CM Convective Model

Recently, Canuto and Mazzitelli (CM 1991) made available, for the purpose of general stellar modeling, a new tool for (locally) evaluating the overadiabatic gradient. The CM model, based on modern phenomenological and theoretical approaches to turbulence, allows the entire spectral distribution of eddies, and assumes as scale length at any given point in the structure the thickness of the turbulent layers at that point. In principle, then, in the CM model there are no free parameters (or quite so, see later) and, when the more reliable and updated physical inputs are included in the computations, the fit to the observed solar radius within better than 1% is achieved, and also a correct *internal stratification* for the solar temperature arises, as shown by helio–seismological comparisons.

Actually, since the scale length corresponds to the whole turbulent region, and not only to the formally convective one (if there is overshooting, it must be included in Λ), the model retains a free parameter: the amount of overshooting, which is not evaluated according to first principles, since also the CM model is a local one. However, since overshooting cannot exceed $0.2 \div 0.3 H_p^{top}$, where

H_p^{top} is the pressure scale height at the top of the convective region, the tuning allowed for the CM model is very narrow; in the solar case, the T_{eff} can be tuned within no more than $\pm\, 1\%$.

The CM model has been successfully applied to a variety of cases, and is a plausible candidate to substitute the MLT in stellar computations. It is however still a *local* treatment, so that it is not yet able to give all the answers about turbulent convection in stars. An application of the CM model, compared to the MLT, is shown in Fig. 1, where the deepening of surface convection in H-rich WDs is shown. With the CM model, the deepening of convection is definitely sharper than with the MLT, and this can be probably related to the observed sharp blue edge of ZZ Ceti pulsators.

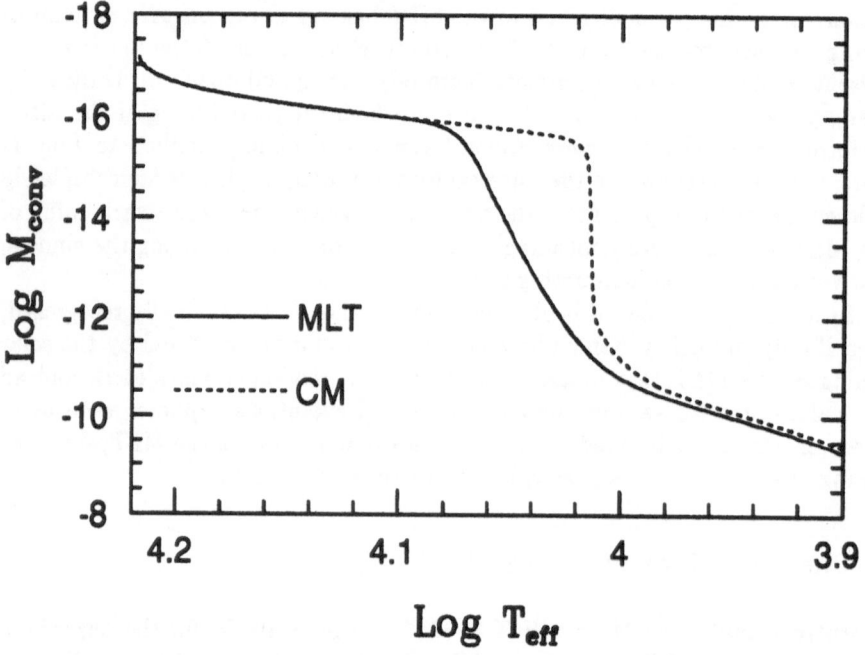

Fig. 1. Deepening of surface convection for a H-rich white dwarf of 0.55 $M_\odot$, with the MLT and the CM convection model. In this last case, deepening is much sharper than with MLT.

7 Overshooting

A local model, like the MLT or the CM one, unavoidably provides velocity and acceleration of the convective elements equal to zero at the formal convective boundaries, defined according to the Schwarzschild criterion. It is then obviously impossible to gather information about convective overshooting from these models.

During the years, a number of models for overshooting have been suggested, based on low order closures of the NSe, or simply on energy or momentum conservation requirements, to be applied to standard, local stellar models. The more widely used of these models is the one by Roxburgh (1989). What seems to appear from these treatments, is that they provide quite large overshootings (usually in the range $1.5 \div 2H_p$), whereas tests performed with parametric overshooting suggest, by comparison with observations (Maeder and Meynet 1991), much lower values ($0.2 \div 0.3H_p$).

No better luck has had an alternative approach to boundary mixing: the so called "semiconvection" (Castellani et al. 1985), which is based upon absolute continuity of the actual temperature gradient at the convective boundaries, and chemically-driven equality between radiative and adiabatic temperature gradients around a convective region. Since both these hypotheses are completely ad hoc, and are not based on any physical ground, semiconvection turns out to be only a numerical recipe, and a quantification of our ignorance about the real occurrences at the convective boundaries.

The case for overshooting is then worse than the one for overadiabatic convection, since we are left with a parametric approach only. The situation will not change until a new generation of stellar models, including full turbulence according to the NSe, will be built up.

8 Numerical Simulations

The possibility of including in stellar modeling the results of full numerical simulations of the NSe (Lydon et al. 1992), has been frequently suggested and, at least in a few cases, some attempts have been performed.

The problem here is that the number N of degrees of freedom of a 3-D turbulent flow is $\propto Re^{9/4}$. For the sun, $N \sim 10^{36}$, to be compared to the 10^9 that represents the current limit for the present generation of computers. So, also these numerical simulations are limited to the peak of the spectral distribution of energy (large eddy simulations), and have to resort either to large viscosities ($Pr \sim 1$, to be compared with $Pr \sim 10^{-10}$ in the sun) and/or to the fit to simple and questionable analytical models (sub-grid models) for the huge range of eddies not directly covered by numerical integration.

Unfortunately, both alternatives have been shown very poor when applied to stars. The large viscosity case is obviously unrealistic, and the sub-grid model approach is, in the best of the cases, qualitatively representative of the occurrences very deep in a convective region, far from the boundaries, and so it has little to do with both overadiabaticity and overshooting, which are boundary phenomena. This situation is not expected to improve very soon.

9 Analytical Models

The last possibility, which seems to be promising enough, is to try to perform some kind of analytic closure of the NSe, based on some sound physical hypotheses, and then integrating the whole set of equations, including the stellar equilibrium ones.

This line has been pursued for the first time by Xiong (1985), who performed a downgrade of the third order momenta in the NSe, and came out with a set of ten differential equations for the stellar structure. He was also able to numerically integrate them, with a Newton-Raphson procedure, in a number of cases.

Unfortunately, the non-locality in the NSe is just in the third order momenta, which have to be retained if overshooting has to be deduced from first principles. Xiong's results, then, still retained some *local* features, depending on an MLT-like formulation.

However, these first results were highly interesting and encouraging, showing not only that largely non-linear systems of this kind can still be numerically integrated with standard techniques, but also for their first physical results. It is for instance worth noting that, due to the large *negative* convective fluxes just beyond the formal convective boundaries, stars can sustain quasi–discontinuities in the gradients (thus contradicting the basic assumption of semiconvection).

A further step ahead was then performed by Canuto (1992) who, in the hypothesis of quasi-normality of the fourth order momenta (Reynolds approximation), was able to retain all the third order momenta, and still provide a network of equations (the exact number depends on some approximations which can or can not be performed, according to the cases to be studied, but ranges between 9 and 12) which, once joined to the four standard equations of stellar equilibrium, should provide a complete description of a star starting from first principles, and spontaneously answering all the questions related to overadiabaticity and overshooting.

The conditional in the above paragraph derives from the formidable non–linearity of the equations which, when applied to stars, have up today resisted any attempt of numerical integration based upon standard techniques, whereas it was possible to integrate them for the case of the terrestrial atmospheric boundary layer, where they have provided satisfactory agreement with the observations. Maybe more sophisticated numerical techniques will have to be applied to the solution of the problem. It is worth explicitly noting that these new sets of differential equations, in spite of their intimidating look, are not expected to require much more computer time to be solved, than the current four–equations set. The problem is then in the software, and not in the hardware, as it is instead for the full numerical simulations.

10 Conclusions

On the basis of the above discussion, the conclusions of the present paper are moderately optimistic. For the most of the physical inputs in stellar modeling, we now have a sufficiently sound and settled understanding as to claim that the results of our computations, at least for a wide class of stellar structures, begin to be *quantitatively*, not only *semi–quantitatively*, reliable. Once solved the numerical integration problems posed by the new networks of differential equations including the NSe, a *second generation* of stellar model, with no parametric dependencies, will greatly increase our understanding of stellar structure and evolution.

References

Alexander, D.R., 1994, private communication

Böhm K.H., 1963, ApJ. 137, 881

Böhm–Vitense E., 1958, Zs. Ap. 46, 108

Canuto V.M., 1992, ApJ 392, 218

Canuto V.M., Mazzitelli I., 1991, ApJ 370, 295

Castellani V., Chieffi A., Pulone L., Tornambe' A., 1985, ApJ 296, 204

D'Antona F., Mazzitelli I., 1994, ApJS 90, 467

Itoh N., Kohyama Y., Matsumoto J., Seki M., 1984, ApJ 258, 758

Itoh N., Mutoh H., Hikita A., 1992, ApJ 395, 622

Kippenhahn R., Thomas H.C., 1970, in Stellar Rotation. D. Reidel Publ. Co., Dordrecht–Holland, 20

Kurucz R.L., 1991, in Stellar Atmospheres: Beyond the Classical Models. L. Crivellari, I. Hubeny, D.G. Hummer eds, NATO ASI series (Dordrecht: Kluwer), 441

Lydon T.J., Fox P.A., Sofia S., 1992, ApJ 397, 701

Maeder A., Meynet G., 1991, A&A 89, 451

Pinsonneault M.H., Kawaler S.D., Sofia S., Demarque P., 1898, ApJ 338, 424

Rogers, F.J., Iglesias, C.A.: ApJS 79, 507

Roxburgh I.W., 1989, A&A 211, 361

Segretain L., Chabrier G., 1993, A&A 271, L13

Van Horn H.M., 1994, in The Equation of State in Astrophysics. IAU Colloquium 147 ed. G. Chabrier and E. Schatzman, Cambridge Univ. Press, 1

Xiong D.R., 1985, A&A 150, 133

Transformation of AGB Stars into White Dwarfs

Thomas Blöcker

Astrophysikalisches Institut Potsdam, D-14473 Potsdam, Germany

1 Introduction

One important aspect of the transformation of AGB stars into white dwarfs is the treatment of mass loss on the AGB and beyond. On the one hand, the AGB mass-loss history determines the star's internal structure reached at the tip of the AGB and therefore the fading speed along the cooling part of the post-AGB evolution. On the other hand, as discussed in Schönberner (1983), the transition times from the AGB to the central-star region depend sensitively on the mass-loss rates employed beyond the AGB.

Based on stellar evolution calculations for initial masses between 1 and $7M_\odot$ of Pop. I $[(Y, Z) = (0.24, 0.021)]$ which consider the whole evolution from the main sequence towards the stage of white dwarfs we will briefly discuss some consequences of the treatment of mass-loss for the post-AGB evolution.

2 Mass loss on the AGB and beyond

Along the Red Giant Branch and during central He-burning we applied Reimers' (1975) rate $\dot{M}_R$ yielding, however, for $M > 2M_\odot$ a total mass loss of only 1-3%. On the AGB we adapted the dynamical calculations of Bowen (1988) getting $\dot{M}_B = 4.8 \cdot 10^{-9} \cdot (L^{2.7}/M_{ZAMS}^{2.1}) \cdot \dot{M}_R$. The resulting final masses agree well with the initial-final mass relation of Weidemann (1987). At the tip of the AGB mass-loss rates of $\approx 10^{-4} M_\odot$/yr or even more are reached (Blöcker 1993, 1994a).

Observations indicate that the mass-loss rates of central stars of planetary nebulae are up to several orders of magnitude below those of the immediately preceding AGB evolution (Perinotto 1989). Therefore, mass-loss has to decrease strongly during the transition between the AGB and the central-star regime. However, up to now it is not known – neither from observations nor from theory – how and at which temperature (range) the strong mass-loss decrease takes place.

We have reduced the AGB mass-loss rates proportional to the (decreasing) radial pulsational period P_0 between $P_0 = 100$ d and 50 d in such a way that the Reimers rate was reached at $P_0 = 50$ d. This point corresponds to our age zero of the central-star evolution. Then, the Reimers rate was kept until the radiation-driven wind theory according to Pauldrach et al. (1988) becomes applicable ($T_{eff} \gtrsim 20000\,K$). The respective mass-loss rates can be adapted like: $\dot{M}_{CPN} = 1.3 \cdot 10^{-15}\,L^{1.9}$. This mass-loss law was used for the remaining part of the post-AGB evolution. The corresponding influence on the evolutionary speed depends on the respective ratio of the burning rate to mass-loss rate and is only important for massive remnants (Blöcker & Schönberner 1990, Blöcker 1993).

The recent post-AGB calculations of Vassiliadis & Wood (1994) consider the results of Pauldrach et al. (1988) in a somewhat different adaption (see Blöcker 1994b). The strong AGB winds were stopped, however, already when the model has moved off the AGB by $\Delta \log T_{eff} = 0.3$ corresponding to $T_{eff} \lesssim 5000\,K$. Since Vassiliadis & Wood (1994) reduced the AGB mass-loss rates abruptly and, particularly, earlier, i.e. closer to the AGB, than we did, their transition times are larger than our ones. Our treatment of mass-loss leads preferably to H-burning models.

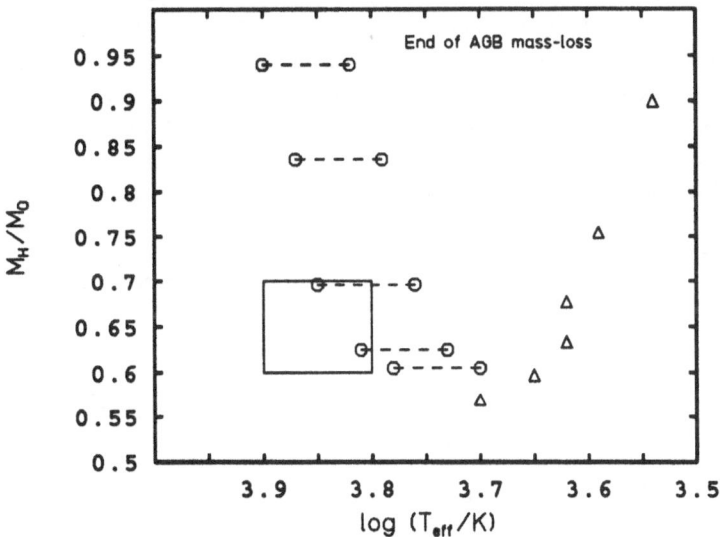

Fig. 1. Core mass ($\approx$ final mass) vs. effective temperature for the end of the strong AGB mass-loss. $\bigcirc$: present work (reduction between $P_0 = 100$ and $50\,d$) — $\triangle$: Vassiliadis & Wood (1994)(abrupt decrease). The rectangle refers to corresponding results of hydrodynamical pulsation calculations (Zalewski 1994).

Hydrodynamical pulsation calculations of Zalewski (1994) indicate that at least for low-mass post-AGB models (e.g. $0.6 M_\odot$) mass-loss should "stop" somewhere between $\log T_{eff} \approx 3.8...3.9$ which nicely coincides with the end of our mass-loss reduction at $P_0 = 50\,d$. Fig. 1 shows our temperatures and core masses at the end of the AGB mass-loss in comparison with the results of Vassiliadis & Wood (1994) and the pulsational calculations of Zalewski (1994).

3 Fading timescales

The seemingly well established statement that remnants fade the faster the more
massive they are has been questioned since a few years. Contrary to the well
known results of Paczynski (1970) and Wood & Faulkner (1986), Blöcker &
Schönberner (1990) showed that a $0.84M_\odot$ can fade *slower* than a $0.61M_\odot$ one.
Whereas their models agree with the initial-final mass relation of Weidemann
(1987) both the models of Paczynski (1970) and Wood & Faulkner (1986) are
based on one initial mass being not consistent with the empirical relation. Later
on, Blöcker (1993) demonstrated that the post-AGB evolution of two massive
models of *equal* remnant mass ($0.84M_\odot$) but *different* initial masses (3 and $5M_\odot$,
resp.) gives completely different fading timescales due to the different internal
structure of the models. This emphasizes the importance of a consistent treatment of the AGB evolution for reliable post-AGB timescales.

Recently, Vassiliadis & Wood (1994) as well as Blöcker (1994b) presented sets
of evolutionary post-AGB tracks. Both calculations are based on models which
agree with the empirical relationship of Weidemann (1987). However, whereas
Vassiliadis & Wood (1994) found that the higher the central-star mass, the lower
is the luminosity on the fading part at a post-AGB age of $\approx 10^4$yrs, the results
of Blöcker (1994b) confirmed that more massive remnants can fade much slower
than hitherto assumed and even slower than less massive ones (see Fig. 2-3).

These contradictory results of evolutionary calculations can be explained in
terms of the stars' internal structure: During the AGB evolution the degeneracy

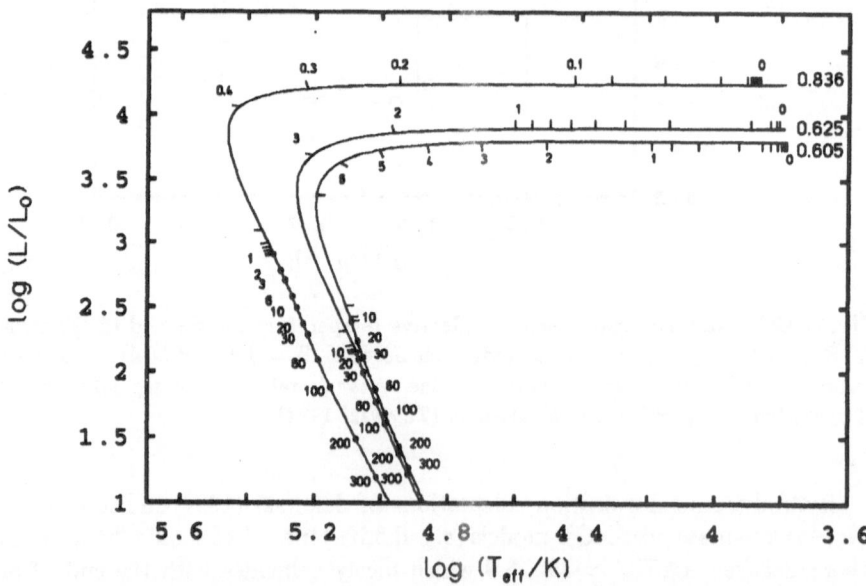

Fig. 2. Tracks for H-burning post-AGB models with $(M_{\mathrm{ZAMS}}, M_{\mathrm{H}}) = (3M_\odot, 0.605M_\odot)$,
$(3M_\odot, 0.625M_\odot)$ and $(5M_\odot, 0.836M_\odot)$. Time marks are in units of 10^3 yrs.

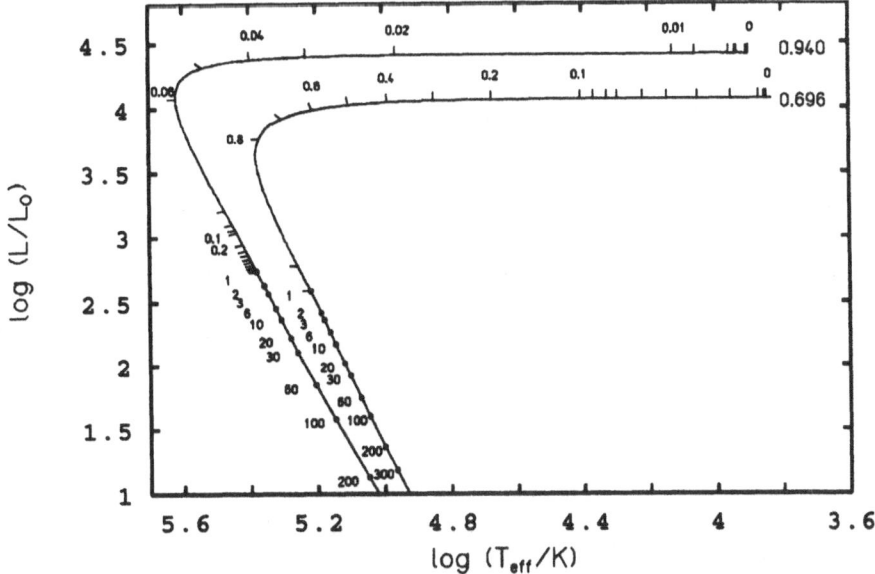

Fig. 3. Tracks for H-burning post-AGB models with $(M_{\text{ZAMS}}, M_{\text{H}}) = (4M_\odot, 0.696M_\odot)$ and $(7M_\odot, 0.940M_\odot)$. Time marks are in units of 10^3 yrs.

of the core increases and an increasing fraction of the energy which is released by contraction is used up by raising the Fermi energy of the electrons, being no longer available for the thermal content of the star. Thus, the more degenerate a model leaves the AGB, i.e. the more compact and cooler the interior, the faster is the final fading. Since for a given initial mass the mean degeneracy increases with increasing AGB duration, i.e. final mass, it is no suprise that the models of Paczynski (1971) and of Wood & Faulkner (1986) fade faster with increasing core mass because they are based on a single initial mass. On the other hand, for a given final mass the models belonging to larger initial masses fade slower since they are hotter and less compact. This explains the different fading times in the case of both $0.84M_\odot$ remnants in Blöcker (1993).

Since Vassiliadis & Wood (1993) employed lower mass-loss rates along the AGB than we did, their remnants spend a longer time on the TP-AGB and are more degenerate for a given initial mass. Thus, their models fade faster than those of Blöcker (1994b). Fig. 4 shows the luminosity vs. core mass ($\approx$ remnant mass) for both evolutionary sets at an post-AGB age of $t = 10^4$ yr. At this age the more massive models of Vassiliadis & Wood (1994) are up to ≈ 0.4 dex less luminous than our ones, and an accumulation of objects can be expected for $\log(L/L_\odot) \approx 2.1...1.9$ instead for $\log(L/L_\odot) \approx 2.5...2.3$.

The position of He-burning models is also shown in Fig. 4. Whereas they evolve much slower than H-burning ones for low-mass stars the situation gets again complex for larger masses. For $t = 10^4$ yr it is hard to distinguish between the He-burning and the slowly fading H-burning model with $0.84M_\odot$. Both have

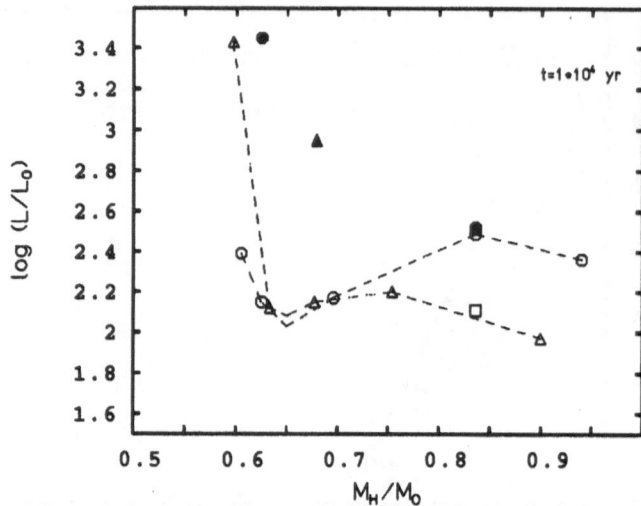

Fig. 4. Luminosity vs. core mass at $t = 10^4$ yr for H-burning (open symbols) and He-burning models (filled symbols). Triangles: Vassiliadis & Wood (1994) [H burner: Z=0.016, He burner: Z=0.008]; circles: this work; square: fast fading $0.84 M_\odot$ remnant belonging to $M_{ZAMS} = 3 M_\odot$ (instead to $5 M_\odot$).

nearly the same luminosity. Thus, in the Vassiliadis & Wood scenario massive and luminous central stars are most likely He-burning objects whereas our results suggest that they can also be explained by H-burning remnants!

Acknowlegdements: This work was supported by the DFG grant Scho 394/1. I thank Jan Zalewski for providing us with the data of his pulsational calculations.

References

Blöcker, T.: 1993, *White Dwarfs*, ed. M.A. Barstow, Kluwer, Dordrecht, p. 59
Blöcker, T., 1994a,b, A&A, in press
Blöcker, T., Schönberner, D., 1990, A&A 240, L11
Bowen, G.H., 1988, ApJ 329, 299
Paczyński, B., 1971, Acta Astron. 21, 417
Pauldrach, A.; Puls, J., Kudritzki, R.P., Méndez, R., Heap, S.R.: 1988, A&A 207, 123
Perinotto, M.: 1989, *Planetary Nebulae*, ed. S.Torres-Peimbert,Reidel,Dordrecht, p.293
Reimers, D., 1975, Mém. Soc. Sci. Liege 8, 369
Schönberner, D.: 1983, ApJ 272, 708
Vassiliadis, E., Wood, P.R., 1993, ApJ 413, 641
Vassiliadis, E., Wood, P.R., 1994, ApJS 92, 125
Weidemann, V.: 1987, A&A 188, 74
Wood, P.R., Faulkner, D.J., 1986, ApJ 307, 659
Zalewski, J., 1994, private communication

The Role of G in the Cooling of White Dwarfs

Enrique García-Berro,[1] *Margarita Hernanz,*[2] *Jordi Isern*[2] *and Robert Mochkovitch*[3]

[1] Dept. de Física Aplicada, Universidad Politécnica de Cataluña, Jordi Girona Salgado s/n, Módul B-4, Campus Nord, 08034 Barcelona, Spain
[2] Centre d'Estudis Avançats (CSIC), Camí de Santa Bàrbara, 17300 Blanes, Spain
[3] Institut d'Astrophysique de Paris (CNRS), 98bis bvd. Arago, 75014 Paris, France

1 Introduction

The value of the constant of gravitation, G, is relatively poorly known. The reason for that is the small strength of gravitational forces and the small relevance of G to the rest of physics. In addition the idea of an expanding Universe has lead to the idea of a gravitational constant evolving with it. Therefore, if G varies, it should change with a time scale proportional to the Hubble time, $\dot{G}/G = \sigma H_0$, where H_0 is the Hubble constant and σ is a parameter depending upon the gravitational theory considered. Of course, in Newtonian and General Relativity theories G is a constant and $\sigma = 0$.

Several constraints to the rate of change of the constant of gravitation have been obtained up to now (see Will (1984) for an extensive review). The best constraint is that obtained from the data of the binary pulsar PSR1913+16 which gives a limit $\dot{G}/G = -(1.10 \pm 1.07) \times 10^{-11}$ yr^{-1} (Damour et al., 1988). The major factor limiting the accuracy of this estimate is the proper motion adopted for the pulsar.

White dwarf stars provide an independent method for measuring a change of G. There are two reasons for that: First, when they are cool enough, their energy is entirely of gravitational and thermal origin and any change of the value of G modifies the energy balance which, in turn, translates into a change of the luminosity. And second, since they are long lived objects, ~ 10 Gyr, even extremely small values of the rate of change of G can become prominent.

2 The role of G in white dwarf cooling

Since we are only interested in the oldest white dwarfs, and in order to avoid the complicated behavior of hot white dwarfs, we will study the evolution of those white dwarfs having luminosities $L \leq 10^{-1} L_\odot$. Therefore, neutrino losses and thermonuclear burning in the outer layers can be neglected and the star luminosity arises only from the thermal and gravitational energy release. Furthermore it is possible to assume, except for the very outer layers, that the star is isothermal. Under these hypothesis, the characteristic cooling time can be easily obtained from the binding energy B (Díaz–Pinto et al., 1993; Hernanz et al., 1994). The change of the gravitational energy of the white dwarf taking into account a possible variation of the gravitational constant, can be derived from the equations of hydrostatic equilibrium, adding a new term to the equation of energy conservation:

$$L = -\frac{dB}{dt} + \frac{\dot{G}}{G}\Omega \qquad (1)$$

which reduces to the usual form when G is taken constant. This equation can be written in a form that is directly related to the luminosity function (Díaz–Pinto et al., 1993; Hernanz et al., 1994):

$$\tau_{cool} = \frac{dt}{d\ln L} = -\frac{\left(\frac{\partial B}{\partial T}\right)_G \frac{dT}{dL}}{[1 + \frac{\dot{G}}{G}\frac{B}{L}(\beta - \frac{\Omega}{B})]} \qquad (2)$$

where τ_{cool} is the characteristic cooling time, dT/dL can be obtained from a model of envelope connecting the isothermal core with the photosphere (see Hernanz et al. (1994) for further details) and β is defined as

$$\beta = \frac{G}{B}\left(\frac{\partial B}{\partial G}\right)_T \qquad (3)$$

Notice that once more this expression reduces to the usual form when G is taken constant. In this paper the envelope models by Wood and Winget (1989) have been adopted.

Figure 1 displays the age of a white dwarf of mass $M_{WD} = 0.6$ versus the luminosity obtained integrating equation (1) with different values of $\dot{G}/G$. When integrating the cooling sequences we have assumed that there is a previous stratification of the C/O mixture (Mazzitelli & D'Antona, 1986) and that the phase diagram of the C/O binary ionic mixture is that of Segretain & Chabrier (1993). The allowance of a variation of G (in the following we will assume $\dot{G} < 0$ for simplicity) accelerates the cooling process, specially at low luminosities. This behavior can be easily understood from the properties of the factor F, defined as:

$$F = 1 + \frac{\dot{G}}{G}\frac{B}{L}(\beta - \frac{\Omega}{B}) \qquad (4)$$

appearing in the denominator of equation (2). Notice that when the white dwarf cools, B remains almost constant whereas the luminosity vanishes. Therefore any possible variation of G will be amplified provided that $\beta \neq \Omega/B$. This can be proved analitically using the virial theorem and taking into account the following contributions to the equation of state: the ideal (Fermi) energy of the electron gas, the ideal thermal energy of ions and the Madelung energy. In the non–relativistic limit we obtain

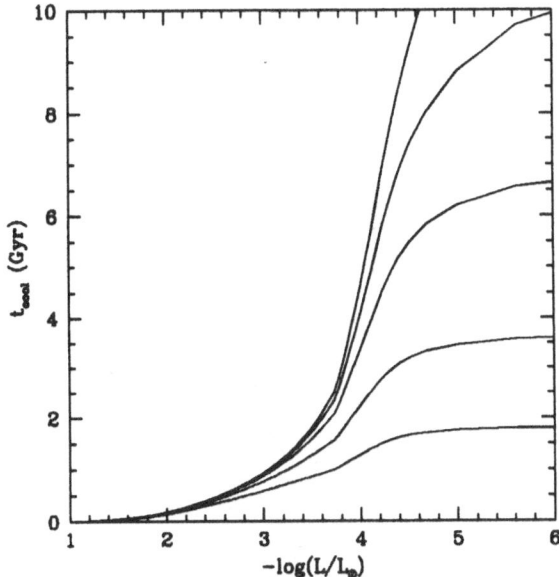

Fig. 1. Cooling times for a 0.6 $M_\odot$ white dwarf for different values of $\dot{G}/G$. From left to right: $\dot{G}/G = 0$, -10^{-11}, $-3\ 10^{-11}$, -10^{-10}, and $-3\ 10^{-10}$ yr^{-1}

$$F = 1 - \frac{\dot{G}}{G} \frac{2E_i}{L} \tag{5}$$

Within this approximation the responsible of the departure of F from zero is the total ionic thermal content. Also noticeable is the absence of effects of the total Madelung energy.

Figure 2 displays the behavior of F for several values of $\dot{G}/G$ for a 0.6 $M_\odot$ white dwarf computed with a detailed equation of state (Segretain et al., 1994).

3 Luminosity functions

With these cooling sequences we have computed the luminosity function of white dwarfs according to Díaz–Pinto et al. (1993) and Hernanz et al. (1994). The source function of white dwarfs, $B_{WD} = \Phi(M)\Psi(t)$, where $\Phi(M)$ is the initial mass function (Salpeter, 1961) and $\Psi(t)$ is the star formation rate per unit volume, has been obtained from a SFR expressed in $M_\odot$ yr^{-1} pc^{-2} and a scale height law: $\Psi(t) = \Sigma_{disk}(t)/h(t)$. For the present calculations the scale height $h(t)$ was that of Mira stars (see Hernanz et al., 1994) and the adopted SFR was that of Clayton (1984). Regarding the cooling sequences and given the exploratory character of the present calculation and the strongly peaked mass distribution of white dwarfs, we have assumed that the cooling times for every white dwarf are those of a 0.6 $M_\odot$ white dwarf.

As we have seen, allowing G to decrease with time accelerates the cooling process and, consequently, the cut–off of the luminosity function moves to the

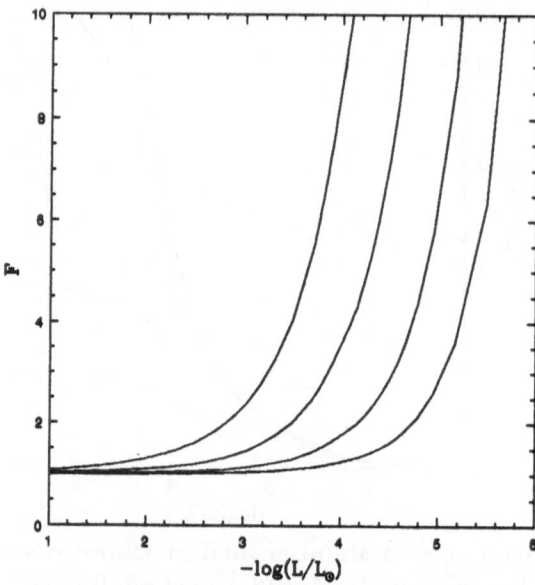

Fig. 2. Correction factor, F, for a 0.6 $M_\odot$ white dwarf for different values of $\dot{G}/G$. From left to right: $\dot{G}/G = -3\ 10^{-10}$, -10^{-10}, $-3\ 10^{-11}$, and -10^{-11}, yr^{-1}

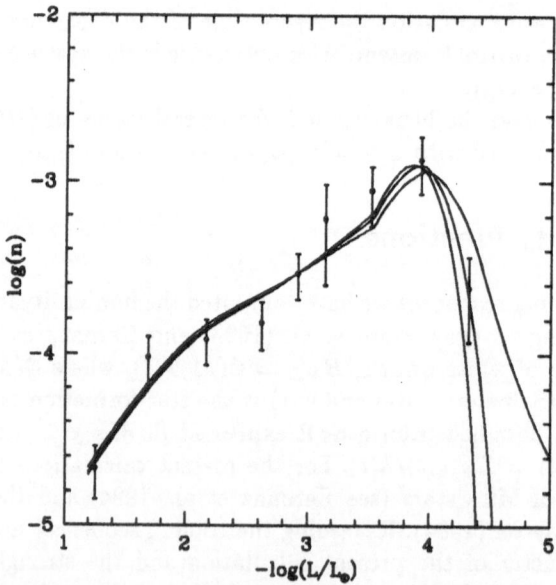

Fig. 3. White dwarf luminosity function for different values of $\dot{G}/G$ and $t_d = 7$ Gyr. From left to right: $\dot{G}/G = 0$, -10^{-11}, and $-3\ 10^{-11}$

region of low luminosities, as can be seen in Figure 3. Therefore allowing G to decrease is equivalent to increase the age of the disk. Since the age of the galactic disk can be determined with independent criteria, it is possible to get an upper bound to the maximum value of $\dot{G}/G$ as a function of the adopted age. The procedure consists in adopting an arbitrary age and adjusting $\dot{G}/G$ to get the best fit to the observational data (Liebert et al., 1988). In particular, if we assume an age of the disk small enough (say 7 Gyr), the position of the cut–off of the luminosity function is determined *only* by the value of $\dot{G}/G$. Conversely, if we fit the observational cut–off by assuming an age of the disk large enough, then the only acceptable result is that G remains constant with time. This can also be seen in Figure 3. Our best guess is, therefore, $\dot{G}/G < -1.0 \times 10^{-11}$ yr^{-1}.

4 Conclusions

We have computed cooling sequences for C/O white dwarfs allowing the gravitational constant to decrease with time. Our results indicate that the cooling of white dwarfs at low luminosities is strongly affected by a non constant G. The cooling process is accelerated if $\dot{G}/G < 0$ which translates into a shift in the position of the cut–off of the theoretical luminosity function, which allows us to derive an independent upper bound to the value of $\dot{G}/G$. This upper limit depends somewhat on the details of the cooling theory of white dwarfs and for the conditions adopted here (i.e.: assuming a previous stratification of C/O in the progenitors of white dwarfs) is $\dot{G}/G < -1.0 \times 10^{-11}$ yr^{-1} comparable to that obtained from the binary pulsar PSR1913+16 which is considered to be the best estimate of an upper bound to $\dot{G}/G$.

References

Clayton, D.D., 1984, *Ap.J*, **285**, 411

Damour, T., Gibbons, G.W., & Taylor, J.H., 1988, *Phys. Rev. Lett.*, **61**, 1151

Díaz–Pinto, A., García–Berro, E., Hernanz, M., Isern J., Mochkovitch, R., 1994, *A&A*, **282**, 86

Hernanz, M., García–Berro, E., Isern, J., Mochkovitch, R., Segretain, L., Chabrier, G., 1994, *Ap.J*, **434**, 652

Liebert, J., Dahn, C.C., Monet, D.G., 1988, *Ap.J*, **332**, 891

Mazzitelli, I., D'Antona, F., 1986, *Ap.J*, **308**, 706

Salpeter E.E., 1961, *Ap.J*, **134**, 669

Segretain, L., Chabrier, G., 1993, *A&A*, **271**, L13

Segretain, L., Chabrier, G., Hernanz, M., García–Berro, E., Isern, J., Mochkovitch, R. 1994, *Ap.J*, **434**, 641

Will, C.M., 1984, in *300 Years of Gravitation*, ed. S.W. Hawking & W. Israel (Cambridge University Press, Cambridge), 80

Wood, M.A., Winget, D.E., 1989, *in IAU Colloq. 114*, *White Dwarfs* (Berlin: Springer-Verlag), 282

Calibrating White Dwarf Cosmochronology

C.F. Claver and D.E. Winget

Department of Astronomy and McDonald Observatory
University of Texas
Austin, TX 78712 USA

1 Introduction

The age and star formation history of the Milky Way is one of the principle pieces of information we have for constraining our ideas about the evolution of spiral galaxies. The comparison of the total age with the Hubble time can teach us about cosmology. Age estimates for the Galaxy are based on the ages of its constituent components – namely the stars. Traditionally, stellar ages have been estimated from isochrone fits to open and globular cluster color-magnitude diagrams, and relative abundances of radioactive isotopes—nucleo cosmochronology.

Within the last seven years white dwarf stars have also been established as reliable chronometers for the stellar population of the galactic disk. These three methods produce a wide spread in age estimates for our Galaxy and its individual stellar populations, ranging between 10 and 20 Gyrs (1 Gyr = 10^9 years). The actual age values quoted are dependent on both technique and technician. For example, using the nucleo–cosmochronology technique Fowler and Meisl (1986) estimate the Galaxy's age at 10 Gyrs, while Thielaman, Cowan, and Truran (1987) obtain an age between 12.4 and 14.7 Gyrs. Similarly, Winget et al. (1987), Iben and Laughlin (1989), and Wood (1992) obtain a Disk age from white dwarfs of 9 Gyrs, while Hernanz et al. (1994 and these proceedings) estimate the white dwarf Disk age to be $\sim 2 - 3.5$ Gyrs older. Demarque et al. (1992) review similar discrepancies found in the stellar isochrone ages. These discrepancies and recent evidence of conflict between Hubble ages and the oldest stellar ages (Jacoby 1994) demand that the various age methods be cross–calibrated. Of the three age techniques the white dwarf ages seem to be most promising. This is because of the underlying physical simplicity of this population of stars. For this reason we have chosen as our standard chronometer the white dwarf cooling ages.

A natural place to begin the cross–calibration process is where we can apply at least two age estimation techniques simultaneously. This points to older star clusters, where an age can be estimated from isochrones, and which are sufficiently evolved to have substantial white dwarf populations. Under the assumption that the cluster is coeval, we have an opportunity to directly compare isochrone ages with white dwarf cooling ages. However, because white dwarfs are intrinsically faint, and given the limitations in current technology, we are limited to examining clusters which are relatively nearby. This essentially eliminates the older, more interesting globular clusters for the time being. Fortunately, there are numerous Galactic (open) clusters whose white dwarf populations are within reach of both ground based and space based observations.

2 Searching the Praesepe for White Dwarf Stars

Recently we have designed, built, and commissioned into service a prime focus $F/3$ CCD imaging system for the McDonald Observatory Boller and Chivens 76–cm Cassegrain telescope (Claver 1994). The instrument contains a Lumigen coated Loral Fairchild 2048 × 2048 CCD with which we can image a $> 46 \times 46$ arcminute field with 1.35 arcsecond resolution over 3000−10000Å. The corrector-telescope combination, is called the Prime Focus Camera (PFC) and is an ideal instrument for surveying Galactic clusters for their white dwarf populations.

While the idea of searching for white dwarf stars in Galactic clusters is not new (Koester and Reimers 1989 and references therein; Anthony-Twarog 1984), the unique qualities of the PFC make new search efforts for white dwarf stars in star clusters potentially very rewarding. One of the better suited clusters to begin the white dwarf–isochrone age calibration is the Praesepe (The Bee Hive; M44; NGC 2362; C0837+201). Although the Praesepe has been searched in the past for white dwarfs (Luyten 1962, 1966; Anthony-Twarog 1982, 1984), our need to determine the low luminosity (age) limit for its white dwarfs requires use of modern tools and techniques which were previously unavailable. The wide field digital imagery of the PFC allows very high precision photometric and astrometric measurements of stars over areas comparable in size to most galactic clusters. In contrast, prior efforts have been with photographic material resulting in lower quality photometry at the faint limits, where precision is needed most to distinguish white dwarfs in the cluster from accidentally aligned field stars.

With the PFC we have surveyed 5 fields in broadband $UBVI$ colors (Bessel 1990) totalling 2.1 square degrees centered on the Praesepe. These data were taken at McDonald Observatory on two runs in February and March 1994. For each filter our total integration times were: $U = 5400$, $B = 1800$, $V = 1080$, and $I = 720$ seconds. These times were divided equally between six shorter exposures to minimize the effects of the bright upper main–sequence stars. We processed and combined the individual images using the tools in IRAF. These images were then evaluated for their stellar content using FOCAS to separate stars from galaxies. Finally, we extracted the stellar photometry using DAOPHOT and transformed these to the standard system using Landolt (1992) standards and IRAF's PHOTCAL.

The combined $UBVI$ photometry from all five fields consists of more than 13,000 stars to a limiting V magnitude of ~ 21. Figure 1 shows the $V, V - I$ color–magnitude diagram for all objects with matched $UBVI$ photometry. We have included three main–sequence isochrones from VandenBerg (1985) scaled to a distance of $m - M_v = 6.2$ (174 pc) indicating a main–sequence age of 0.85 ± 0.10 Gyrs. Also shown is a mean white dwarf cooling track determined from Monet *et al.* (1992) trigonometric parallax. On the basis of their photometric colors and magnitudes we have selected 25 objects consistent with white dwarfs at the Praesepe's distance. Using a digitized portion of the POSS survey as our second epoch (1954.9), we examined each of these photometric candidates astrometrically for proper motions consistent with the Praesepe's 0.035 arcseconds per year (1.48 arcseconds between these epochs). Of the 25 photometrically selected white dwarf candidates, four have proper motions consistent with cluster membership. Two objects fell outside the area scanned on the POSS plate and could not be tested for kinematic membership. Our sample consists of these six objects plus four objects from the literature which fell outside our survey area. We believe it is significant that no objects below $V = 19$ were found, indicating that we have successfully reached the low luminosity limit for the Praesepe's white dwarfs.

3 Discussion of Results

In Figure 2 the objects from Table 1 are plotted in the $V, V - I$ color– magnitude plane along with calculated white dwarf isochrones at the Praesepe's distance. We calculated the white dwarf isochrones using evolution models from Wood (1994, these proceedings). We used the DA bolometric correction procedure in Liebert, Dahn, and Monet (1988) to transform the model luminosities to the absolute visual magnitude scale. For the effective temperature to color transformation, we computed synthetic broadband colors using Bergeron model DA atmospheres and the $UBVI$ response functions of Bessel (1990). We set the color zero points to Vega using its flux distribution as measured by Hayes (1985). From these isochrones we estimate the *"white dwarf"* age for the Praesepe to be ~ 1.0 Gyr, with an estimated uncertainty of 0.3 Gyr. Within the estimated errors, there is no measurable difference between the main sequence and white dwarf ages for the Praesepe.

The white dwarf age is dependent on the oldest identified white dwarf. Therefore, if we are to have any confidence in this age method we must have some estimate of the completeness of our white dwarf sample relative to the number of white dwarfs thought to have formed in the cluster. We can do this by differencing the estimated number of stars formed in the cluster above the main sequence with the number of stars observed. In Figure 3 we have normalized the Taff (1974) mean luminosity function (LF, solid points) for Galactic clusters to the observed LF (open points) from Jones and Stauffer (1991). The number of stars predicted in the Praesepe above the $M_v = 1.63$ turnoff is 29 ± 5. From our observations and the photometry lists of Johnson (1952) and Klei–Wassink (1927) the number of stars observed brighter than $M_v = 1.63$ is 22 ± 4. From these estimates we find there are 7^{+9}_{-4} *"evolved"* stars missing above the turnoff — the white dwarfs — in the Praesepe. Our list contains 10 objects, which is consistent with our predicted number. Therefore, we have some degree of confidence that we have not missed a significant number (if any) of white dwarfs in

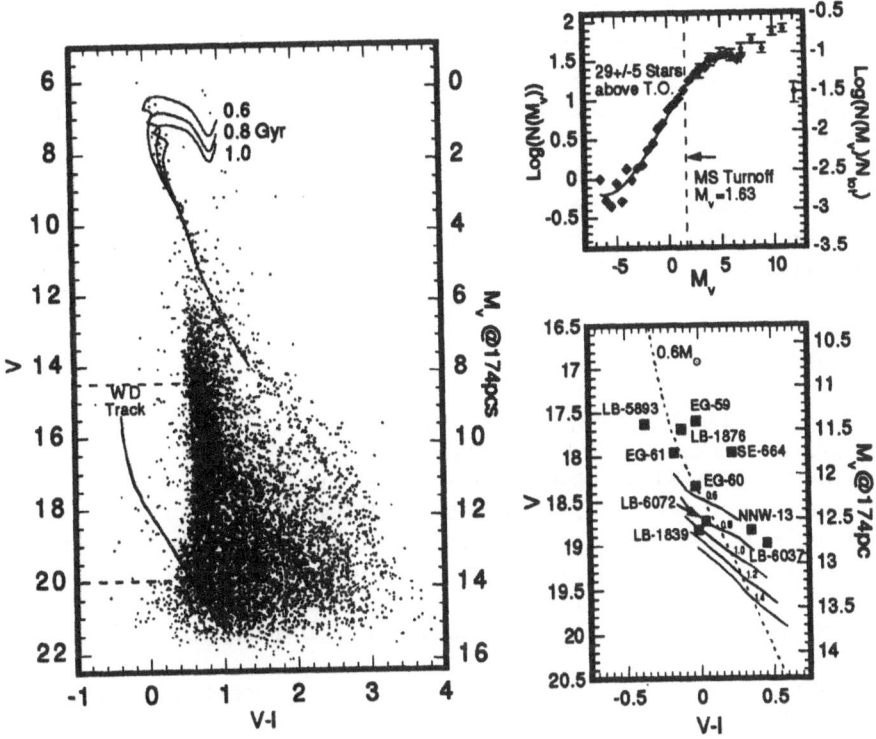

Fig. 1. The V versus V-I color–magnitude diagram (left) for all stars with *UBVI* photometry in the 5 Praesepe survey fields. **Fig. 2.** White dwarf and candidate stars for the Praesepe (bottom right) with white dwarf isochrones for 0.6, 0.8, 1.0, and 1.2 Gyrs. **Fig. 3.** (top right) From the Taff (1976) mean Galactic cluster LF normalized to the observations of Jones and Stauffer (1991) we estimate there are 29 ± 5 *evolved* stars above the mainsequece turn–off at $M_v = 1.63$ in the Praesepe.

the cluster.

It is clear from this work and similar work presented at this meeting by von Hippel that on timescales between 1 and 3 Gyrs there is substantial agreement between white dwarf cooling and main-sequence stellar ages. It is essential that we extend our calibration to clusters with ages greater than 3 Gyrs because the model input physics for both white dwarf cooling and main-sequence evolution dramatically changes. In white dwarf models crystallizing begins shortly after 3 Gyrs and theoretical uncertainties regarding phase separation of carbon and oxygen become important (Wood, 1994, Segretain et al. 1994). Similarly, the main-sequence model 3 Gyrs represents the beginning of the transition from CNO bi–cycle to pp–chain hydrogen burning as the dominant energy source. By the time the age of the disk is reached the turnoff mass is near one solar mass and more than 95% of the stellar luminosity is generated by pp–chain burning.

The uncertainties in the white dwarf ages at low luminosity are due phase-separation of carbon and oxygen. They will be removed independently by studying the shape of the luminosity function for the disk at the cool end. Thus the most critical calibration remaining will be NGC 188. This will resolve lingering questions about the uncertainty in the nuclear reactions, and allow us to tie in the age estimates for the globular clusters. With this in hand, we can compare the Hubble time with the age of our galaxy, thereby moving forward our understanding of cosmology.

References

Anthony–Twarog, B. A.1984, *A. J.*, 89, 276.

Bessell, M. S.1990, *P. A. S. P.*, 102, 1181.

Claver, C. F.1994, Ph. D. Dissertation, The University of Texas at Austin.

Demarque, P., Green, E. M., and Guenther, D. B.1992, *A. J.*, 103, 151.

Fowler, W. A., and Meisl, C. C.1986, In *Cosmological Processes*, eds. W. D. Arnett, C. J. Hansen, J. W. Truran, and S. Tsuruta, Utrecht: VNU Science Press.

Hayes, D. S.1985, In *IAU Symposium 111: Calibration of Fundamental Stellar Quantities*, ed. D. S. Hayes, L. E. Pasinetti, and A. G. D. Phillip, Dordrecht: D. Reidel Pub. Co.

Hernanz, M., Garcia–Berro, E., Isern, I, Mochkovitch, R., Segretain, L.,and Chabrier, G. 1994, *Ap. J.*, 434, 652.

Iben, Jr., I., and Laughlin, G.1989, *Ap. J.*, 341, 312.

Jacoby, G. H.1994, *Nature*, 371, 741.

Johnson, H. L.1952, *Ap. J.*, 116, 640.

Jones, B. F., and Stauffer, J. R.1991, *A. J.*, 102, 1080.

Klein–Wassink, K. N.1927, *Pub. Kaptayn Ast. Lab.*, No. 41.

Koester, D., and Reimers, D.1989, *A. A.*, 223, 326.

Landolt, A. U.1992, *A. J.*, 104, 340.

Liebert, J., Dahn, C. C., and Monet, D. G.1988, *Ap. J.*, 332, 891.

Monet, D. G., Dahn, C. C., Veba, F. J., Harris, H. C., Pier, J. R., Luginbuhl, C. B., and Ables, H. D. 1992, *A. J.*, 103, 638.

Segretain, L., Chabrier, G., Hernanz, M., Garcia–Berro, E, and Mochkovitch, R.1994, *Ap. J.*, 434, 641.

Taff, L. G.1974, *A. J.*, 79, 1280.

Thielemann, F. -K., Cowan, J. J., Truran, J. W.1987, In *13th Texas Symposium on Relativistic Astrophysiscs*, Teaneck, NJ: World Scientific Pub. Co.

VandenBerg, D. A.1985, *Ap. J. Supp. Ser.*, 58, 711.

Winget, D. E., Hansen, C. J., Liebert, J., Van Horn, H. M., Fontaine, G.,Nather, R. E., Kepler, S. O., and Lamb, D. Q. 1987, *Ap. J. Let.*, 315, L77.

Wood, M. A.1992, *Ap. J.*, 386, 539.

Wood, M. A.1994, In *IAU Colloquium 147: The equation of State in Astrophysics*, eds. G. Chabrier and E Schatzmann, Cambridge: Cambridge University Press.

Thermodynamic Quantities for Evolutionary and Pulsational Calculations

Werner Stolzmann[1] and Thomas Blöcker[2]

[1] Institut für Astronomie und Astrophysik der Universität, D-24098 Kiel, Germany
[2] Astrophysikalisches Institut Potsdam, D-14473 Potsdam, Germany

1 Introduction

In order to understand the behaviour and the evolution of stars it is necessary to know the thermodynamics of interacting many particle systems over a wide region of densities and temperatures. An important aspect for the modelling of the equation of state (EOS) is the nonideality according to the interactions between the charged particles. Astrophysical applications, e.g. the calculation of stellar evolution or pulsation, require an accurate determination of exchange and correlation interactions valid for density-temperature conditions from the center to the surface of the star. A powerful method to consider the complex physics over a large region of stellar densities and temperatures is the Padé approximant technique (cf. Baker and Gammel 1970). The aim of this contribution is to present results for the Helmholtz free energy and the EOS of interacting fully ionized systems consisting of electrons and ions, which are described by means of Padé approximants.

2 Helmholtz free energy and pressure

Starting with the Coulombic parts of the Helmholtz free energy for a fully ionized plasma

$$F^{\text{coul}} = F^{\text{x}}_{\text{ee}} + F^{\text{c}}_{\text{ee}} + F^{\text{c}}_{\text{ii}} + F^{\text{cq}}_{\text{ii}} + F^{\text{c}}_{\text{ie}} \tag{1}$$

we obtain for the pressure (EOS) equivalent contributions. We calculate the exchange [x] - correlation [c] terms which are generally determined by time consuming numerical calculations, by means of Padé approximants. For the non-relativistic exchange energy F^{x}_{ee} we used the formula from Perrot and Dharma-wardana (1984) and for the correlation energies of the electrons [ee], ions [ii], and ions-electrons [ie] are applied Padé formulas. The ionic quantum correction can be described by the theory from Nagara et al. (1987). Details for the Coulomb interaction terms expressed by Padé approximants are given by Ebeling (1990) and Stolzmann and Blöcker (1995). In the following section we are interested to demonstrate by figures the validity and to give comparisons with other theories or "closed-form" parametrizations.

3 Numerical results

In Figs. 1 and 2 the single parts of the Coulombic Helmholtz free energy (1)
normalized to NkT are shown, which are given by Padé formulas. Fig. 1 illustrates the dependence from the degeneracy ($\Theta = kT/kT_{\mathrm{F}}$) in the case of strong
particle coupling ($\Gamma = 3$). The Coulomb coupling parameter is defined by

$$\Gamma = \langle Z \rangle^{1/3} \langle Z^{5/3} \rangle \frac{e^2}{kT} \left(\frac{4\pi}{3} \sum_i n_i \right)^{1/3} , \qquad (2)$$

with the charge averages $\langle Z^k \rangle$ and the ion particle density n_i. The classical regime
corresponds to $\log \Theta = 2$. The electrons are highly degenerate at $\log \Theta = -2$.
The dependences from the coupling strength Γ at an intermediate degeneracy
($\Theta = 1$) are shown in Fig. 2. Both the limiting law for the weakly coupled
particles (Debye-Hückel) and for the strongly coupled particles (Wigner-Seitz)
are exactly given by the ionic correlation f_{ii}. This contribution is determined
by a classical expression for ionic mixtures in the fluid and solid phase (Brami
et al. 1979, Ebeling 1990). Comparisons with other approaches as from Pokrant
(1977), Dandrea et al. (1986), Ichimaru et al. (1987), Saumon and Chabrier
(1989) are presented in Stolzmann and Blöcker (1993).

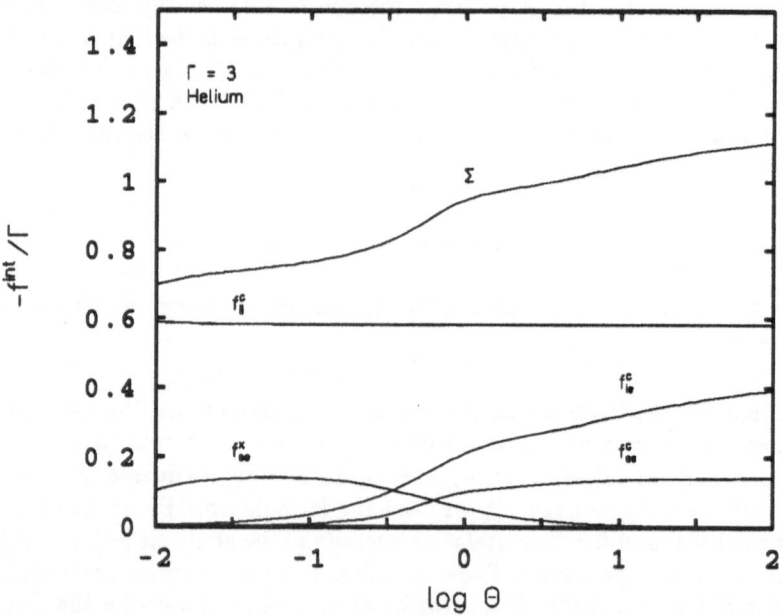

Fig. 1. Helmholtz free energy contributions of the electron-electron, ion-ion, and
ion-electron interaction vs. the degeneracy parameter $\Theta = kT/kT_{\mathrm{F}}$ for helium at $\Gamma = 3$.
The labels denote the the interaction free energy terms f^{int}

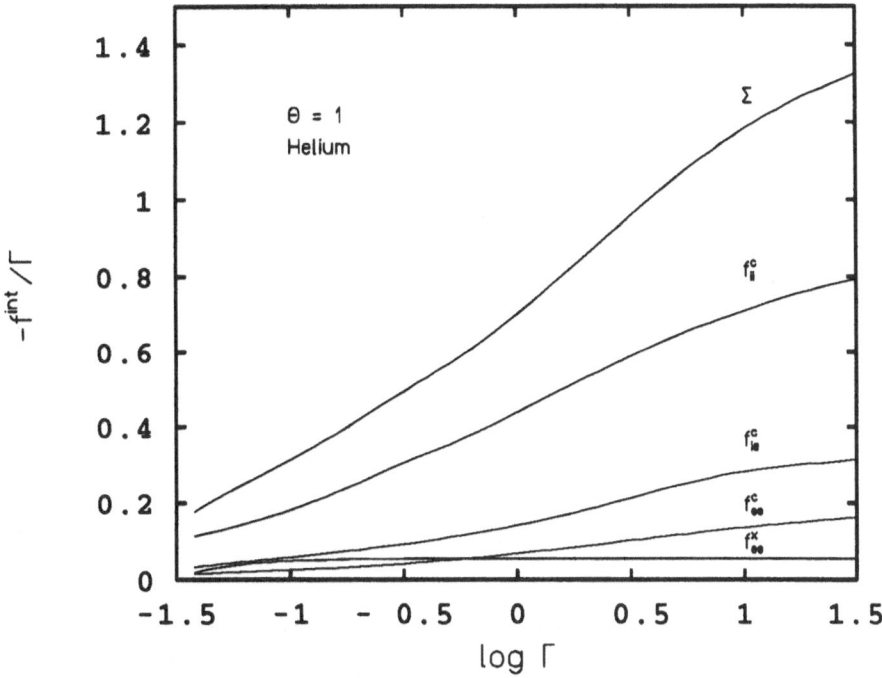

Fig. 2. Helmholtz free energy contributions of the electron-electron, ion-ion, and ion-electron interaction vs. the Coulomb coupling parameter Γ for helium at $\Theta = 1$. The labels denote the the interaction free energy terms f^{int}

Figs. 3 and 4 show the different Coulombic pressure terms normalized to the ideal pressure of the electrons and ions. In Fig. 3 we present the interaction terms for a mixture of carbon and oxygen as a function of the density at $T = 10^7\mathrm{K}$. For high densities we have to include relativistic correction for the ideal and exchange pressure. Fig. 4 compares the various interaction contributions for pure helium at a fixed density as a function of the temperature. We remark that in the highly degenerate region the Coulomb interaction is dominated by the electron exchange and the ion correlation. In the nondegenerate region (weakly coupled regime) all interacting terms can be neglected. Fig. 5 shows the relative pressure (ratio of total to ideal pressure) for a hydrogen-helium plasma at a fixed mean electron distance r_s as a function of the helium mass-fraction Y. The mean electron distance in units of Bohr radius, r_s, has been chosen to 1. The respective lines refer to different degeneracy parameters $\Theta = kT/kT_F$. The Coulomb interaction increases with the mean charge number Z of the ions. Moreover, this effect is even stronger for low Θ, i.e. at high degeneracy. For $\Theta = 0.01$ we have added the data of the EOS of Hubbard and DeWitt (1985).

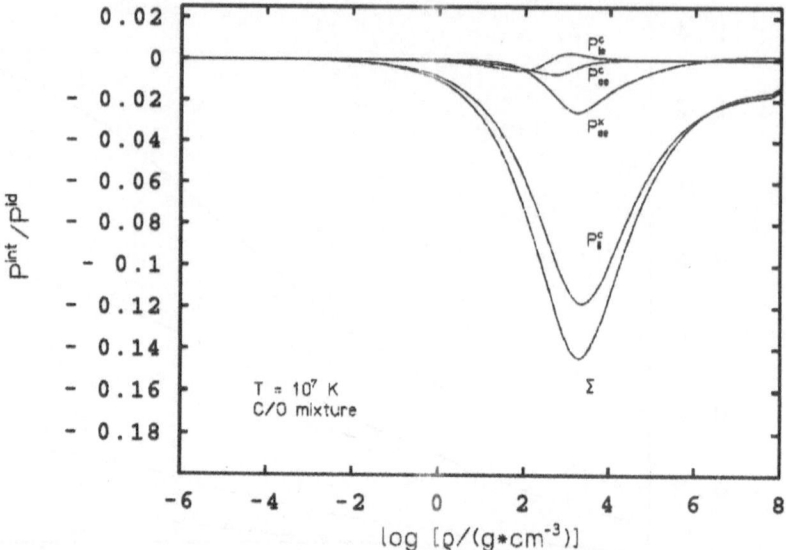

Fig. 3. Relative pressure contributions of the electron-electron, ion-ion, and ion-electron interaction vs. density for a carbon-oxygen mixture with C/O = 0.5/0.5 [mass fractions] at $T = 10^7$ K. The labels denote the the interaction pressure terms P^{int}

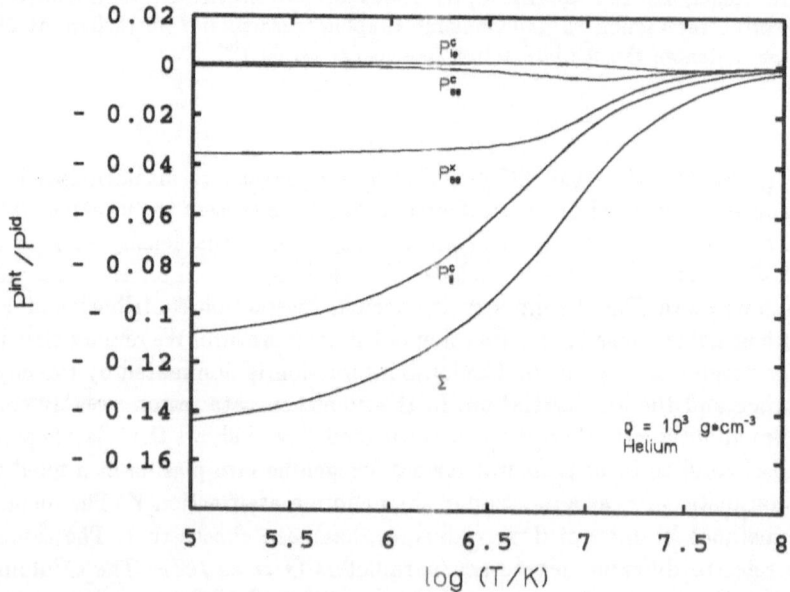

Fig. 4. Relative pressure contributions of the electron-electron, ion-ion, and ion-electron interaction vs. temperature for helium at $\rho = 10^3 g \cdot cm^{-3}$. The labels denote the the interaction pressure terms P^{int}

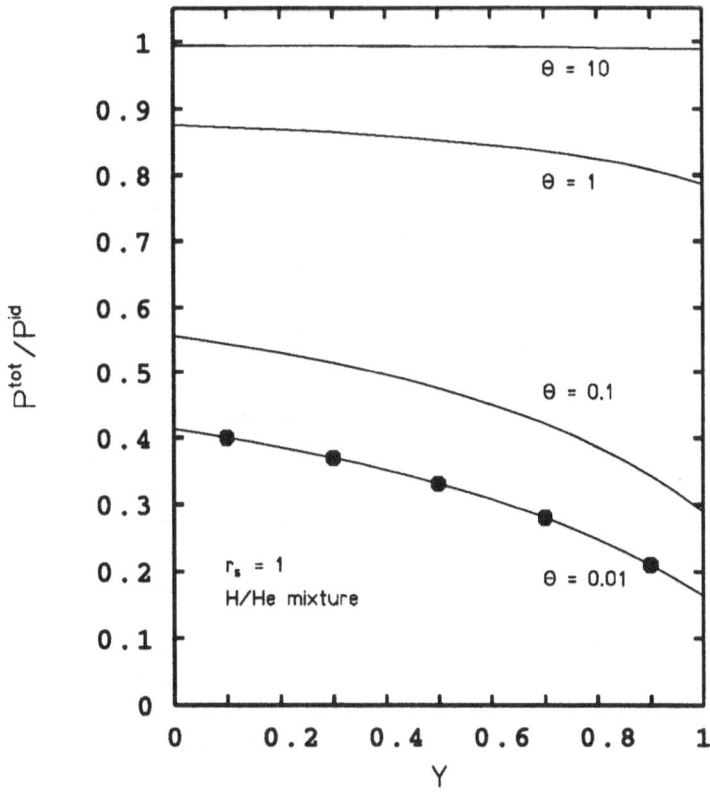

Fig. 5. Relative pressure vs. mass fraction Y of helium for a H/He mixture at $r_s = 1$ ($\rho/\mu_e = 2.6754$) for different degeneracy parameters Θ. The dots are the results from Hubbard and DeWitt (1985)

References

Baker Jr., G.A., Gammel, J.L. 1970, *The Padé Approximant in Theoretical Physics*, Academic Press, New York

Brami, B., Hansen, J.P., Joly, F. 1979, Physica A 95, 505

Dandrea, R.D., Ashcroft, N.W., Carlsson, A.E. 1986, Phys. Rev. B34, 2097

Ebeling, W. 1990, Contr. Plasma Phys. 30, 553

Hubbard, W. B., DeWitt, H. E. 1985, ApJ 290, 388

Ichimaru, S., Iyetomi, H., Tanaka, S. 1987, Phys. Rep. 149, 91

Nagara, H., Nagata, Y., Nakamura, T. 1987, Phys. Rev. A 36, 1859

Perrot, F., Dharma-wardana, M. W. C. 1984, Phys. Rev. A 30, 2619

Pokrant, M.A. 1977, Phys. Rev. A 16, 413

Saumon, D., Chabrier, G. 1989, Phys. Lett. A 134, 2397

Stolzmann, W., Blöcker, T. 1993, Contrib. Plasma Phys. 33, 391

Stolzmann, W., Blöcker, T. 1995, A&A, in preparation

Numerical Simulations of Convection and Overshoot in the Envelope of DA White Dwarfs

Bernd Freytag[1], Matthias Steffen[1], Hans-Günter Ludwig[2]

[1] Institut für Astronomie und Astrophysik der Universität, D-24098 Kiel, Germany
[2] Max-Planck-Institut für Astrophysik, D-85740 Garching, Germany

Abstract: We present results of realistic 2D numerical radiation hydrodynamics calculations, simulating the surface convection zones of DA white dwarfs in the range of effective temperatures from 14 200 K down to 11 400 K. Comparison with mixing length theory (MLT) yields a conflicting picture: The dynamics of convection is not governed by up- and downflowing bubbles which dissolve after travelling some characteristic distance – but by the formation, advection, merging, and disruption of fast narrow downdrafts in a slowly upstreaming surrounding. MLT tremendously underestimates the depth of the region where material is mixed. Nevertheless, it turns out that a mixing length model with $\alpha = 1.5$ gives a good fit of the *photospheric temperature structure* ($T_{\text{eff}} = 12\,600$ K) and that a 1D temperature stratification suffices to reproduce the mean spectrum of the 2D simulations, indicating that the photospheric temperature inhomogeneities are negligible for spectroscopic analysis. In *deeper layers* the temperature stratification of our hydrodynamical models corresponds to larger values of α. Introducing our envelope models into nonadiabatic pulsation calculations results in a blue edge of the ZZ Ceti instability strip near $T_{\text{eff}} = 12\,400$ K at $\log g = 8.0$.

1 Introduction

Hydrogen-rich DA white dwarfs with effective temperatures near the ZZ Ceti instability strip show a surface convection zone in their envelope due to partial ionization of hydrogen. This convection zone extends far into the photosphere and is able to influence the emergent spectrum. In the context of MLT the photospheric temperature stratification depends sensitively on the adopted mixing length parameter α. This results in uncertainties in the determination of effective temperatures. At the same time the depth of the model convection zone (and therefore its ability to mix underlying material into surface layers) also depends on α (cf. Bradley and Winget 1994). Unfortunately, the spectroscopically determined value of α is in conflict with that derived from pulsational calculations reproducing the blue edge of the ZZ Ceti instability strip. One way to overcome the problems and contradictions of the MLT approach is the use of parameter-free numerical simulations based on the constitutive equations.

2 The Radiation Hydrodynamics Code

The computer program used for our simulations (for details see Ludwig, Jordan, and Steffen 1994) solves the equations of nonlinear hydrodynamics and non-local radiative transfer in 2D Cartesian geometry. Compressibility effects are fully taken into account which is essential for modelling sonic convective flows in a stratified medium. The computational box covers only the uppermost (non-degenerate) layers of the star but is nevertheless more than ≈ 10 pressure scale heights deep. We use a realistic equation of state. For DA white dwarfs it takes into account ionization of hydrogen but neglects non-ideal effects which are unimportant in the surface layers. The energy equation includes a detailed treatment of radiation transport with realistic opacities (grey or frequency-dependent approximation). Rotation and magnetic fields are ignored. Our grid of stellar models so far covers

- Surface layers of solar type convection zones with $3.54 \leq \log g \leq 4.44$,
- Main sequence A stars with $7\,500\,\mathrm{K} \leq T_{\mathrm{eff}} \leq 9\,500\,\mathrm{K}$,
- DA white dwarfs with $11\,400\,\mathrm{K} \leq T_{\mathrm{eff}} \leq 14\,200\,\mathrm{K}$, $\log g = 8.0$.

3 Dynamics of the Convective Pattern

A simulation is started with an initial condition taken from a previous run or with a radiative stratification after applying a small random velocity field as a perturbation. At the beginning we see a linear growth of the most unstable convective mode with symmetric up- and downflows. When the amplitude of the velocities becomes larger, nonlinear effects produce an asymmetry between narrow downdrafts and broad upflow regions. The downdrafts are advected horizontally, merge and increase the asymmetry. Such merging events may be accompanied by the production of waves and often lead to deep-reaching plumes. In relatively quiet regions with small vertical velocities the convective instability produces new downdrafts which move sideways and initiate a new 'cycle'.

Our simulations suggest that the diameter of downdrafts is related to the wavelength of the most unstable mode and that a jet can be considered as a 'wave packet'. The mean distance between downdrafts (the size of the 'granules') is the result of the competition between convective instability creating new downdrafts and horizontal advection sweeping them into established downflows.

The properties of convection depend on effective temperature: The hotter models (e.g. Fig. 1) show *weak* convection with moderate velocities and convective fluxes not exceeding a few percent. With decreasing T_{eff} convection starts to carry a significant fraction of the total energy flux and becomes *vigorous*. The velocity field in the downdrafts and in parts of the photosphere becomes supersonic (see Figs. 2 and 3) and the evolution of the flow pattern becomes more and more complex. At still lower temperatures the convection becomes *efficient*, dominating the energy transport in a deep, adiabatic, solar-like convection zone (simulations of this type are not yet available for white dwarfs).

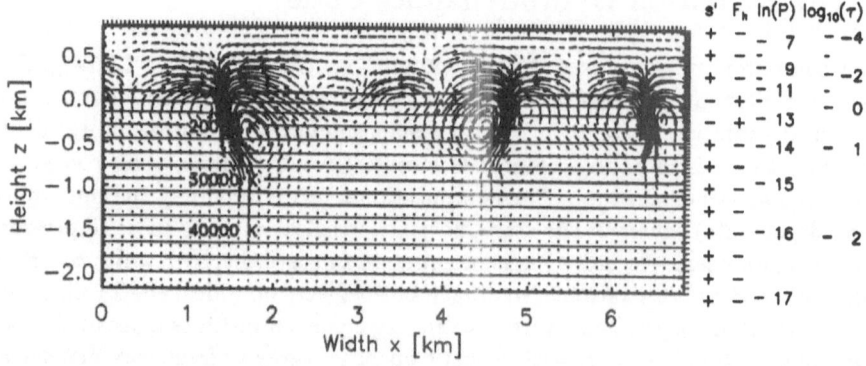

Fig. 1. Snapshot from a pure hydrogen DA white dwarf model (zt134g80n1) with $T_{eff} = 13\,400\,\mathrm{K}$, $\log g = 8.0$, as an example of *weak convection*. The velocity field is indicated by pseudo streamlines followed over $t_{int} = 0.2\,\mathrm{s}$. The maximum velocity is $v_{max} = 4.1\,\mathrm{km\,s^{-1}}$ at this instant. Contour lines indicate the temperature field, which is nearly plane-parallel here. Closely spaced tick marks at the top and the right show the positions of the computational grid points. Four columns at the right show the sign of the vertical mean entropy gradient s', the sign of the enthalpy flux F_h ($F_{h,max} = 0.8\%$), the natural logarithm of the mean pressure $\ln(P)$, and the logarithm of the optical depth $\log_{10}(\tau_{Ross})$. A positive entropy gradient defines a convectively stable region and a negative one an unstable zone. The geometrical height scale is given on the left side, with zero height corresponding to optical depth unity. Top and bottom are closed, the lateral boundaries are periodic. Although there is only a very thin convectively unstable zone, the velocity field is non-zero down to the bottom of the model. The descending material is concentrated in narrow downdrafts which are surrounded by broad upflows. Note that the maximum upward velocity is usually not found in the center of the upflows but near the downdrafts.

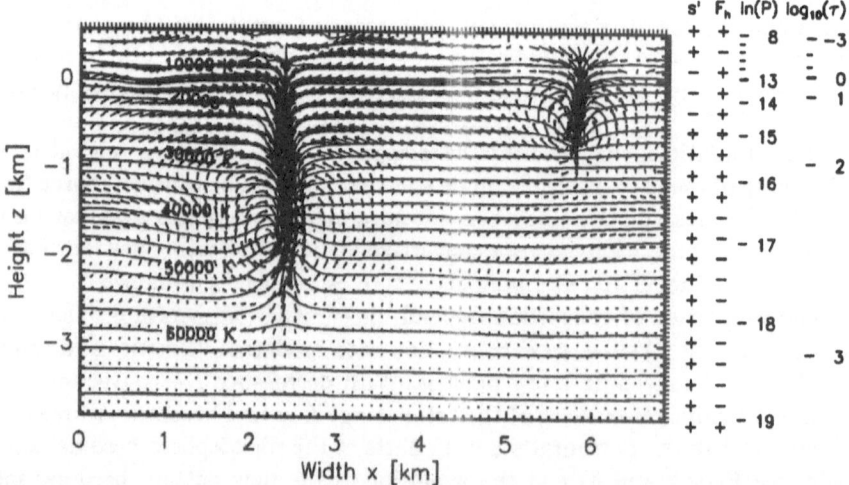

Fig. 2. Snapshot from a DA white dwarf model (zt120g80n4) with $T_{eff} = 12\,000\,\mathrm{K}$, $\log g = 8.0$, $t_{int} = 0.05\,\mathrm{s}$, $v_{max} = 20.8\,\mathrm{km\,s^{-1}}$, $F_{h,max} = 41\%$. Temperature fluctuations, velocities and overshoot region are much larger than in Fig. 1, convection is *vigorous*.

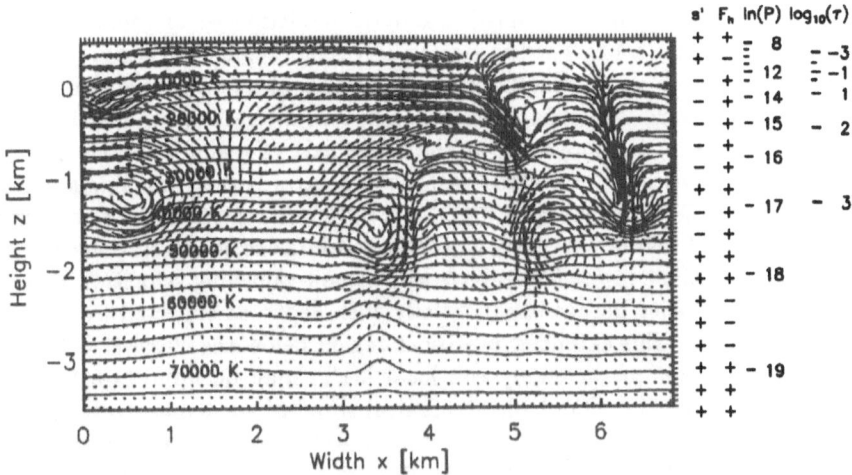

Fig. 3. Snapshot from a DA white dwarf model (zt114g80n2) with $T_{eff} = 11\,400$ K, $\log g = 8.0$, $t_{int} = 0.03$ s, $v_{max} = 21.0$ km s^{-1}, $F_{h,max} = 100\%$. Sometimes there are deep-reaching jets as in Fig. 2 but very often they are torn to pieces as shown here.

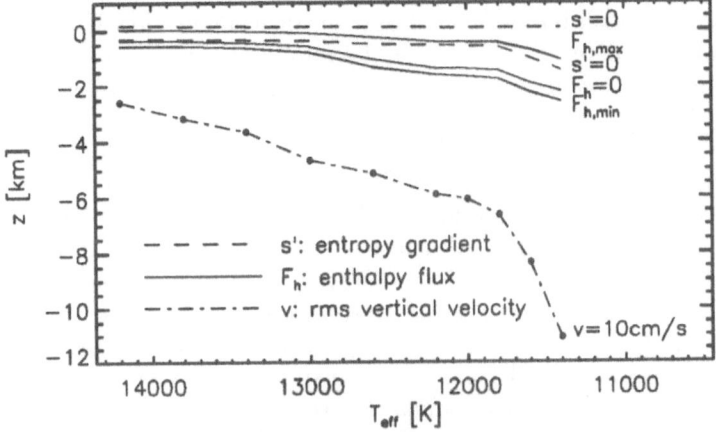

Fig. 4. Boundaries of the convection zone versus T_{eff} as derived from our hydrodynamics models. Extrapolating the convective vertical velocity outside our models, we determine the level where the velocity falls below 10 cm/s and mixing becomes inefficient. It lies far below the convectively unstable layers.

4 Overshoot and the 'Boundaries' of a Convection Zone

As can be seen from Figs. 1 to 4 there are several possible criteria for defining the position of a convection zone. Some possible definitions are:

1. $s' < 0$: negative entropy gradient, convective instability,
2. $F_h > 0$: positive enthalpy flux,
3. $F_h \neq 0$: nonvanishing enthalpy flux,
4. $v_{conv} \neq 0$: nonvanishing convective velocities.

In local MLTs all definitions coincide but in the models based on radiation hydrodynamics the criteria 1 to 4 define increasingly larger regions. In region 1 ($s' < 0$)
small disturbances grow, leading to the formation of jets. Throughout region 2
($F_h > 0$) downward velocities are correlated with a temperature deficit, resulting in acceleration by buoyancy. In the lower part of region 3 (where $F_h < 0$)
downward velocities are correlated with a temperature excess, causing a deceler-
ation by buoyancy. Below region 3 ($v_{conv} \neq 0$, $F_h \approx 0$) there are no temperature
fluctuations and no buoyancy work is done, but large-scale pressure fluctuations
extending into these layers from above lead to non-vanishing velocities. Here the
rms vertical velocity decays exponentially with depth as expected for unstable
linear g^--modes. This extended overshoot region below the classical convection
zone, which is not predicted by the local MLT, may be important for diffu-
sion/mixing problems. Our hydrodynamical simulations demonstrate that the
mass in the convectively mixed region may exceed the mass in the unstable zone
by a factor of 10 to 100, depending on effective temperature (see Fig. 4).

5 Synthetic Spectra, Pulsation Calculations

Our 2D convection models have also been used for the computation of synthetic
spectra. The main results of this study may be summarized as follows.

1. No significant errors are introduced in the spectroscopic analysis of DAs
near the blue edge of the ZZ Ceti instability strip by neglecting thermal inhomogeneities (cf. poster abstract by Ludwig and Steffen, this proceedings).

2. For a DA white dwarf model with $T_{eff} = 12\,600\,\mathrm{K}$ and $\log g = 8.0$ the
spectrum based on our HD calculations is almost indistinguishable from that
of a MLT $\alpha = 1.5$ model in the whole spectral region from the UV to the red
(Ludwig, Jordan, and Steffen 1994). But a larger α is necessary to account for
the temperature gradient in deeper, subphotospheric layers. No choice of α can
reproduce the velocity and flux profile of the hydrodynamical convection model.

In another investigation radiation hydrodynamics models of 4 DA white
dwarfs of different T_{eff} have been combined with hydrostatic interior models
for calculations of linear nonadiabatic pulsations. In this way a parameter-free
determination of the position of the blue edge of the ZZ Ceti instability strip has
become possible for the first time. Preliminary results presented in the poster
abstract by Gautschy and Ludwig (this proceedings) indicate that the blue edge
is located somewhere between $T_{eff} = 12\,600\,\mathrm{K}$ and $12\,200\,\mathrm{K}$ at $\log g = 8.0$.

References

Bradley, P. A., Winget, D. E.: ApJ. **421** (1994) 236-244
Gautschy, A., Ludwig, H.-G.: The blue edge of the ZZ Ceti instability strip. Poster
    abstract (this proceedings)
Ludwig, H.-G., Jordan, S., Steffen, M.: A&A **284** (1994) 105-117
Ludwig, H.-G., Steffen, M.: Spectroscopic effects of temperature inhomogeneities in the
    atmospheres of DA white dwarfs. Poster abstract (this proceedings)

The Formation of Massive White Dwarfs and the Progenitor Mass of Sirius B

Francesca D'Antona[1] and Italo Mazzitelli[2]

[1] Osservatorio Astronomico di Roma, Via dell'Osservatorio 1,
  I-00040 Monte Porzio (Roma) Italy
[2] Istituto di Astrofisica Spaziale C.N.R.,
  Frascati, Italy

1 Scientific Rationale

Single DA white dwarfs present a mass distribution sharply peaked at $< M > \sim$ $0.56 - 0.58 M_\odot$ (Bergeron *et al.* 1992, Weidemann & Koester 1984) holding important constraints on the earlier phases of stellar evolution through mass loss and planetary nebula ejection. A small secondary peak is perhaps present at $\sim 0.9 \pm 0.1 M_\odot$ (Bergeron *et al.* 1992, Weidemann 1988). Low and intermediate mass stars ($M \lesssim 5 M_\odot$) ascend the Asymptotic Giant Branch (AGB), and become Carbon Stars, in the "Thermal Pulse" (TP) phase, during which the white dwarf to be C–O core grows in mass while the envelope is reduced by wind mass loss. The absence of Carbon stars at $M_{bol} \lesssim -6$ is regarded as one of the indications that AGB evolution is finished at relatively small core masses for the bulk of these stars.

For more luminous stars, the discovery of Lithium in the atmospheres of red giants having $-7 < M_{bol} < -6$ in the LMC and SMC (e.g. Plez *et al.* 1993) led to suspect that these objects are descendants of initially more massive ($M \gtrsim 5 M_\odot$) stars, which go ahead along AGB evolution until meeting "hot bottom burning" (Scalo *et al.* 1975) at the basis of the convective envelope. In these conditions, the $^3He +^4 He =^7 Be$ nuclear reaction can proceed, *Be* is transported by turbulent diffusion to the stellar surface and decays into Lithium, which goes back to the convective bottom where it will be ultimately destroyed. This process can go ahead only for a short while, until all the ^3He is destroyed, but ^7Li will survive long enough to be detected. If the above interpretation is correct, the fact that in the LMC and SMC *there are no giants without Lithium* at even larger luminosities ($M_{bol} \lesssim -7$) leads us to suspect that the AGB phase is terminated by mass loss very soon after the envelope becomes Lithium rich (D'Antona & Matteucci 1991). *It is then of interest to compute at which core masses these stars begin experiencing hot bottom burning, since these core masses will be* **very close** *to the final WD masses.*

2   New Computations

We decided to map again the evolution of masses undergoing the second Dredge
Up (DU, Becker & Iben 1979). The program contains many updates in physical inputs. Complete results will be given elsewhere (D'Antona & Mazzitelli
1995). Canuto and Mazzitelli (1991, 1992) model for overadiabatic convection is
adopted. Main results are:

1) For population I, all masses $M > 4.5 M_\odot$ undergo the second DU;

2) All masses undergoing the second DU **spontaneously** *achieve hot bottom
burning already at the first TPs* **with CM convection.** *To get the same result
with the MLT one needs $\alpha = l/H_p \gtrsim 2.5$, whereas the solar fit requires $\alpha = 1.5$.*

*3) After the second DU, the stars do not follow a canonical core mass – lumi-
nosity relation, but luminosity increases very fast with core mass, in agreement
with the finding by Blöcker and Schönberner (1991);*

*4) The lower the metal abundance, the lower is the minimum mass for the
second DU and Hot Bottom Burning.*

*We conclude that population I stars with $M \gtrsim 4.5 M_\odot$ release their envelope
and become WDs of total mass* **only slightly larger than the 2nd DU core
mass,** *so, according to our results, we expect a secondary peak in the WD mass
distribution around $0.9 \div 1 M_\odot$ (see Mazzitelli 1988).*

These results can be applied to the determination of Sirius B progenitor
mass. As the WD mass is $M = 1.05 \pm 0.028 M_\odot$ (Gatewood & Gatewood 1978),
we find $M_{initial} \sim 6.5 \div 7 M_\odot$. In these range of progenitor masses, $\sim 470 R_\odot$ is
the maximum pre–WD radius of Sirius B allowed to avoid Roche lobe overflow
to A and common envelope evolution (D'Antona 1982). This is consistent with
an (extrapolated) beginning of the AGB for a $7 M_\odot$ around $T_{eff} \sim 3.6$.

References

Becker, S.A., Iben, I. Jr. 1979: ApJ 232, 831
Bergeron, P., Saffer, R. A., Liebert, J.: 1992, ApJ 394, 228
Blöcker, T., Schönberner, D. 1991: ApJ 244, L43
Canuto, V. M., & Mazzitelli, I. 1991: ApJ 370, 295.
Canuto, V. M., & Mazzitelli, I. 1992: ApJ 389, 724.
D'Antona, F. 1982: A&A 114, 289
D'Antona, F., Matteucci, F. 1991: A&A 248, 62
D'Antona, F., Mazzitelli, I. 1995: A&A to be submitted
Gatewood, G. D., Gatewood, C. V. 1978, ApJ 225, 191
Mazzitelli, I. 1988: in "White Dwarfs", Proc. IAU Colloquium 114, ed. G. Wegner, p.
    29 (Springer–Verlag)
Plez, B., Smith, V.V., Lambert, D.L. 1993: ApJ 418,812
Scalo, J. M., Despain, K. H., Ulrich, R. K. 1975: ApJ 196, 805
Weidemann, V. 1988: in "White Dwarfs", Proc. IAU Colloquium 114, ed. G. Wegner,
    p. 1 (Springer–Verlag)
Weidemann, V., Koester, D. 1984: A&A 132, 195

HST Observations of White Dwarf Cooling Sequences in Open Clusters

Ted von Hippel[1,2], Gerard Gilmore[1], D.H.P. Jones[3]

[1] Institute of Astronomy, Madingley Road, Cambridge CB3 0HA
[2] WIYN Telescope, NOAO, PO Box 26732, Tucson, AZ 85726-6732, USA
[3] Royal Greenwich Observatory, Madingley Road, Cambridge CB3 0EZ

The white dwarf luminosity function of a stellar cluster will have a sharp truncation at a luminosity which is determined by the time since formation of the first white dwarfs in that cluster. Calculation of the dependence of this limiting luminosity on age is based on relatively well-understood physics, and is independent of stellar evolutionary models. Thus, measurement of the termination of the white dwarf luminosity function provides an independent method to determine the age of a cluster, and thereby to calibrate stellar evolutionary ages. We have obtained HST WFPC2 V- and I-band photometry in the open clusters NGC 2420 and NGC 2477, identifying the white dwarf sequences, and proving the feasibility of this approach.

Effects of Updated Neutrino Rates in Evolutionary White Dwarf Models

Paul A. Bradley

Los Alamos National Laboratory
Los Alamos NM 87545

I use the white dwarf evolution code ($\equiv$ WDEC) of Lamb & Van Horn (1975) to evolve realistic models of C/O white dwarfs suitable for use in computing white dwarf luminosity functions (Wood 1992) and for pulsation analysis and asteroseismology (Bradley 1993, 1994). Recently, a complete set of updated by Kohyama et al. (1993) and Itoh et al. (1989, and references therein) became available. Here, I describe the differences between these neutrino rates and the older ones of Beaudet, Petrosian, & Salpeter (1967, hereafter BPS).

In my hottest models, the Itoh et al. neutrino luminosities are about 95 % of the BPS luminosities, due mainly to Itoh et al.'s inclusion of neutral current effects in the plasmon neutrino rates. Between $\log L/L\odot \sim -0.2$ and $\log L/L\odot \sim -2.1$, the Itoh et al. model neutrino luminosities are up two times greater. During this phase, the Itoh models are up to 10 % younger. Between $\log L/L\odot \sim 1.5$ and $\log L/L\odot \sim -1.0$, the neutrino luminosity exceeds the photon luminosity by up to a factor of four. At nearly all luminosities, the Itoh et al. Bremsstrahlung and recombination neutrino emission rates are less than half of those of BPS. The Itoh models are less than 1 % older than the BPS models by the time $T_{\text{eff}} \sim 12,000$ K, corresponding to the location of the ZZ Ceti instability strip. At this point, the Itoh neutrino luminosities are about half the BPS values. There is a sharp dropoff in the Itoh neutrino luminosities near $\log L/L\odot \sim -2.6$ due to the core cooling to below 10^7 K, where the fitting formulae are invalid; we switch off the neutrino rates at this point.

References

Beaudet, G., Petrosian, V., & Salpeter, E. 1967, ApJ, 150, 979
Bradley, P.A. 1993, Ph. D. Thesis, Univ. Texas
Bradley, P.A. 1994, ApJ, submitted
Itoh, N., Adachi, T., Nakagawa, M., Kohyama, Y., & Munkata, H. 1989, ApJ, 339, 354; erratum 1990, ApJ, 360, 741
Kohyama, Y., Itoh, N., Obama, A., & Mutoh, H. 1993, ApJ, 415, 267
Lamb, D.Q., & Van Horn, H.M. 1975, ApJ, 200, 306
Wood, M.A. 1992, ApJ, 386, 539

Part III
Atmospheres I

Non-LTE Line Blanketed Model Atmospheres of Hot, Metal-Rich White Dwarfs

Ivan Hubeny and Thierry Lanz

Universities Space Research Association,
NASA Goddard Space Flight Center, Greenbelt, MD 20771, USA

1 Introduction

At first sight, the title of this paper may seem to be a complete misnomer. First, white dwarfs are not viewed as typical objects for which one needs or requires a non-LTE description. And, second, DA white dwarfs do not seem to be the typical objects for which one invokes metal line blanketing. Yet, as we will show in this paper, both above mentioned effects are not only important, but may in fact become *crucial* for our understanding of hot white dwarfs. The aim of this paper is to explain why this is so, and what we may do about it, i.e. how we construct appropriate model atmospheres.

Let us first set out the stage by specifying the terminology. The acronym LTE stands for local thermodynamic equilibrium, and the term non-LTE (or NLTE) refers, somewhat loosely, to any description allowing for some kind of departure from LTE. In practice, one usually means that populations of selected energy levels of some atoms and ions deviate from the Saha-Boltzmann statistics. For a general discussion, see e.g. Mihalas (1978). The term line blanketing describes an influence of thousands to millions of spectral lines on the atmospheric structure and predicted emergent spectrum.

As it is well known, the DA white dwarfs were originally thought to be of pure hydrogen composition, or to have a stratified hydrogen-helium structure. Consequently, simple pure-hydrogen, or hydrogen-helium model atmospheres were deemed to be sufficient. However, it was convincingly demonstrated in recent years that the atmospheres of at least some hot DA's are appreciably contaminated by heavier elements. Weak lines of C IV, Si IV, and N V were first detected in the *IUE* high-resolution spectra by Dupree & Raymond (1982) and Bruhweiler & Kondo (1983) in two stars, Feige 24 and G191-B2B, which later became the prototypes of hot, metal rich DA white dwarfs. The above authors have presented evidence for their photospheric origin, and suggested that these elements are supported against the downward diffusion by radiative levitation.

Additional evidence came from X-ray observations. Kahn et al. (1984) first demonstrated that the observed data require the presence of some X-ray opacity in the photosphere. They discussed the rôle of "metals", specifically C, N, and O.

However, theoretical diffusion calculations carried out by Vauclair et al. (1979) suggest that these elements cannot be supported. Kahn et al. therefore concluded that it is most likely helium which provides the missing opacity. Subsequent studies either elaborated this idea (e.g., Paerels & Heise 1989), or refined it by considering H-He stratified models (Koester 1989; Vennes & Fontaine 1992).

However, this was not the end of the story. It soon became clear that even the sophisticated stratified H-He models do not provide a satisfactory explanation of the observed EUV and X-ray flux. The most dramatic demonstration of this effect was given by Barstow et al. (1993) who have analyzed *ROSAT* fluxes for about 30 DA white dwarfs and have shown that virtually all DA's with effective temperature hotter than about 40,000 K exhibit substantially lower EUV flux than predicted by pure-hydrogen models. They also convincingly demonstrated that helium alone cannot provide sufficient EUV opacity, and that the heavier elements have to be included. Originally, only C, N, O, Si, (Mg), were considered as "heavy" elements (Paerels et al. 1986; Vennes et al. 1989; Vennes et al. 1991, Vennes 1992, Wilkinson et al. 1992). Later, even heavier elements were detected, namely iron (Sion et al. 1992; Vennes et al. 1992), and very recently nickel (Holberg et al. 1994). Finally, several extremely metal rich DA's were discovered (Holberg et al. 1993). In other words, the presence of heavy metals in the hot DA white dwarf atmospheres seems now to be the rule rather than the exception.

On the modeling side, the assumption of LTE was traditionally adopted in view of the high surface gravity, and therefore high atmospheric densities, in the atmospheres of white dwarfs. However, a high atmospheric density does not necessarily guarantee the validity of the assumption of LTE. For very high temperatures the radiation field is very strong and consequently the radiative rates of many atomic transitions may dominate over the corresponding collisional rates – a signature that the atmosphere may be susceptible to NLTE effects. Auer & Shipman (1977) first discussed this problem in the context of analyzing EUV spectrum of the hot DA star HZ 43, and concluded that NLTE effects are negligible. Wesemael et al. (1980) calculated an extensive grid of pure-hydrogen model atmospheres and demonstrated that the NLTE effects are indeed very small for the UV and visible continua. Nevertheless, they found some NLTE effects in the line centers of the first members of the Lyman and Balmer series. However, it is important to note that they computed the so-called NLTE/C models, wherein the lines are assumed to be in detailed radiative balance, and therefore their results for the hydrogen line profiles should be used with caution.

The demonstration of Wesemael et al. (1980) that NLTE effects are not very important in the atmospheres of hot DA white dwarfs has convinced most workers in the field that it is safe to use LTE models. Subsequent refinements in modeling thus proceeded in the direction of developing adequate approaches for computing models with self-consistent diffusion and helium stratification (Jordan & Koester 1986; Vennes & Fontaine 1992), rather than in investigating NLTE effects.

Once it became clear that the heavier elements must absorb a large fraction of the EUV flux in the very region where the pure-H models predict the maximum flux, it also became clear that this blocked flux must be redistributed to longer-wavelength portions of the spectrum. The assumption that the heavy elements do not influence the atmospheric structure thus became questionable. This point is not only of academic interest; instead, it may be crucial in view of the need for accurate determinations of the basic stellar parameters (effective temperature, surface gravity, chemical abundances), for instance for studying late stages of stellar evolution. Moreover, hot DA white dwarfs, like G191-B2B, are being used as radiometric standards for UV spectrophotometry, for instance with *IUE* and *HST/FOS*. Therefore, reliable model atmospheres of DA white dwarfs are very important in a broad astrophysical context.

Since most of the important opacity sources are located near or shortward of the flux maximum, the corresponding ions which provide the opacity may be significantly influenced by NLTE effects. This may be expected from the analogy from other stellar classes, as for instance the main-sequence early A-type stars. For this stellar class the NLTE effects for hydrogen, and consequently for the predicted visible and near-UV continua, were found to be negligible, while the far-UV opacity sources like C I are significantly influenced by NLTE effects.

The only way how to determine whether the metals influence the atmospheric structure, and whether the metal lines and continua are in turn influenced by the NLTE effects, is to calculate a set of fully consistent NLTE models where the metals are allowed for explicitly; in other words, to calculate NLTE line-blanketed model atmospheres. We note that the LTE line-blanketed model atmospheres for the DAO star Feige 55 were recently presented by Bergeron et al. (1993); they found that the effects of line blanketing on the atmospheric structure and the predicted Balmer line profiles are indeed quite important. However, as we demonstrate later on, the LTE line-blanketed models may overestimate the effects of EUV metal opacity, and therefore these results should be viewed with caution.

The only previously calculated NLTE metal line-blanketed model atmospheres for white dwarfs are those by Dreizler & Werner (1992, 1993). They used the accelerated lambda iteration (ALI) method, applied first to model atmosphere construction by Werner (1986) and later refined by Werner (1989). Werner and collaborators then used the method extensively to compute model atmosphere for the PG1159 stars, which are believed to be precursors of white dwarfs (for a review, see Werner 1992).

Here, we will present several representative NLTE model atmospheres for hot DA white dwarfs. These models are not meant to be used for a theoretical analysis of any particular star. Instead, we will briefly discuss the two following important methodological questions: First, under what conditions can the so-called "trace elements" be left out of the construction of the model atmosphere and included only in the spectrum synthesis. Second, whether LTE is a viable approximation, or, analogously, when does a detailed NLTE description become necessary. A more detailed discussion of the two above questions are presented in a separate paper (Lanz & Hubeny 1995).

2 Numerical Method and Assumptions

We have calculated several model atmospheres under the assumptions of a plane-parallel geometry, and the hydrostatic, radiative, and statistical equilibrium. For the present exploratory purposes, we adopt a relatively simple chemical composition, namely, an atmosphere composed of hydrogen, carbon and iron. We vary the carbon and iron abundances all the way from zero to relatively large values. In all the models we assume an homogeneous chemical composition.

We use the new numerical method developed by Hubeny & Lanz (1993, 1995), called the hybrid complete linearization/accelerated lambda iteration (CL/ALI) method. It combines the advantages of its both constituents; its convergence rate (i.e., the number of iterations required to reach a given accuracy) is virtually as high as for the standard CL method, while the computer time per iteration is almost as low as for the standard ALI method. The method formally resembles the standard complete linearization (Auer & Mihalas 1969; Hubeny 1988, Hubeny & Lanz 1992); the only difference being that the radiation intensity at selected frequency points is not explicitly linearized; instead, it is treated by means of the ALI approach. The method can be applied to calculate metal line-blanketed NLTE model atmospheres, by using the idea of "superlevels" and "superlines" introduced originally by Anderson (1989). For a detailed discussion of the method, the reader is referred to Hubeny & Lanz (1995).

We have computed several model atmospheres for the parameters $T_{\text{eff}} = 60,000$ K and $\log g = 7.5$ that are representative of hot DA white dwarfs with trace metal lines in their spectrum. These models have either a pure-hydrogen composition or they include carbon and/or iron with various abundances. Hydrogen is treated essentially exactly, as described by Hubeny et al. (1994). The first 8 lowest levels are classical NLTE levels, and all higher levels are lumped into a merged level within the occupation probability formalism. We have adopted a model atom for carbon with 12 NLTE levels of C III, 12 NLTE levels of C IV plus one level of C V. Iron is treated as described in detail by Hubeny & Lanz (1995). We consider four ions of iron, Fe IV to Fe VII. Fe IV is represented by 21 superlevels and 109 superlines, which represent 7897 genuine lines; Fe V is represented by 19 superlevels and 82 superlines, which represent 3670 genuine lines; Fe VI is represented by 11 superlevels and 31 superlines, which similarly represent 1100 genuine lines. Fe VII is considered as a one-level ion. The photoionization cross-sections are taken from the detailed Opacity Project tables (Sawey & Berrington 1992).

3 Theoretical Spectra

3.1 UV Continuum and the Hydrogen Lines

In agreement with previous studies (e.g., Wesemael et al. 1980), we found the effects of NLTE to be small for UV and visible continua, as well as for most hydrogen lines. There are some NLTE effects in the center of Lyα, but they are always masked by interstellar absorption (for details, refer to Lanz & Hubeny

1995). In that paper, it was found that the influence of metals in determining the atmospheric structure is more important than NLTE effects for metal abundances around or above 10^{-5}. In such cases, spectroscopic analyses based on pure-hydrogen models overestimate the effective temperature. The only striking NLTE effect for pure-H models is a central emission in Hα, discussed already by Werner (1988 – private communication) and Vennes et al. (1991), who have shown that the central emission is a consequence of a NLTE temperature rise at the stellar surface. We have shown (Lanz & Hubeny 1995) that the height of the central emission increases with increasing metallicity, in spite of the cooling effect of metal lines.

3.2 Carbon Lines

The carbon lines were first detected in the spectra of DA white dwarfs in the high-resolution *IUE* spectra obtained by Dupree & Raymond (1982) and Bruhweiler & Kondo (1983). Henry et al. (1985) have calculated a grid of LTE equivalent widths for a number of lines, including the C III $\lambda1176$ and C IV $\lambda1549$ lines. These predicted equivalent widths were later used to derive carbon abundances for several DA stars (e.g., Vennes et al. 1991, Sion et al. 1992), but the authors found serious discrepancies between the abundance derived from the C III and C IV lines. This was interpreted either as an argument in favor of a non-photospheric origin of carbon, or as a consequence of NLTE effects.

Figure 1 shows the theoretical curve of growth for the C III $\lambda1176$ and the C IV $\lambda1549$ lines. We see that while the equivalent widths of the C IV lines are virtually unaffected by NLTE effects, the predicted C III $\lambda1176$ multiplet is much weaker in NLTE. Consequently, the LTE abundance determinations based on these lines will systematically underestimate the carbon abundance by about an order of magnitude (for $T_{\text{eff}} \approx 60,000$ K). Therefore, we conclude that NLTE effects are *very* important for describing the carbon resonance line formation, and are able to explain previously found discrepancies in abundance determination based on the C III and C IV lines.

3.3 UV Iron Lines

We present in Fig. 2 synthetic spectra in the region between $\lambda1269$ and 1274 Å. We have selected this region because it is rich in iron lines, and moreover contains lines of three ionization degrees – Fe IV, Fe V, and Fe VI. The upper panel shows that the differences between the NLTE and LTE predictions are dramatic. The Fe VI lines are significantly stronger because the NLTE ionization balance favors Fe^{5+}. The Fe V lines are weaker, again due to the NLTE ionization shift; and the Fe IV line at λ 1270.8 Å almost disappears, because for Fe^{3+} the NLTE underpopulation is even greater. For details, see Lanz & Hubeny (1995).

The lower panel of Fig. 2 compares the predictions for the H-Fe and the H-C-Fe models, both assuming NLTE iron line formation. While the Fe VI lines are not influenced by the presence of carbon because Fe^{5+} is the dominant ionization degree, the Fe V lines are stronger for the H-C-Fe model because the

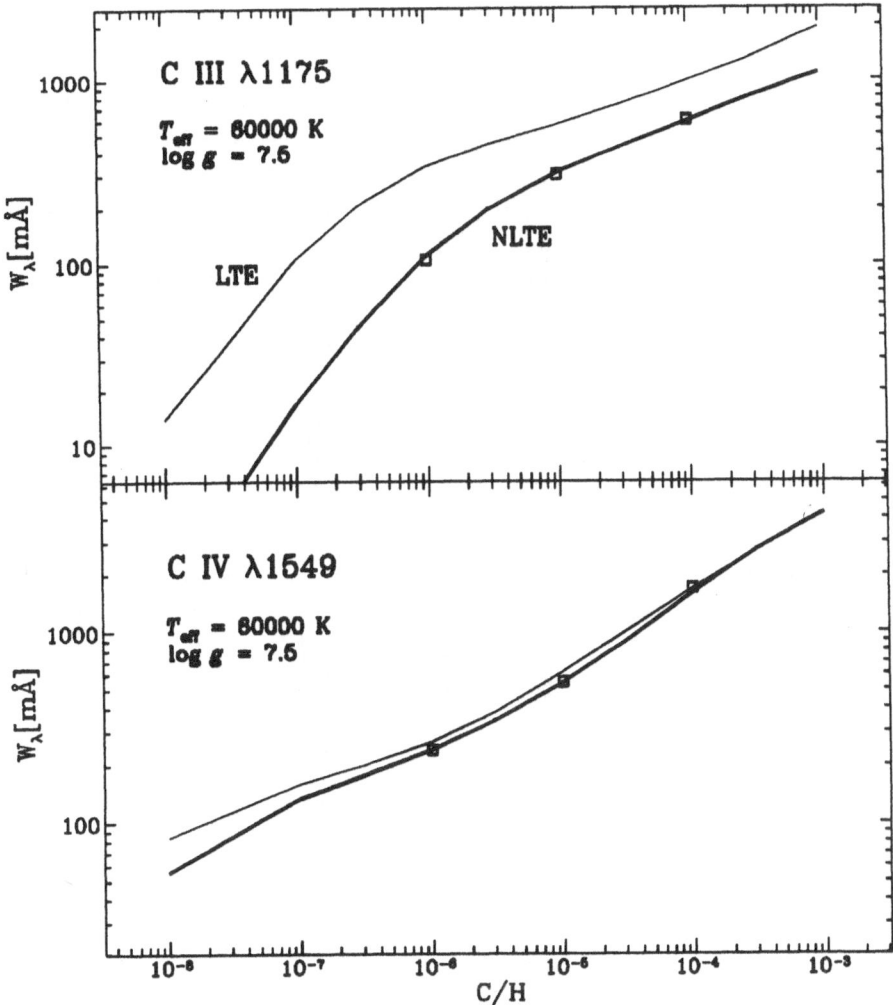

Fig. 1. Theoretical curve of growth of the C III λ1176 (upper panel), and the C IV λ1549 (lower panel) multiplets. Heavy lines are calculated for the NLTE H-C models; thin lines are calculated for the LTE pure-H model atmosphere, with carbon included only in the spectrum synthesis. Open squares represent the results for H-C-Fe NLTE models.

ionization balance is closer to LTE when the carbon EUV opacity is considered. This result dramatically demonstrates that the usual strategy, namely computing first a model atmosphere with a simplified chemical composition (typically pure-H), and then determining abundances of individual elements, may be quite inaccurate because varying the abundance of one element (in this case carbon) influences the ionization balance, and therefore the predicted line strengths, of another element, such as iron.

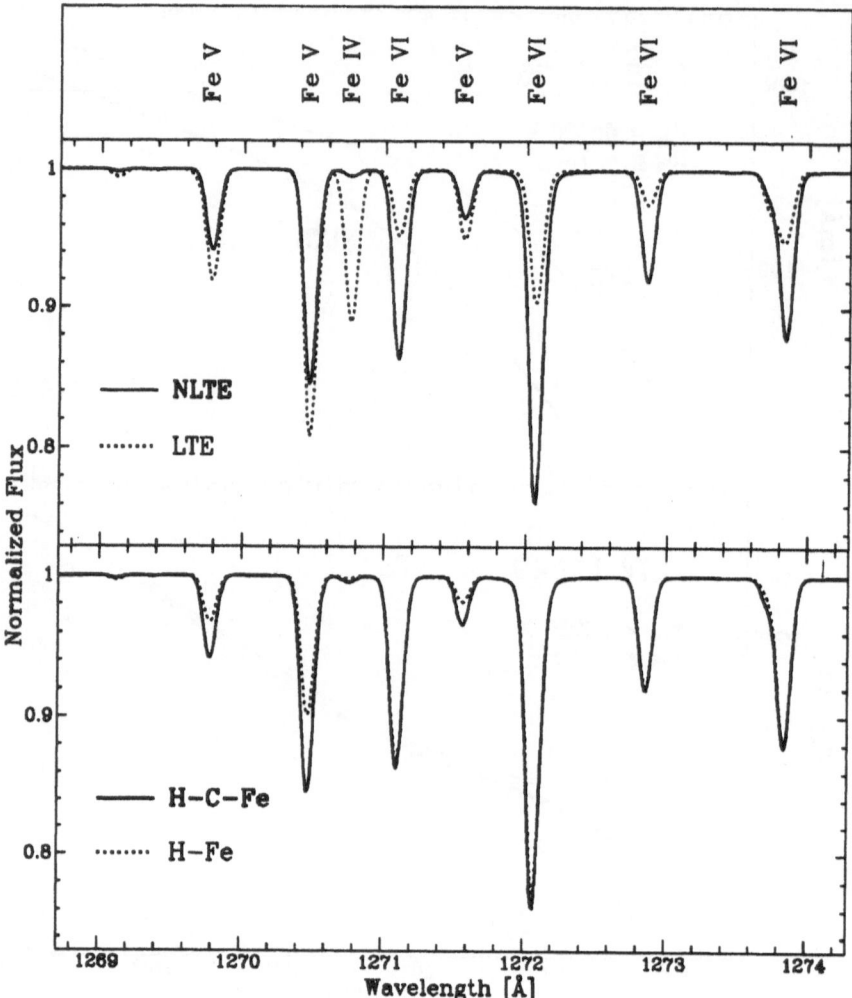

Fig. 2. Predicted iron lines in a typical UV spectral region for various model atmospheres. Upper panel – a comparison of NLTE (full line) and LTE (dotted line) predictions. LTE profiles are calculated for the pure-H LTE model, with iron included only in the spectrum synthesis; Fe/H = 10^{-4}. Lower panel – an influence of carbon on the iron line formation. Full line: H-C-Fe model; dotted line; H-Fe model; C/H = Fe/H = 10^{-4}. The spectra are convolved with a gaussian profile with FWHM = 0.1 Å.

3.4 EUV Spectrum

Finally, we present preliminary results for the most interesting spectral region – the EUV spectrum. Figure 3 shows the influence of iron on the predicted EUV spectrum for several H-Fe models with different iron abundance, together with a pure-H NLTE model. The effects of iron opacity, as well as the influence of NLTE, are quite dramatic. For $N(\text{Fe})/N(\text{H}) = 10^{-4}$, the iron contin-

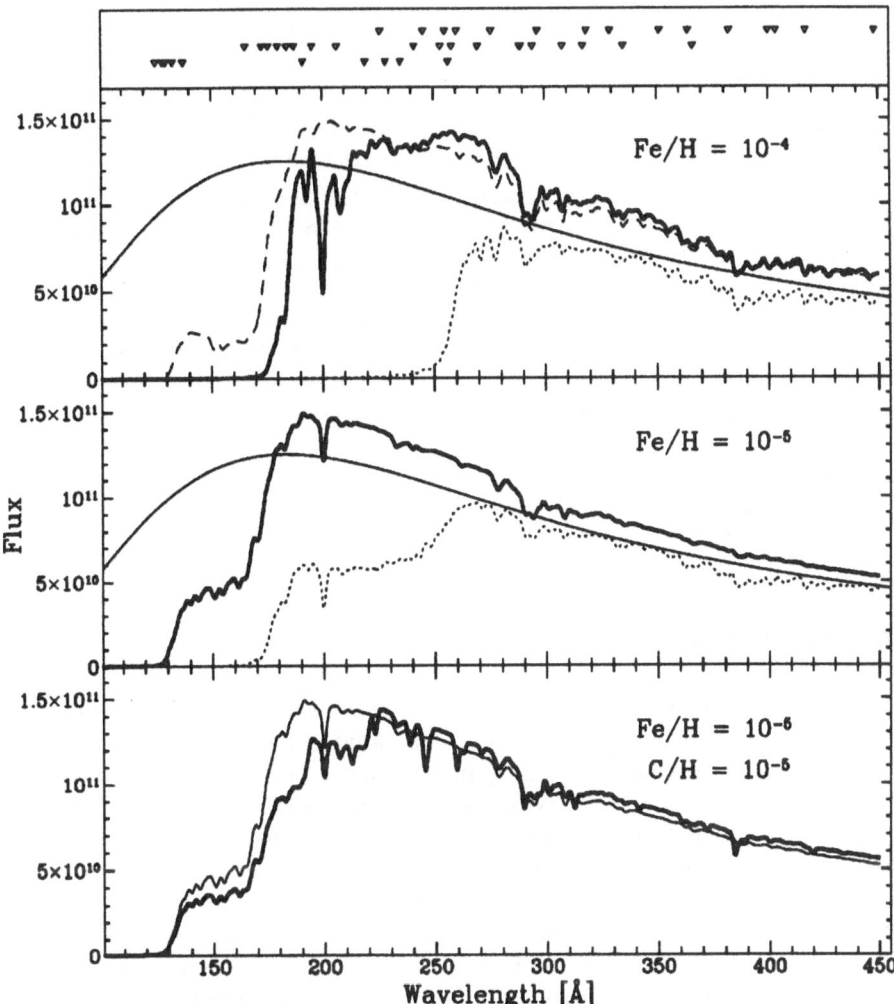

Fig. 3. Predicted EUV spectrum. Thick lines: NLTE models, dotted line: LTE models. Upper panel, dashed line: NLTE model with hydrogenic cross-sections for iron. Thin full line: pure-H model. Upper and middle panel: $Fe/H = 10^{-4}$, and 10^{-5}, respectively. Lower panel: A comparison of NLTE H-C-Fe models (heavy line), and NLTE H-Fe models (thin line), for $C/H = Fe/H = 10^{-5}$. Small triangles at the top indicate (from top to bottom) the positions of edges for the superlevels of Fe IV, Fe V, and Fe VI.

uum opacity removes practically all the flux for $\lambda \leq 180$ Å, where many ionization edges for low-lying states of Fe^{4+} are located. Strong line-like features at $\lambda = 190, 200, 208$ Å, etc., correspond to strong autoionization resonances in the iron cross-sections. The resonance features are obviously missing in the spectrum computed assuming the hydrogenic cross-sections. Both the shape of the spectrum, as well as the detailed predicted features, are significantly different for $\lambda < 230$ Å, which shows that the use of the accurate (e.g., Opacity Project) photoionization cross-sections is mandatory for interpreting the EUV observations. We also note that the position of the flux maximum is displaced to longer wavelengths with increasing metallicity.

Differences between LTE and NLTE predictions in the EUV region are also dramatic. LTE predicts much lower flux, in particular for $\lambda < 260$ Å. This wavelength represents the position of several edges from mildly excited levels of Fe^{4+} and low-lying levels of Fe^{3+}, whose LTE populations are much larger than the NLTE ones. The lower flux for $\lambda > 260$ Å follows from the fact that the LTE spectrum is calculated for a pure-H model structure, which has a lower temperature in the continuum-forming layers. This example shows the inadequacy of treating iron as a trace element. Self-consistent LTE line-blanketed models would solve this particular problem, but the resulting blanketing effect would be exaggerated due to an incorrect ionization balance of iron.

The lower panel of Fig. 3 shows that the effect of carbon on the predicted EUV flux, for abundances Fe/H = C/H = 10^{-5} (due to the C^{3+} edges at $\lambda = 220$ and 190 Å, in particular) is significant, moving the flux maximum to longer wavelengths.

4 Conclusions

The aim of this paper was to discuss two important issues, *i)* for which features are the NLTE effects important; and *ii)* to what extent do the trace elements influence the atmospheric structure; in other words, whether the usual strategy of computing first a simple (typically pure-hydrogen) model atmosphere, and considering metals only in the spectrum synthesis, is reliable. In other words, are NLTE metal line blanketed model atmospheres really necessary for spectroscopic studies of these stars?

We have found, in agreement with previous studies, that the NLTE effects are rather small in the context of pure-hydrogen models. The only exception is a central emission in Hα. In contrast, NLTE effects are crucial in predicting the profiles of the C III and C IV resonance lines. Likewise, NLTE effects on the iron ionization balance are quite dramatic. In addition, we have shown that the iron ionization balance is significantly influenced by the opacity of light elements, as for instance carbon, if its abundance $N(C)/N(H) \geq 10^{-5}$.

The general conclusion of this study is that for low abundances of the "trace elements", i.e. below 10^{-6} relative to hydrogen by number, it is a reasonable approximation to calculate first a pure-hydrogen model, and then include trace elements in the spectrum synthesis. However, for higher abundances, it is increasingly important to calculate self-consistent, NLTE line-blanketed model atmospheres. This procedure becomes mandatory if the abundances of the trace elements are above 10^{-5}. Finally, we have found that the accurate photoionization cross-sections (including autoionization resonances) are very important for spectrum synthesis.

This work was supported in part by NASA grants NAGW-3025 and NAGW-3834.

References

Anderson, L. S. 1989, ApJ, 339, 588

Auer, L. H., & Mihalas, D. 1969, ApJ, 158, 641

Auer, L. H., & Shipman, H.L. 1977, ApJ, 211, L103

Barstow, M.A., et al. 1993, MNRAS, 264, 16

Bergeron, P., Wesemael, F., Lamontagne, R., & Chayer, P. 1993, ApJ, 407, L85

Bruhweiler, F.C., & Kondo, Y. 1983, ApJ, 269, 657

Dreizler, S., & Werner, K. 1992, in The Atmospheres of Early Type Stars, Lecture Notes in Physics, Vol. 401, ed. U. Heber & and C.S. Jeffery (Berlin: Springer), 436

Dreizler, S., & Werner, K. 1993, A&A, 278, 199

Dupree, A.K., & Raymond, J.C. 1982, ApJ, 263, L63

Henry, R.B.C., Shipman, H.L., & Wesemael, F. 1985, ApJS, 57, 145

Holberg, J.B., et al. 1993, ApJ, 416, 806

Holberg, J.B., Hubeny, I., Barstow, M.A., Lanz, T., Sion, E.M., & Tweedy, R.W. 1994, ApJ, 425, L105

Hubeny, I. 1988, Comput. Phys. Commun., 52, 103

Hubeny, I., Hummer, D. G., & Lanz, T. 1994, A&A, 282, 157

Hubeny, I., & Lanz, T. 1992, A&A, 262, 501

Hubeny, I., & Lanz, T. 1993, in Peculiar versus Normal Phenomena in A-Type and Related Stars, eds. M.M. Dworetsky, F. Castelli, & R. Faraggiana, IAU Coll. No. 138, ASP Conf. Ser. 44 (ASP: San Francisco), 98

Hubeny, I., & Lanz, T. 1995, ApJ, 439 (in press)

Jordan, S., & Koester, D. 1986, A&AS, 65, 367

Kahn, S.M., et al. 1984, ApJ, 278, 255

Koester, D. 1989, in White Dwarfs, IAU Coll. 114, ed. G. Wegner (Berlin: Springer), 206

Lanz, T., & Hubeny, I. 1995, ApJ, 439 (in press)

Mihalas, D. 1978, Stellar Atmospheres (San Francisco: Freeman)

Paerels, F.B.S., et al. 1986, ApJ, 309, L33

Paerels, F.B.S., & Heise, J. 1989, ApJ, 339, 1000

Sawey, P.M.J., & Berrington, K.A. 1992, J. Phys. B, 25, 1451

Sion, E.M., Bohlin, R.C., Tweedy, R.W., & Vauclair, G. 1992, ApJ, 391, L32

Vauclair, G., Vauclair, S., & Greenstein, J.L. 1979, A&A, 80, 79

Vennes, S. 1992, ApJ, 390, 590 1989, ApJ, 336, L25

Vennes, S., Chayer, P., Fontaine, G., & Wesemael, F. 1989, ApJ, 336, L25

Vennes, S., & Fontaine, G. 1992, ApJ, 401, 288 1989, ApJ, 336, L25

Vennes, S., Thejll, P., & Shipman, H.L. 1991, in White Dwarfs, eds. G. Vauclair & E. Sion (Dordrecht: Kluwer), 235

Vennes, S., et al. 1992, ApJ, 392, L27

Werner, K. 1986, A&A, 161, 177

Werner, K. 1989, A&A, 226, 265

Werner, K. 1992, in The Atmospheres of Early Type Stars, Lecture Notes in Physics, Vol. 401, eds. U. Heber & C.S. Jeffery (Berlin: Springer), 273

Wesemael, F., Auer, L.H., Van Horn, H.M., & Savedoff, M.P. 1980, ApJS, 43, 159

Wilkinson, E., Green, J.C., & Cash, W. 1992, ApJ, 397, L51

New He I Line Profiles for Synthetic Spectra of DB White Dwarfs

A. Beauchamp[1], F. Wesemael[1], P. Bergeron[1], R.A. Saffer[2], and James Liebert[3]

[1] Département de Physique, Université de Montréal, C.P. 6128, Succ. Centre Ville, Montréal, Québec, Canada, H3C 3J7
[2] Space Telescope Science Institute, 3700 San Martin Dr., Baltimore, MD 21218, USA
[3] Steward Observatory, University of Arizona, Tucson, AZ 85721, USA

1 Introduction

The DB spectral type is assigned to those helium-atmosphere white dwarfs cool enough to display predominantly (or exclusively) the lines of neutral helium in their optical spectra. At low effective temperatures ($T_{\text{eff}} \sim 11,000$ K), they connect to the featureless DC stars, while at high temperatures, there appears to be a gap between the hottest DB star ($T_{\text{eff}} \lesssim 29,000$ K) and the coolest DO star ($T_{\text{eff}} \sim 45,000$ K). The existence of such a gap appears in contradiction with spectral evolution schemes which postulate a parallel and independent evolution for hydrogen-atmosphere and helium-atmosphere white dwarfs. On the other hand, the presence of at least two types of white dwarfs at high effective temperatures raises important questions about the identity and properties of their respective progenitors. Not unexpectedly, our understanding of the fundamental atmospheric properties of both hydrogen-rich and helium-rich white dwarfs remains a key ingredient in our groping with these problems.

Among several methods available for determining atmospheric parameters of white dwarfs, one of the most direct ones is the comparison of the observed line spectrum with the predictions of model atmosphere calculations, parametrized by T_{eff}, $\log g$ and — when appropriate — composition. Much progress has been made recently in spectroscopic studies of DA stars by using the large sensitivity of Balmer lines profiles to the atmospheric parameters, and a mass distribution for DA stars has been derived on that basis for a homogeneous sample of more than one hundred objects (Bergeron, Saffer, & Liebert 1992). In comparison, progress has been slower for DB stars, despite important earlier spectroscopic studies by Koester, Schulz, & Wegner (1981), Wickramasinghe (1983), Wickramasinghe & Reid (1983), Koester et al. (1985), and Wegner & Nelan (1987), as well as the spectrophotometric analysis of Oke, Weidemann, & Koester (1984).

With this backdrop in mind, we initiated, a few years ago, a spectroscopic study of DB stars which is similar in many respects to the Bergeron et al. (1992) study of DA stars. Our aim is to derive the atmospheric parameters of a sample of nearly 50 DB stars through a detailed analysis of their optical spectra (3700-5200 Å) based on a new grid of synthetic spectra. To supplement the optical data, secured at the Steward Observatory 2.3m telescope, we also have in hand ultraviolet spectrophotometry for 25 of those objects from the IUE archives, as well as a handful of spectra, already discussed by Beauchamp et al. (1994), in the 5200-7100 Å range. To complement this observational material, we have conducted, in parallel, a major reexamination of the broadening theories for several neutral helium lines, some features of which will be discussed below.

2 Spectroscopic Determinations of Atmospheric Parameters of DB Stars

2.1 Problems Inherent to the DB Stars

The atmospheric parameters of DB stars are, in general, less well constrained than those of DA stars. There are several reasons for this state of affairs. Firstly, a spectroscopic determination of the atmospheric parameters of DB stars depends sensitively on our understanding of the helium atom. While improvements have been made recently to the continuum opacity of that atom (Koester et al. 1985; Seaton et al. 1992), the treatment of line opacities has not been improved for nearly 20 years. Secondly, the treatment of convection represents a significant uncertainty in the atmospheric structure. Of course, this is a problem both in DA (Bergeron, Wesemael, & Fontaine 1992; Ludwig, Jordan, & Steffen 1994) and DB models, but — in contrast to the situation of the DA stars — almost all the DB stars have a convective atmosphere. To render this problem even more vexing, we find that, in contrast to the findings of Thejll, Vennes, & Shipman (1991), an important fraction of the observed bright DB stars are in the effective temperature range in which the derived atmospheric parameters do depend on the convective efficiency chosen. Thirdly, as a result of their relative rarity, DB stars are, on the whole, rather faint objects.

2.2 The Behavior of He I Lines in DB Spectra

The variation of the line strength in synthetic spectra of DB stars underscores the need for an accurate prediction of the He I line spectrum. It is well known that, in cool DB stars, the line strength increases rapidly with effective temperature. Stars below 11,000 K are essentially featureless DC stars, and the line strength peaks near 20,000 K. At temperatures above this peak, the strength decreases, but only slowly, with effective temperature. The determination of accurate effective temperatures on the basis of the optical He I line spectrum in the hottest DB stars thus requires that detailed broadening calculations be available for as many He I lines as possible.

The gravity determination of DB stars is less of a problem, since many of the He I lines are quite sensitive to log g. Some of these sensitive lines are near the dominant optical He I line, the 2^3P-4^3D $\lambda4471$ transition, and have been widely used before as gravity indicators. However, as has been demonstrated more than 15 years ago (Wickramasinghe 1979), many additional lines in the 3700-4200 Å range are also very sensitive to gravity. Nevertheless, little use has been made in the past of this spectral region, since detailed broadening calculations were not available for the vast majority of these lines (the $2^3P - 5^3D$ transition $\lambda4026$ being the exception).

2.3 Stark Broadening Theories for He I Lines

The major source of line broadening for DB stars hotter than $\sim 15,000$ K is Stark broadening by ions and electrons. Two theories are available for the calculation of Stark profiles incorporating these perturbers. The first, for isolated lines, simply describes the profile as a Lorentzian, and is derived under the twin assumptions of quadratic Stark broadening and impact regime for the electrons. This Lorentzian profile may be quite different from a more accurately-computed profile, because a transition from quadratic to linear Stark broadening may occur near the core for some lines, while the transition from impact to quasi-static regime for electrons typically occurs a few tens of Angströms from the core. In contrast, the theory of Barnard, Cooper and Smith (1974), for overlapping lines, includes the transition from quadratic to linear Stark broadening, the transition from the impact to the quasi-static regime for electrons, and furthermore accounts for the presence of forbidden components. These improvements tend to have some consequences for the asymptotic behavior of the profile in the far wings, among other things.

Since reliable profiles computed on the basis of the second theory were available for five neutral helium lines only, we have generated our own tables of detailed profiles for all those observable lines in the spectra of DB stars which required such a treatment, about twenty of them altogether. Details are provided by Beauchamp (1994).

3 A New Generation of Synthetic Spectra

These improved broadening data have been incorporated in our calculations of a new generation of LTE model atmospheres and synthetic spectra for DB stars. Among additional improvements, our models also include the effects of the He_2^+ molecule (Stancil 1994), which is one of the main sources of continuum opacity in the coolest DB stars, as well as the Hummer & Mihalas (1988) equation of state along with the treatment of pseudo-continuum opacity from Däppen, Anderson, & Mihalas (1987). Additional details are given elsewhere (Beauchamp 1994).

These new ingredients have a dramatic influence on the profiles of some He I lines. Figure 1 compares two normalized synthetic spectra, both computed from the atmospheric stratification of a $T_{\rm eff}$=24,000 K, log g=8.0 model. The top spectrum is taken from our new grid of synthetic spectra, while the bottom one has been calculated with the broadening data available before this study was undertaken. For the five lines with existing tabulated profiles (i.e, He I $\lambda4026$, $\lambda4388$, $\lambda4471$, $\lambda4922$, and $\lambda5016$), changes are only significant in the red wing of $\lambda4026$. Among the other lines, three transitions namely $2^3P - 7^3D$ $\lambda3705$, $2^3P - 6^3D$ $\lambda3820$, and $2^1P - 6^1D$ $\lambda4144$, all gravity sensitive, appear very different in these spectra. For transitions to such excited levels, linear Stark broadening operates over the whole profile, which is almost hydrogenic. Strong forbidden components are blended with the main permitted ones and must be taken into account in those transitions.

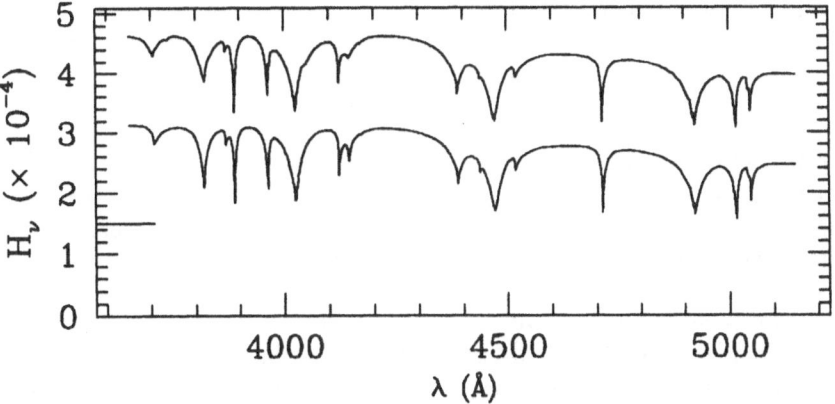

Fig. 1. Synthetic spectra without (bottom) and with (top) the new broadening data

Figure 2 is a preliminary fit to the optical spectrum of the brightest DB star, GD 358, obtained on the basis of our new grid of synthetic spectra. Each line, or group of overlapping lines, has been suitably rectified. The thick line indicates the spectrum corresponding to our current value of the atmospheric parameters of GD 358, namely T_{eff}=25,200 K, log g=7.9, a result compatible with those of Koester et al. (1985) and Thejll, Vennes, & Shipman (1991).

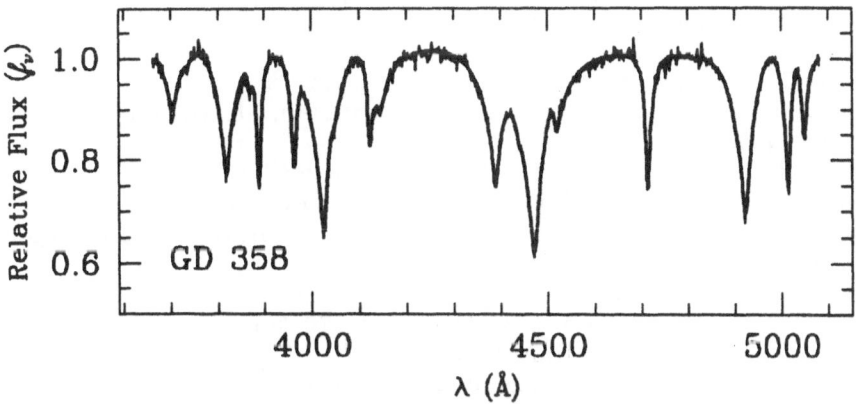

Fig. 2. A preliminary fit to the DB star GD 358

As could be anticipated from Figure 1, a good match in the redder part of the spectrum, above 4300Å, was already achievable on the basis of the limited data available previously, since the lines in that region either are well described by a Lorentzian shape, or have tabulated profiles. By contrast, the quality of the fit has been greatly improved in the blue, where almost all lines required a detailed treatment. Furthermore, the red wing of He I λ4026, which provided a poor fit when studied with the older broadening data (Koester et al. 1985), is now well matched in this latest analysis.

4 Concluding Remarks

Our preliminary, and most encouraging, tests on the spectrum of GD 358 suggest that it is now possible to fit simultaneously the complete line spectrum of DB stars in the optical range, as long as the dominant broadening mechanism remains Stark broadening. These results thus open the door to a systematic study of our much larger sample of DB spectra, as well as to our addressing several important issues related to DB stars: what is their mass distribution, and how does it compare with that of DA stars? What are the spectroscopic boundaries of the instability strip of variable DB stars? What does the luminosity function of helium-rich degenerate dwarfs look like, and what are its implications for our current understanding of the spectral evolution of white dwarf stars?

This work was supported in part by the NSERC Canada, by the Fund FCAR (Québec), and by the NSF grant AST 92-17961.

References

Beauchamp, A. 1994, Ph.D. thesis, Université de Montréal
Beauchamp, A., Wesemael, F., Bergeron, P., Liebert, J. 1994, ApJL, submitted
Bergeron, P., Saffer, R.A., Liebert, J. 1992, ApJ, **394**, 228
Bergeron, P., Wesemael, F., & Fontaine, G. 1992, ApJ, **387**, 288
Däppen, W., Anderson, L., Mihalas, D. 1987, ApJ, **319**, 195
Hummer, D.G., Mihalas, D. 1988, ApJ, **331**, 794
Koester, D., Schulz, H., & Wegner, G. 1981, A&A, **102**, 331
Koester, D., Vauclair, G., Dolez, N., Oke, J.B., Greenstein, J.L., Weidemann, V. 1985, A&A, **149**, 423
Ludwig, H.-G., Jordan, S., & Steffen, M. 1994, A&A, **284**, 105
Oke, J.B., Weidemann, V., & Koester, D. 1984, ApJ, **281**, 276
Seaton, M.J., Zeipper, C.J., Tully, J.A., Pradhan, A.K., Hibbert, A., & Berrington, K.A. 1992, Rev. Mex. Astr. Af., **23**, 19
Stancil, P.C. 1994, ApJ, in press
Thejll, P., Vennes, S., & Shipman, H.L. 1991, ApJ, **370**, 355
Wegner, G., & Nelan, E.P. 1987, ApJ, **319**, 916
Wickramasinghe, D.T. 1979, in White Dwarfs and Variable Degenerate Stars, eds. H.M. Van Horn & V. Weidemann (Rochester: University of Rochester Press), p. 35
Wickramasinghe, D.T. 1983, MNRAS, **203**, 903
Wickramasinghe, D.T. & Reid, N. 1983, MNRAS, **203**, 887

Line Broadening in Hot Stellar Atmospheres

Thomas Schöning

Institut für Astronomie und Astrophysik der Universität München,
Scheinerstr. 1, D-81679 München

1 Introduction

In the past decade outstanding improvements in the development of new techniques and instruments (e.g. CCD detectors, satellite telescopes) have enabled the observation of optical and UV spectra with excellent resolution and high signal to noise ratio. This, in turn, has impelled the development of powerful theoretical methods for spectral diagnostics, among them the sophisticated applications of NLTE line formation theory to the quantitative spectroscopy of hot stellar atmospheres (see Kudritzki & Hummer 1990). However, the quality of the synthetic spectra depends above all on the accuracy of the atomic data. Whereas nowadays the calculation of radiative data represents a minor problem (e.g. Seaton et al. 1994), collisional data, particularly line broadening parameters, are more difficult to obtain. We note here that at the atmospheric conditions of hot stars only Doppler and Stark broadening are relevant. Although computer programs that make the calculation of reliable Stark profiles practical do exist, accurate line broadening data are still lacking for many line transitions of interest, particularly those in multi-electron atoms and ions. These deficiencies are considered to be a large source of uncertainty in the spectral analysis of subdwarf stars (see Werner 1992). In the following sections we briefly review the available Stark broadening data for one- and multi-electron systems and present the results of recent calculations.

2 Hydrogen and Singly Ionized Helium

Hydrogenlike systems are especially sensitive to Stark broadening since the energy degeneracy of the angular momentum eigenstates (neglecting relativistic effects) leads to a linear Stark effect if an electric microfield interacts with the radiating atom. With regard to hydrogen, the unified theory (Vidal et al. 1973) is the standard theory for the calculation of complete Stark profiles. This theory has been extended to lines in the He II spectrum (Schöning & Butler 1989a, 1989b). Apart from the line centres the unified theory profiles usually coincide well with experimental measurements but show much more structure in the core. Although the motion of the perturbing ions plays an important rôle in this part of the Stark profile, it has been neglected in the unified theory. However, we have shown that ion dynamical effects are masked by thermal Doppler broadening for the plasma conditions prevailing in the line forming depths of hot white dwarf stars (Schöning 1994a). As expected, within the framework of a systematic investigation of line formation calculations for hot stellar atmospheres ($T_{eff} \geq$ 20 000 K) Napiwotzki & Rauch (1994) have confirmed the insignificance of ion dynamical effects for the analyses of O,B subdwarfs. Nevertheless, Stehlé (1994) has performed extensive calculations of Stark profile functions using the model microfield method (Brissaud & Frisch 1971) which allows for ion dynamics.

3 Multi-electron Atoms and Ions

3.1 Isolated Lines

The initial and final states of isolated line transitions (e.g. resonance lines) in multi-electron systems are energetically well separated from neighbouring levels and experience merely a quadratic Stark effect in the presence of perturbing electric fields. Apart from a few exceptional cases (e.g. Seaton 1988) fully quantum-mechanically calculated line broadening parameters are not yet available. On the other hand, using semiclassical perturbation methods for impact broadening, Dimitrijević and Sahal-Bréchot (1994 and references therein) have expended considerable efforts on the compilation of extensive tables of Stark broadening parameters for multiplets in the spectra of neutral and ionized emitters of astrophysical interest. Nevertheless, data are lacking for numerous transitions, e.g. in Fe ions, and one has to fall back on less accurate semiempirical formulae for electron impact broadening (Griem 1968, Cowley 1971). We note, however, that even with the semiclassical methods an accuracy of no better than approximately 20 % can be achieved (Dimitrijević & Sahal-Bréchot 1991).

3.2 Overlapping Lines

The spectra of very hot objects show overlapping lines arising from transitions between highly excited levels which are close to degeneracy (e.g. N v $n = 6 \rightarrow n = 7$). In typical subdwarf O,B star atmospheres the densities of slowly moving perturbing particles (ions) are sufficiently large that the interaction energies are of the order of the corresponding energy level separations. Consequently adjacent angular momentum eigenstates effectively mix with each other and the broadening ranges between the quadratic and linear Stark effect regimes. The line shapes may then exhibit complicated structures with overlapping allowed and forbidden components (Fig. 1).

Obviously for overlapping lines one has to include the broadening by slow ion collisions for which the impact theory is not applicable. Therefore it is clear that line broadening parameters calculated with the semiclassical or semiempirical impact methods (Sect. 3.1) fail to reproduce the line shapes. Unfortunately only a few tabulated Stark profiles for selected line transitions in the He I (Gieske & Griem 1969, Barnard et al. 1974) and C IV (Schöning 1993b) spectra have been published. Thus Werner et al. (1991) have introduced an approximate formula for ions with a single valence electron which presumes broadening of hydrogenic line transitions according to Holtsmark's statistical theory of pressure broadening. In order to compensate for the assumption of a linear Stark effect the widths are adjusted by empirical correction factors. The approximate method still overestimates the Stark widths as can be seen from a comparison with the results of more accurate theories (Fig. 1).

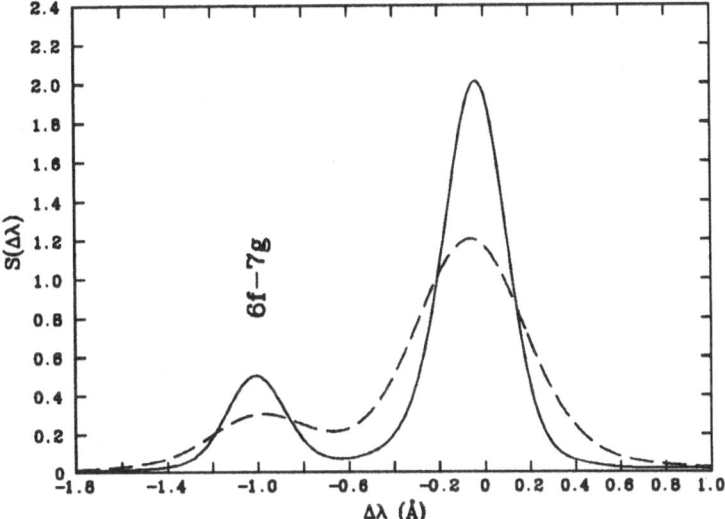

Fig. 1. N v 6f,g,h–7f,g,h,i 4945 Å. Comparison of the Stark profile (dashed) according to the hydrogenic approximation of Werner et al. (1991) with the result (full) of our quantum mechanical calculation (Schöning 1994c). The plasma conditions (electron density $N_e = 1 \times 10^{15} \mathrm{cm}^{-3}$ and electron temperature $T_e = 80\,000$ K) are typical for the line forming depths in hot subdwarf O-star atmospheres. On the left hand the 6f-7g line is partially overlapping with the main component (6g,h–7f,g,h,i). All line shapes are area normalized and the theoretical profiles have been convolved with a thermal Doppler and radiative damping profile.

Fortunately our new quantum mechanical method (Schöning 1993a, 1994b) has proved to be especially useful for the computation of complete Stark profiles for overlapping lines in the spectra of multi-electron atoms and ions. This technique is based on powerful methods developed in recent years for the efficient handling of structure and scattering problems in atomic physics. In particular, the radiator wavefunctions are accurately calculated in the LS coupling approximation using configuration interaction methods and make it possible to simulate the gradual transition from quadratic to linear Stark broadening. Further we treat the electron broadening in the impact approximation and apply the R-matrix version of the close-coupling theory (Berrington et al. 1987) for the quantum mechanical solution of the electron scattering problem on complex targets. The ion broadening is included through an approximate implementation of the model microfield method (Brissaud & Frisch 1971) allowing for the contributions of fast and slow ion collisions.

Presently we are performing extensive numerical computations of Stark profile functions for selected N v lines which will cover the useful range of plasma parameters and chemical compositions for hot star atmospheres (Schöning 1994c). As a first test we have performed line formation calculations for a hot subdwarf O-star atmosphere using the new Stark profiles. Fig. 2 shows the comparison between our results and emerging model line profiles obtained with Stark profiles according to the approximate formulae of Werner et al. (1991). The deviations could be significant for the quantitative high resolution spectroscopy of subdwarf stars but detailed investigations are necessary to clarify this point. Future work will be performed for He i lines involving transitions to highly excited levels and lines in other complex atoms (ions) of astrophysical interest.

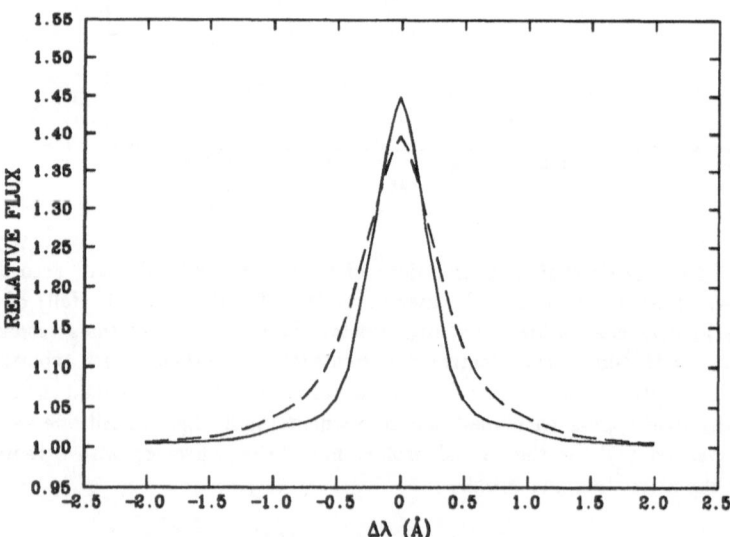

Fig. 2. N v 6f,g,h–7f,g,h,i 4945 Å. Comparison of model line profiles calculated with our new Stark profiles (full) and the hydrogenic approximation (dashed) of Werner et al. (1991). The model parameters are $\log g = 5.3$, $T_{\text{eff}} = 130\,000$ K and $\log (N/He) = -3$.

References

Barnard A.J., Cooper J., Smith E.W., 1974, JQSRT, 14, 1025

Berrington K.A., Burke P.G., Butler K., et al., 1987, J. Phys. B, 20, 6379

Brissaud A., Frisch U., 1971, JQSRT, 11, 1767

Cowley C.R., 1971, Observatory, 91, 139

Dimitrijević M.S., Sahal-Bréchot S., 1991, J. Physique IV, 1, C1

Dimitrijević M.S., Sahal-Bréchot S., 1994, A&AS, 107, 349

Gieske H.A., Griem H.R., 1969, ApJ, 157, 963

Griem H.R., 1968, Phys. Rev., 165, 258

Kudritzki R.P., Hummer D.G., 1990, ARA&A, 28, 303

Napiwotzki R., Rauch T., 1994, A&A, 285, 603

Schöning T., 1993a, J. Phys. B, 26, 899

Schöning T., 1993b, A&A, 267, 300

Schöning T., 1994a, A&A, 282, 994

Schöning T., 1994b, J. Phys. B, 27, 4501

Schöning T., 1994c, A&AS, submitted

Schöning T., Butler K., 1989a, A&A, 219, 326

Schöning T., Butler K., 1989b, A&AS, 78, 51

Seaton M.J., 1988, J. Phys. B, 21, 3033

Seaton M.J., Yu Yan, Mihalas D., Pradhan A.K., 1994, MNRAS, 266, 805

Stehlé C., 1994, A&AS, 104, 509

Vidal C.R., Cooper J., Smith E.W., 1973, ApJS, 25, 37

Werner K., 1992, Evolved Stars. In: Heber U., Jeffery C.S (eds.) The Atmospheres of Early-Type Stars. Springer, Berlin, p. 273

Werner K., Heber U., Hunger K., 1991, A&A, 244, 437

Diffusion in PG 1159 Stars

K. Unglaub and I. Bues

Remeis Sternwarte Bamberg, Sternwartstr. 7, 96049 Bamberg, Germany

1 Introduction

The results of new diffusion calculations will be presented for models with an effective temperature $T_{\text{eff}} = 120000\,\text{K}$ and two different surface gravities: $\log g = 6.0$ and 7.0, respectively. Most of the stars in this region of the $T_{\text{eff}} - \log g$ diagram are PG 1159 stars with atmospheres dominated by the elements He, C, O. (For a review see e.g. Werner (1993)). How do this abundances come about? Have these stars lost not only their H- rich envelope but also most of their He-rich mantle surrounding the C-O core as suggested by Werner et al. (1991)? Or can this atmospheric chemical composition be explained by the interplay of gravitational settling and selective radiative forces? To check this second possibility the chemical composition of the atmosphere will be predicted by taking into account the feedback effects between chemical composition and atmospheric structure. Because of the high abundances expected this is a necessary improvement in comparison to previous calculations (e.g. Chayer et al., 1993, Vauclair, 1989). The elements He, C, N, O, Ne are taken into account. The abundance distribution of these elements will be investigated in the outer regions (comprising about 10^{-7} stellar masses) where they are not yet fully ionized and thus are subject to selective radiative forces due to line and continuum absorption of photons.

The bulk of the results will be published elsewhere (Unglaub and Bues, 1995). The computational method and the physical assumptions will be described there in detail as well. The model parameters used in the present paper have been chosen in view of the spectral analysis of the hot helium- rich white dwarf KPD 0005+5106 which yields $T_{\text{eff}} = 120000\,\text{K}$, $\log g = 7.0$, (Werner et al., 1994). Although the model parameters are typical for PG 1159 stars, however, the derived abundances of C, O are lower by about a factor of 1000 in comparison to the PG 1159 objects.

2 Results

In Figs. 1, 2 the results are shown for $T_{eff} = 120000$K and $\log g = 6.0, 7.0$, respectively. As predicted by stellar evolution theory the C-O core may be surrounded by a thick helium layer ($\approx 10^{-2}M_*$) with C, N, O, Ne as trace elements. The outer regions of the helium layer are assumed to be in a stationary, time-independent state. The physical assumptions used are similar to those in Unglaub and Bues (1993). The radiative energy transport is treated in the diffusion approximation. Continuum absorption cross sections for non-hydrogen-like ions have been taken from Clark et al. (1986), occupation numbers and ionization equilibria have been obtained in LTE. The momentum balance equations for the various species of particles do not provide a simple equilibrium between radiative and gravitational accelerations, but in addition take into account the concentration gradient. Thus they can be solved only if the chemical composition is known somewhere in the atmosphere. We assume number ratios C/He = N/He = O/He = Ne/He = 10^{-3} at the lower boundary of the layers considered here. In the presence of strong radiative forces the surface composition is decoupled from the internal composition and therefore the choice of the inner boundary conditions has little influence on the results. This decoupling has already been mentioned by Vennes et al. (1988) for the case of hydrogen and helium. The Eddington approximation ($T_{(\bar{\tau}=0)}^4 = \frac{1}{2}T_{eff}^4$) is used as outer boundary condition. So we are confronted with a two-point boundary value problem. This has been solved by a special iteration method which will be explained in Unglaub and Bues (1995).

From Fig. 1 we can see that for a surface gravity $\log g = 6.0$ the atmosphere for $\bar{\tau} > 1$ is dominated by heavy elements, helium is predicted to be a trace element in wide regions. This result is due to the momentum transfer by photons to matter via continuum absorption as well as the steepening of the temperature gradient because of the high opacity of the elements oxygen, neon and nitrogen. For $\log g = 7.0$ (see Fig. 2) the atmosphere is dominated by helium . The carbon number fraction reaches a value of about 0.1 in the lower atmosphere where carbon has hydrogen- like configuration, towards the stellar surface it decreases below 10^{-4}. The predicted abundances are lower than in PG1159 stars (log C/He = -0.30, log O/He = -1.0). They may rather apply to KPD 0005+5106 (log C/He = -3.5, log N/He = -4.0, log O/He = -4.0). At $\bar{\tau} = \frac{2}{3}$ we obtain log C/He = -4.4, log N/He = -2.3, log O/He = -3.0, log Ne/He = -2.2. But the comparison with observational results seems to be difficult. E.g. the number ratios of the elements N and O vary within a factor of ten in the optical depth range $1 \leq \bar{\tau} \leq 10$. Similar variations occur for $\bar{\tau} \leq 1$. These layers are probably more important for the line formation, but the results are not so reliable there.

These results may be compared with the most recent results of Chayer et al. (1994). On the one hand these authors neglect the feedback effects between chemical composition and temperature structure and the contribution of the bound-free transitions to the total radiative acceleration, on the other hand their computation of the contribution of the bound-bound transitions is more accurate. They included more transitions using more reliable atomic data and

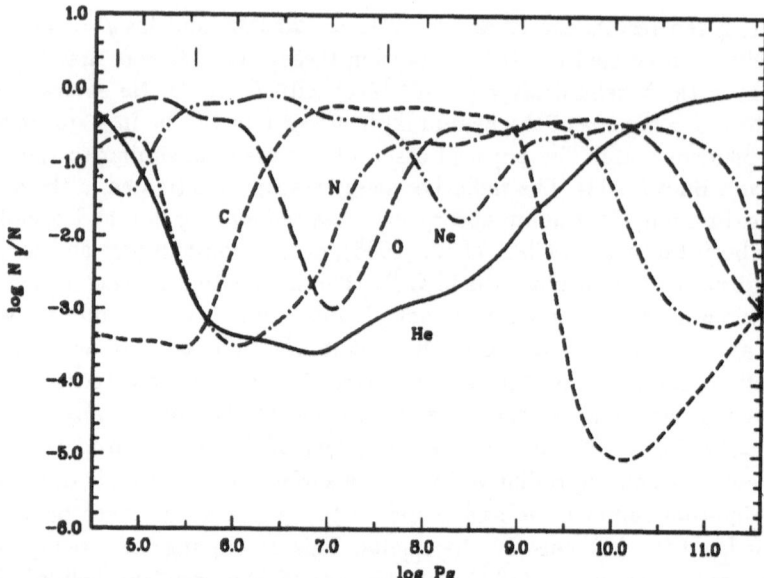

Fig. 1. Number fractions of the elements He, C, N, O, Ne as a function of the gas pressure for a model with $T_{\text{eff}} = 120000\text{K}$ and $\log g = 6.0$. The gas pressure is in SI units! The number fractions are given on a logarithmic scale and are defined as the ratio of all particles of an element over all heavy particles. The four tick marks in the upper part of the figure show where the Rosseland mean optical depth is $\bar{\tau} = 1, 10, 100, 1000$, respectively.

a more realistic Voigt function instead of a simple Lorentzian law for the line profile. According to our experience a comparison of the contributions of bound-bound and bound- free transitions to the total radiative acceleration yields the tendency that at low densities and high abundances the contribution of the bound-free transitions dominates. It must not be neglected for effective temperatures above 100000 K and gravities lower than $\log g = 7.0$. The opposite is true at high densities and low abundances: here the contribution of bound-bound transitions dominates and the feedback effects are negligible. The photons are effectively trapped in the broad line wings, which increases the radiative acceleration due to bound-bound transitions.

For the model $T_{\text{eff}} = 120000$ K, $\log g = 7.0$ our carbon abundance at $\bar{\tau} = \frac{2}{3}$ is lower by a about a factor of eight in comparison to Chayer et al. (1994). We clearly have $g_{\text{rad,L}} > g_{\text{rad,C}}$ for carbon (see Table 1) , therefore the difference is probably due to different input physics in computing $g_{\text{rad,L}}$.

Our nitrogen abundance is higher of about a factor of five, which is probably due to the dominating contribution of $g_{\text{rad,C}}$ to the total radiative acceleration. For oxygen we obtain an abundance lower than about a factor of ten, although $g_{\text{rad,C}} > g_{\text{rad,L}}$ according to our computations. $g_{\text{rad,C}}$ has been maximized by the

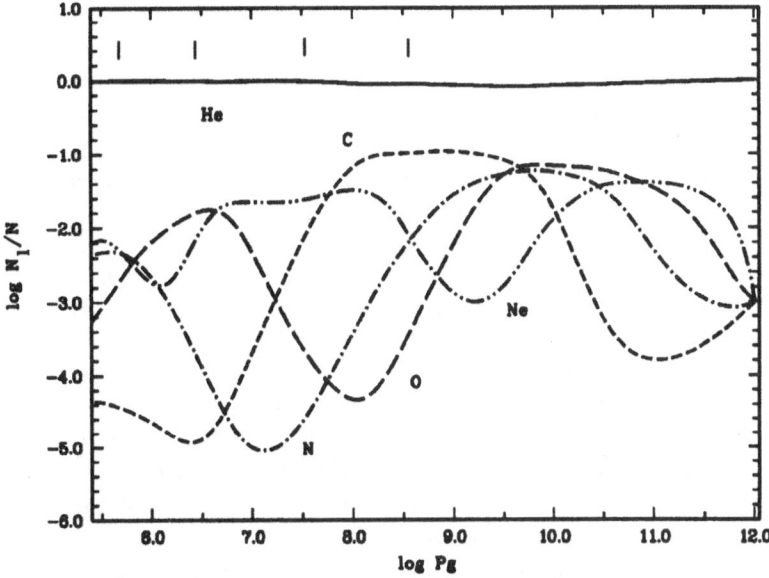

Fig. 2. The same as Fig. 1 for $T_{\rm eff} = 120000$K and $\log g = 7.0$

Table 1. Comparison of the contributions of bound-bound transitions $g_{\rm rad,L}$ and bound-free transitions $g_{\rm rad,C}$ with the total radiative acceleration at $\bar{\tau} = 0.9$ for $T_{\rm eff} = 120000$ K and $\log g = 7.0$.

	He	C	N	O	Ne
$\log g_{\rm rad,L}$	5.96	6.62	5.90	6.11	6.59
$\log g_{\rm rad,C}$	5.83	6.19	6.65	6.43	6.51

assumption that all the photon momentum is transferred to the heavy particle in a bound-free transition. In comparing the various results it is important to know that deviations in the radiative accelerations of, say a factor of two, may lead to deviations in the "equilibrium abundances" of a factor of 20, as will be explained in Unglaub and Bues (1995).

3 Discussion

The results obtained here cannot explain the high carbon- and oxygen abundances observed in the PG1159 stars. The predicted abundances are by far too low. But according to the evolutionary tracks of Wood and Faulkner (1986) the PG1159 stars have passed through regions in the HRD with effective temperatures above 100000 K and surface gravities below $\log g = 7$. In these regions helium is expected to sink and heavy elements to accumulate in the atmosphere. So a helium rich atmosphere could be transformed into a metal rich atmosphere by diffusion, provided that the diffusion time scales are short compared to the time scales of stellar evolution. If the atmosphere is rich enough in heavy elements, probably convection will set in. This will smear out the abundance distribution predicted by diffusion theory and may prevent the heavy elements from sinking back until the star has evolved into the PG1159 region in the HRD. In this scenario the helium layer surrounding the C- O core need not necessarily be extremely thin, but it could as well be thick as predicted by standard evolution theory. However, the low nitrogen abundances found in all PG1159 stars can hardly be understood then. Finally we note that a hydrogen rich atmosphere can never be transformed by diffusion into a helium rich or a metal rich one, as test calculations have shown.

The results indicate that for $T_{\text{eff}} > 100000$ K a simultaneous solution of the radiative transport- and the diffusion equations are necessary to obtain consistent results. Because of the weak coupling between surface composition and internal composition it should be possible to obtain synthetic spectra directly in dependence of T_{eff} and $\log g$ with an iteration method similar to the one used here. Some problems are still to be solved concerning the momentum exchanges due to collisions between the various species of particles (thermal and ambipolar diffusion, for a review see e.g. Michaud (1992)). They are due to the deviations of the particle distribution functions from a Maxwellian one. In difference to the method proposed by Montmerle and Michaud (1976) we assumed in this work that these effects have no influence on the chemical composition gradient.

References

Chayer P., Pelletier C., Fontaine G., Wesemael F. 1993, in White Dwarfs: Advances in Observation and Theory, ed. M.A. Barstow, NATO ASI Ser. C, Vol. 403, p. 261
Chayer P., Fontaine G., and Wesemael F. 1994, Astrophys. J. (submitted)
Clark R., Cowan R., Bobrowicz F. 1986, Atomic Data and Nuclear Tables 34, 415
Michaud G. 1992, Lecture Notes in Physics Vol. 401, The Atmospheres of Early-Type Stars, eds. U. Heber and C.S. Jeffery, Springer Berlin, p.189
Montmerle T., Michaud G. 1976, Astrophys. J. Suppl. Ser. 31, 489
Unglaub K., Bues I. 1993, in White Dwarfs: Advances in Observation and Theory, ed. M.A. Barstow, NATO ASI Ser. C, Vol. 403, p. 221
Unglaub K., Bues I. 1995, Astron. Astrophys. (in preparation)
Vauclair G. 1989, Proc. IAU Colloquium 114: White Dwards, ed. G. Wegner, Springer Berlin, p. 176
Vennes S., Pelletier C., Fontaine G., Wesemael F. 1988, Astrophys. J. 331, 876
Werner K., Heber U., Hunger K. 1991, Astron. Astrophys. 244, 437
Werner K. 1993, in White Dwarfs: Advances in Observation and Theory, ed. M.A. Barstow, NATO ASI Ser. C, Vol. 403, p. 67
Werner K., Heber U., Fleming T. 1994, Astron. Astrophs. 284, 907

Pure Hydrogen in the Spectrum of GD229

Dieter Engelhardt and Irmela Bues

Dr.Remeis-Sternwarte Bamberg, Sternwartstr. 7, D-96049 Bamberg

1 Introduction

The spectra of some white dwarfs with very strong magnetic fields, indicated by strong linear and circular polarization ($>$ 1 per cent), show absorption features, which positions are not understood. Observers and theoreticians have been tackled for a long time by the peculiar nature of the spectrum of GD229, one of these extreme objects with the largest linear polarization observed. It has a very simple characteristic: There are eight very broad features (half width up to 100 Å) in the visible and the IUE -spectrum, which are nearly equally spaced in terms of frequency. The effective temperature is about 20kK (Schmidt et al. 1990).

Another extreme object, Grw+70°8247, could be analyzed successfully with a pure hydrogen model atmosphere by Jordan (1992) at a polar field strength of $3.2 \cdot 10^4$ Tesla, thanks to the extensive calculation of the hydrogen atom in fields up to 10^8 Tesla of the Tübingen group (Rösner et al. 1984). Despite of the simplicity of the spectrum of GD229 all attempts to model the spectrum or even to compare the line positions to these data up to a field of $2 \cdot 10^5$ Tesla failed (Schmidt et al. 1990, Östreicher et al. 1992).

The simplicity of the spectrum of GD229 and the strong linear polarization might indicate a larger magnetic field. Thus we use a new approach: The hydrogen atom within very strong magnetic fields can be investigated in terms of the Dirac equation. The results are compared to line positions.

2 Dirac's Hydrogen Atom and the Adiabatic Approximation

From the investigation of Dirac's hydrogen atom in a strong magnetic field beyond the critical field strength ($B > B_c = 4.7 \cdot 10^5$ Tesla) we draw the conclusion, that no bound states exist in direction of a homogeneous magnetic field (Fig. 1a). If the energy has an extremum perpendicular, quasi-Landau resonances will be shifted to the visible spectrum, illustrated by O'Connell's semi-classical model (1974), introduced for the explanation of the experimentally observed quasi-Landau spacing of the energy eigenvalues close to the continuum limit.

The adiabatic approximation, introduced by Schiff and Snyder (1939) to explain experimental results of Jenkins and Segrè (1939) for the quadratic Zeeman-effect of Na and K, is appropriate to give accurate results within some percent beyond 10^6 Tesla. Within a classical picture the motion of the electron in the direction of a homogenous magnetic field and perpendicular may be separated.

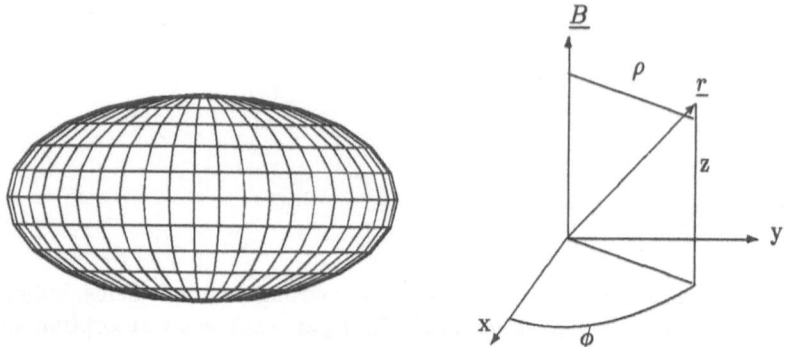

Fig. 1: The elliptical electron cloud of the Dirac hydrogen atom for $B > B_c$, indicated in terms of an elliptical grid (1a, left). Cylindrical coordinates (ρ, z, ϕ) related to cartesian coordinates (x, y, z) (1b).

At first look it is obvious, that within a strong field lines in the optical region may be only due to transitions of an electron in direction of the magnetic field (z), because perpendicular the energy difference of two electron states is of the order of the cyclotron energy.

Thus an investigation of the hydrogen atom in the direction of the magnetic field in terms of the Dirac equation (definition of cylindrical coordinates see Fig.1b) seems reasonable. We start with the extreme case of an arbitrary strong magnetic field. The Dirac equation of the one-dimensional hydrogen atom may be written as

$$\begin{pmatrix} 2\epsilon_z - \alpha/|z| + 1/\alpha & i\sigma_1(\partial/\partial z) + c. \\ i\sigma_1(\partial/\partial z) + c. & 2\epsilon_z - \alpha/|z| - 1/\alpha \end{pmatrix} \Psi = 0, \qquad (1)$$

where ϵ_z represents the energy (normalized to the Rydberg energy) in z-direction, α the fine structure constant , $c.$ is a constant, with respect to z, 2×2 matrix, σ_1 is a Pauli matrix and Ψ is a four component wave function. The ρ dependence of the Coulomb potential is neglected.

Equation (1) can be solved in terms of a Frobenius-expansion $\Psi_i = z^s \sum_\nu a_\nu z^\nu$ $i = 1, .., 4$ ($z > 0$). From the characteristic equation, which reflects the asymptotics $z \to 0$, s must equal $\pm i\alpha$. The complex value of s indicates, that Ψ does not fulfill the completeness relation. But if this usual Ansatz is used, the resulting recurrence relation implies, that no coefficient of the expansion is equal to zero. Thus the expansion is an infinite sum and ϵ_z is continuous. This problem may be solved by a sophisticated mathematical treatment. An easy solution is obtained, if a cut-off procedure similiar to Landau-Lifschitz (1985, III, 112) is applied, to avoid the problem of the inner boundary condition:

$$\epsilon_z = 4log(\sqrt{B_0})^2 \approx 4(\sqrt{B_0} - 1)^2 \approx 4B_0, \tag{2}$$

where $B_0 = B/B_c$ is the magnetic field, normalized to the critical field strength. The difference between the model in terms of the Schrödinger equation (Hasegawa et al. 1961) and the prediction of the Dirac equation for hydrogen is, that in the latter the cut-off procedure from above is needed for all states concerning the z-direction.

The difficulties of the treatment of the hydrogen atom in direction of the magnetic field do not occur in the perpendicular direction.

3 Quasi-Landau Resonances

In the model of O'Connell (1974), the Landau-radius of the electron within a magnetic field (without Coulomb-potential) is simply inserted into the Coulomb-potential. To calculate the energy of an optical transition, the energy is diffentiated with respect to the Landau quantum number n. For $n \rightarrow \infty$ the result must correspond to the classical energy of the system in accordance with the Bohr-Sommerfeld quantization rule. The total energy is:

$$\epsilon_z = 4B_0(n - \frac{1}{2\sqrt{B_0 n}}) + \epsilon_z, \tag{3}$$

where n is the nth-Landau quantum number. The dependence of the Coulomb-potential on z is neglected. The energy difference is derived with respect to n:

$$\frac{\partial \epsilon}{\partial n} = \frac{3}{2}4B_0 - \frac{\epsilon}{2n} + \frac{\epsilon_z}{2n}. \tag{4}$$

If $(3/2)4B_0 - \epsilon/2n = 0$ (corresponds to $\partial(\epsilon - \epsilon_z)/\partial n = 0$) , then

$$\partial \epsilon/\partial n = 2B_0/n. \tag{5}$$

Therefore the optical transitions are spaced equally in terms of the frequency space, if the energy $\epsilon - \epsilon_z$ has an extremum with respect to n.

4 Application and Results

To compare the prediction of the equation (5) with the spectrum of GD229, we used the Fourier-technique. An algorithm, based on the 'Fast Fourier Transform', extended to non equally spaced data (Lomb Algorithm, Numerical Recipes 1992) was used. First we analyzed the IUE spectra, flux versus energy, of Grw+70°8247. The spectral features at 1600 Å, 2100 Å and 3200 Å may be regarded as Landau resonances, with quantum numbers $n = 2$ and a modulation factor one, $n = 1$ and 1.5, and $n = 1$ and 1, respectively. The highest Landau resonance of the Fourier spectrum indicates a field strength of $3.2 \cdot 10^4$ Tesla. The combined IUE spectra of GD229 (SWP7550, LWR8110), flux versus wavelength,

were analyzed. In addition, from the spectrum of Schmidt et al. the positions of
the main features were used to calculate a simple spectrum, with the assumption
that the depressions correspond to Lorentz profiles. The results are plotted in
Fig. 2 and 3.

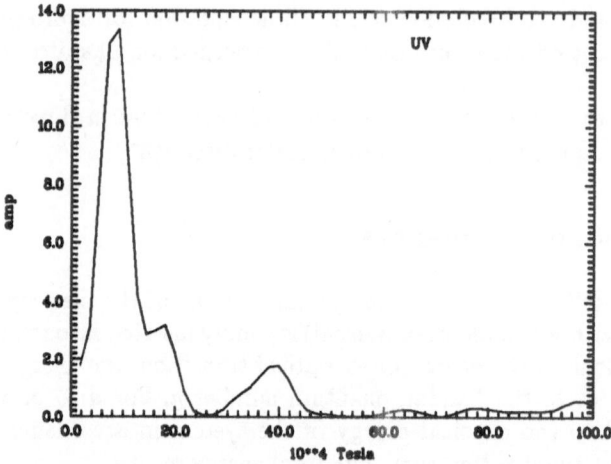

Fig. 2. Fourier amplitude *amp* versus magnetic field strength. Fourier-transform applicated to the combined (F_λ, λ) IUE spectra swp7550, lwr8110 of GD229.

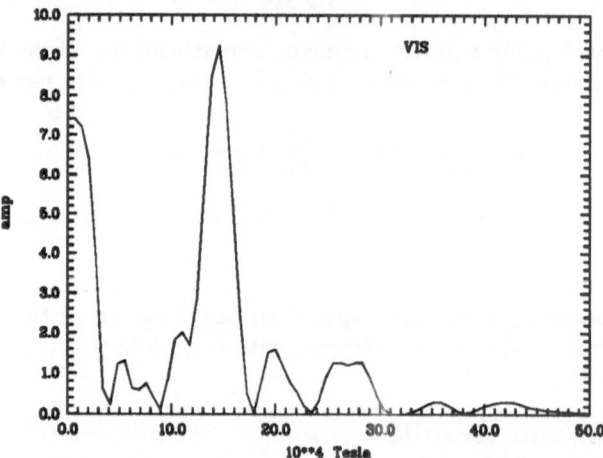

Fig. 3. Fourier- transform applicated to the (F_λ, λ) spectrum of Schmidt et al.

The sum of the Fourier amplitude (*amp*) over all calculated points of the
Fourier spectrum is normalized to 100. From the abzissa, the magnetic field may
deduced, if one assumes that the maximum of the curve in Fig.2 and 3 is due to
the cyclotron energy times $1/n$ (Formula (5) implies a field two times stronger).
With a numerical significance of 100 per cent the lines in the spectrum of GD229
are due to the energy spacing $\epsilon \propto 1/n$.

In Fig.4 formula (5) is compared with the spectrum of Greenstein and Boksenberg (1978, scanned by hand). The field strength is $2.45 \cdot 10^5$ Tesla. The

differences between prediction and observation are about 50 Å and 150 Å.

The simplicity of the spectrum of GD229 and the existence of linear polarization in excess of circular polarization suggest a magnetic field stronger than in the magnetic white dwarf PG1031+234 and might be even larger than the critical field strength of $B_c = 4.7 \cdot 10^5$ Tesla.

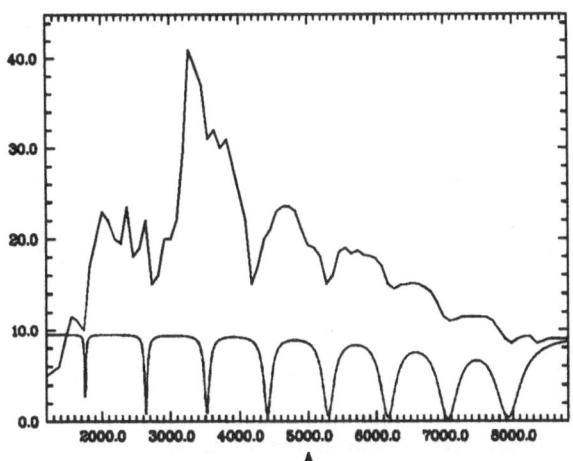

Fig. 4. Flux (a.u.) versus frequency in Å. Model compared to spectrum of Greenstein and Boksenberg (1978) and IUE spectrum (upper curve).

The investigation of the Dirac hydrogen atom within a very strong magnetic field leads to the conclusion, that the solution in direction of the magnetic field, which is stronger than the critical field $4.7 \cdot 10^5$ Tesla, in terms of an usual Frobenius expansion implies, that the energy is continuous. We have shown, that Landau transitions occur in the region of the optical spectrum for a field strength of $2.5 \cdot 10^5$ Tesla, if the energy has an extremum within the plane defined by $z = 0$. Thus the Fourier-analysis of the optical and the IUE spectrum have confirmed the simple spectroscopic rule $\Delta \epsilon = 2B_0/n$, where n is the Landau quantum number. This model is valid for a one electron atom. The temperature of GD229 indicates, that hydrogen only contributes to the absorption features.

References

O'Connell, R., F.: 1974, Astrophys. J., **187**, 275

Greenstein, J.L., Boksenberg, A.: 1978, Mon. Not., R., Astron. Soc., **185**, 823

Hasegawa, H., Howard, R.E.: 1961, J.Phys.Chem. Solids, **21**, 179

Jenkins, F.A., Segrè, E.: 1939, Phys.Rev., **55**, 52

Jordan, S.: Astron.Astrophys., **265**, 570

Östreicher, R., Seifert, W., Friedrich, S., Ruder, H., Schaich, M., Wolf, D., Wunner, G.: 1992, Astron.Astrophys., **257**, 353

Press, W.H., Teukolsky, S.A., Vetterling, W.T., Flannery, B.P.: 1992, 'Numerical Recipes', Cambridge

Rösner, W., Wunner, G., Herold,H., Ruder,H.: 1984, J.Phys.B, **17**,29

Schiff, L.I., Snyder, H.: 1939, Phys. Rev., bf 55, 59

Schmidt, G.D., Latter, W.B., Foltz, C.B.: 1990, ApJ, **350**, 758

Spectroscopic Effects of T-Inhomogeneities in the Atmospheres of DA White Dwarfs

Hans-Günter Ludwig[1] and Matthias Steffen[2]

[1] Max-Planck-Institut für Astrophysik, D-85740 Garching, Germany
[2] Institut für Astronomie und Astrophysik der Universität, D-24098 Kiel, Germany

From detailed 2-dimensional numerical radiation hydrodynamics calculations of time-dependent compressible convection (cf. Freytag, this volume) we have obtained the thermal structure of the convective surface layers of DA white dwarfs with effective temperatures near the blue edge of the ZZ Ceti instability strip (T_{eff}=12 200 K, 12 600 K; $\log g$ = 8.0). Synthetic line profiles of Lα and Hβ computed from the inhomogeneous hydrodynamical models are compared with the spectra resulting from corresponding static plane-parallel model atmospheres constructed to have the same radiative flux as the respective hydrodynamical models at all depths. Mixing-length theory is not needed in this differential approach. We find that in the investigated range of effective temperature no significant errors are introduced in the spectroscopic analysis by neglecting the thermal inhomogeneities generated by photospheric convection.

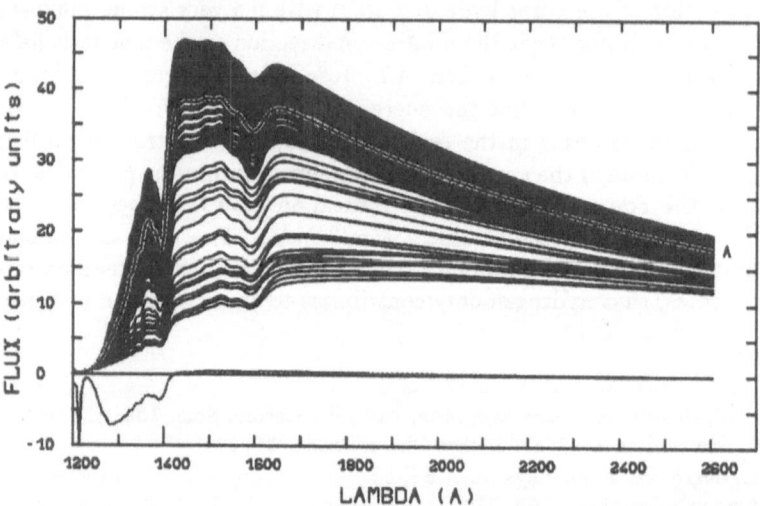

Fig. 1. Lα: 136 flux spectra originating in different parts of the photosphere of a 2D convection model of a DA white dwarf (T_{eff}=12200 K, log g = 8.0). The horizontally averaged (observable) spectrum of the 2D model (labelled 'A') is almost indistinguishable from the spectrum of the equivalent 1D model (not shown) although the surface flux at λ 1500 Å varies by a factor of 5 from the cool downdrafts to the hot upflows. The lower part of the figure shows $(F_{1\mathrm{D}} - F_{2\mathrm{D}})/F_{2\mathrm{D}}$ in %. The flux difference attains a maximum of $\approx -7\%$ near λ 1300 Å and is less than 0.5 % for wavelengths at the red side of the 1400 Å feature. Similar results hold for the 12600 K model.

The π Line Polarization in Magnetic White Dwarfs *

*N. Achilleos[1]** and D. T. Wickramasinghe[2]*

[1] Département de Physique, Université de Montreal, C. P. 6128, Succ. A, Montréal, Québec H3C 3J7
[2] Dept. of Mathematics / Astrophysical Theory Centre, Australian National University, G. P. O. Box 4, Canberra ACT 2601, Australia

The π sub-components of hydrogen line transitions only absorb light which is linearly polarized parallel to a local magnetic field. In the spectra of magnetic white dwarfs, the circular polarization in these π lines is usually expected and often observed to be reduced in magnitude relative to the neighbouring continuum. However, observations of some strongly magnetized white dwarfs show some π lines which have circular polarization equal to that of the adjacent continuum. We show that this is a consequence of the effect of magnetic anomalous dispersion (magneto-optical effects) on radiative transfer.

We do this by considering :

(i) Analytical models of magnetized atmospheres with source functions linear in optical depth.

(ii) More detailed numerical models which include realistic atmospheric structures and opacities.

The equality between continuum and π line polarization is obtained, in theory, at adequately high fields and for special field geometries. We apply our modeling technique, with some success, to the strongly magnetized white dwarfs Grw+70°8247 and PG 1031+234. The former has surface field strengths in the range 160–320 MG, and shows the π effect in the 2s0 − 3p0 hydrogen line. The latter has the highest field observed to date for a white dwarf (up to ∼ 900 MG), but does not show the π effect.

* To be published in "Monthly Notices of the Royal Astronomical Society"
** Current postal address: Dept. of Physics and Astronomy, University College London, Gower Street, London, U. K. WC1E 6BT

Compton Scattering in Stellar Atmospheres at T_{eff} Below 100 000 K

Jerzy Madej

Warsaw University Observatory, Al. Ujazdowskie 4, 00-478 Warszawa, Poland

This abstract reports results of the LTE model atmosphere computations which include noncoherent Compton scattering. Compton scattering represents physically correct description of photon–electron collision, in which both particles can partly exchange their initial momenta and energies. The corresponding LTE equation of transfer in plane-parallel geometry reads

$$\mu \frac{\partial I_\nu}{\partial z} = \kappa_\nu (B_\nu - I_\nu) - \sigma_T I_\nu + \sigma_T J_\nu + \sigma_T Y_\nu \,, \tag{1}$$

where

$$Y_\nu = \frac{kT}{mc^2} \nu^2 \frac{\partial^2 J_\nu}{\partial \nu^2} + \frac{h\nu - 2kT}{mc^2} \nu \frac{\partial J_\nu}{\partial \nu} + \frac{h\nu}{mc^2} J_\nu - \frac{h\nu}{mc^2} \frac{c^2}{h\nu^3} J_\nu \left(J_\nu - \nu \frac{\partial J_\nu}{\partial \nu} \right) \,. \tag{2}$$

Electron scattering coefficient σ_T has been set to the classical Thomson value. The equation of radiative equilibrium resulting from (1) contains explicitly noncoherent scattering terms.

Pure hydrogen model atmospheres were computed at $T_{eff} = 5 \times 10^4, 7 \times 10^4$, and 1×10^5 K, respectively, and $\log g$ ranging from 5.0 to 7.0. Such values of T_{eff} and $\log g$ correspond to the hottest white dwarfs and subdwarf O stars. Compton scattering models were always compared to the twin models computed with the account of the coherent Thomson electron scattering. Comptonized atmospheres exhibit both the increase of gas temperature in the uppermost layers, and flux depression in the soft X-ray domain below $\lambda = 60$ angstr. Effects of Comptonization are most prominent at the higher T_{eff} and at the lowest $\log g$ investigated here. In case of $T_{eff} = 1 \times 10^5$ K and $\log g = 6.0$ the inversion of gas temperature generates emission core in the theoretical profile of the Balmer Hα line.

Results of this research demonstrate the importance of Compton scattering in hot stellar atmospheres with low $\log g$ approaching the Eddington limit. This process can significantly influence both the temperature distribution in the uppermost layers and the emerging X-ray radiation.

Hydrogen Absorption in High Field Magnetic White Dwarfs *

N. Merani, J. Main and G. Wunner

Ruhr-Universität Bochum, Lehrstuhl für Theoretische Physik I

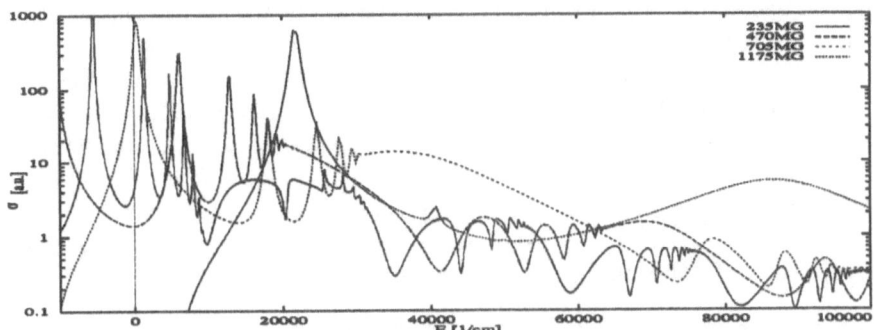

Fig. 1. Photoabsorption cross section of the $|2p0\rangle \longrightarrow |0^+\rangle$ -transition at 235, 352, 470, 705, 1175 MG. Cross section is plotted as a function of energy.

Recent observations of magnetic white dwarfs have revealed the existence of very high field strengths (Tab.1). We have tried to meet the need of accurate opacity data for these objects.

Object	B[MG]	Observer	publ.
PG 1031+234	500	Schmidt et al.	ApJ 86
PG 1015+014	100	Wickramasinghe	MNRAS
		& Cropper	88
G 227-35	130	Cohen et al.	ApJ 93
SBS 1349+5434	760	Liebert et al.	ApJ 94
PG0945+245	670	Liebert et al.	ApJ 93
		Glenn et al.	PASP 94

Table 1. List of recent high field observations.

B[MG]	E[Ry]	λ[A]
235	-0.22482	2 806.9781
352	-0.19982	2 606.3595
470	-0.17037	2 461.7762
705	-0.10301	2 262.3935
1 175	$+0.05048$	2 028.3012

Table 2. Ionization edges and energies for several high field strenghts of the $|2p0\rangle$-state.

A major source of opacity is absorption by atomic hydrogen. Data for b-b transitions can be found in the recent book by Ruder et al. (Springer 1994). There used to be no accurate data for *bound-free* absorption, but the zero-order approximation by Lamb/Sutherland (North Holland publ. 1974) has recently been improved by our group. The theory, proper references, further details etc. concerning these first high field data including photoionization are presented in: Merani et.al., *Balmer and Paschen bound-free opacities for hydrogen in a strong white dwarf magnetic fields*, A&A, 1994, in press.

* Supported by the Deutsche Forschungsgemeinschaft

LTE for the Analysis of White Dwarfs?

Ralf Napiwotzki

Dr. Remeis-Sternwarte, Sternwartstr. 7, 96049 Bamberg, Germany

Analysis of white dwarf atmospheres is traditionally the domain of LTE techniques. Usually deviations from LTE are kept small by the high densities of white dwarf atmospheres. However, it will be shown that for very hot white dwarfs ($T_{eff} \approx 60000\,\mathrm{K}$) the LTE assumption is only valid with important limitations.

Hydrogen and helium composed model atmospheres were calculated with the NLTE code developed by Werner (1986; A&A 161, 177). For details cf. Napiwotzki & Rauch (1994; A&A 285, 603). LTE models were calculated with the NLTE code by drastically enhancing the collisional rates ($\times 10^{20}$) between the atomic levels. An a posteriori check guaranteed that the Saha and Boltzmann equations are fulfilled in every atmospheric layer.

Balmer lines of hot DA/DAO white dwarfs: In a recent paper Bergeron et al. (1994; ApJ 432, 305) reported a strong influence of small traces of helium ($n_{He}/n_H = 10^{-4} \ldots 10^{-5}$) on the Balmer line profiles. The reason for this behavior is the strong He II absorption edge at 228 Å, which cause a heating of the atmosphere. Our comparison ($T_{eff} = 60000\,\mathrm{K}$, $\log g = 7.5$) showed, that this effect is an artifact caused by the LTE assumption. The reason for this different behavior is the dramatic overionization of helium in NLTE. The Balmer lines are much less affected by traces of helium than predicted by LTE calculations. Thus the best that can be done with LTE model atmospheres for the analysis of DA white dwarfs is the use of pure hydrogen models for the Balmer line fits.

The agreement of LTE and NLTE is better for DAO compositions ($n_{He}/n_H = 10^{-3}$ and 10^{-2}). However, there exists a slight trend of the LTE profiles of Hα and Hβ to be too weak. This trend is expected to be larger for lower gravities ($\log g \leq 7.0$) more typical for DAO white dwarfs (Bergeron et al. 1994; Napiwotzki 1993; PhD thesis, Univ. Kiel). Thus for LTE analyses of DAO white dwarfs the correct helium abundance should be used. However, one should be aware of deviations caused by NLTE.

The HeII 4686 Å line of DAO white dwarfs: The He II 4686 Å profiles displayed moderate NLTE effects. This can lead to systematic errors of the order 0.5 dex, if a LTE analysis is performed. For more accurate results the use of NLTE calculations is necessary.

DO white dwarfs: Even for cool DO white dwarfs ($T_{eff} \geq 50000\,\mathrm{K}$) large differences between LTE and NLTE models are recognized for many important lines, e.g. He I 4471 Å (30%), He II 4686 Å (20%), and 4540 Å (10%). The numbers in brackets indicate the deviations of W_λ for the $T_{eff} = 55000\,\mathrm{K}$, $\log g = 7.5$ model. Thus, we strongly discourage the use of LTE model atmospheres for the analysis of DO stars.

Spherically Symmetric and Plane Parallel NLTE Model Atmospheres for Hot High Gravity Stars

Jiří Kubát

Astronomický ústav, Akademie věd České republiky, 251 65 Ondřejov, Czech Republic

Abstract: Theoretical NLTE Hα line profiles of a pure hydrogen atmosphere ($T_{\text{eff}} = 100000K$ and $\log g = 8.0$) are calculated for spherical and planar geometry. Small differences between Hα profiles were found.

The new version of a recently developed computer code ATA that enables calculation of both static spherical and plane-parallel NLTE model stellar atmospheres in radiative equilibrium is applied to pure hydrogen hot high gravity atmospheres ($T_{\text{eff}} = 100000K$ and $\log g = 8.0$). The temperature structures of the model atmospheres differ less than 300K, i.e. 0.5 %. Although there are no differences between continuum flux and Lα profiles, there are small differences between Hα lines. The plane-parallel model predicts higher temperature of the outer layers where the central emission originates. Consequently, it affects Hα line profile. The differences between spherical and plane-parallel models of the hot high gravity atmospheres are very small. Nevertheless, we have found slightly more pronounced differences in the Hα line profiles caused by small differences of the temperature run for our particular model. More detailed analysis of the sphericity effects in the atmospheres of very hot stars will be published in A&A. This work was partially supported by the integral grant of the Academy of Sciences of the Czech Republic No.303401.

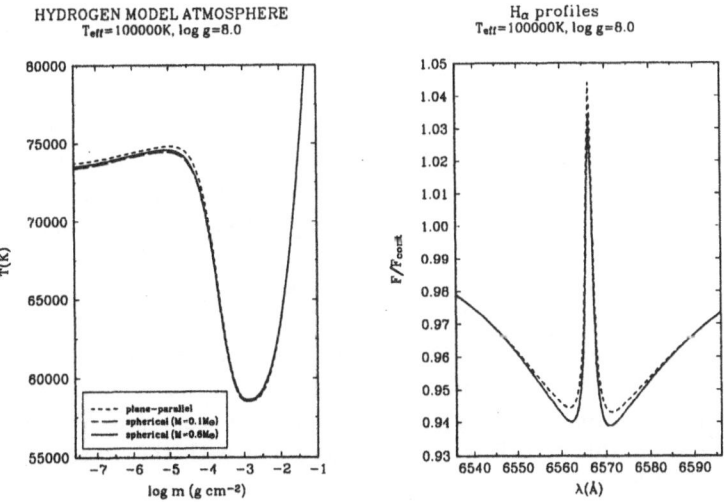

Fig. 1. Temperature structure (left) and Hα profiles (right) in plane parallel and spherical atmospheres.

The Influence of the Bound-Free Opacity on the Radiation from Magnetic DA White Dwarfs

Stefan Jordan[1], Nikil Merani[2]

[1] Institut für Astronomie und Astrophysik der Universität, D-24098 Kiel, Germany
[2] Ruhr-Universität Bochum, Lehrstuhl für Theoretische Physik I, Germany

About 3 % of all white dwarfs show surface magnetic fields with strengths B between a few KG to several hundred MG (Schmidt 1987, IAU Coll. 95, p. 377; Wesemael et al. 1993, PASP 105, 761). As in the case of non-magnetic white dwarfs most of the stars show only hydrogen in their spectra (spectral type DA).

At the moment the highest known field strengths are about 320 MG (Grw+70°8247) and 1000 MG (PG1031+234). Data for the bound-bound transitions of hydrogen in arbitrary magnetic fields have been calculated by groups in Tübingen and Baton Rouge (Forster et al. 1984, J.Phys. 17, 1301; Rösner et al. 1984, J.Phys. 17, 29; Henry and O'Connell 1984, ApJ 282, L97, and 1985, PASP 97, 333). For several objects these data have been used by our group to interpret the spectra and polarization of magnetic white dwarfs (e.g. Jordan 1989, in White Dwarfs, Lecture Notes in Physics 328, p. 333; Jordan et al. 1991, A&A 242, 206; Jordan 1992, A&A 265, 570; Jordan 1993, in White Dwarfs: Advances in Observation and Theory, p. 333; Reimers et al. 1993, A&A 285, 995; Putney & Jordan, submitted to ApJ). However, rather crude extrapolations had to be used for the photoionization cross-sections, which may partly explain discrepancies between the observed polarization and the predictions from the model atmospheres.

Complex energy eigenvalues and oscillator strengths for bound-free transitions are presently calculated (Merani et al. 1995, A&A in press, and these proceedings). We have performed test calculations with a first set of data for six different field strengths (47, 64, 117, 164, 235, and 352 MG). Up to now the grid of photoionization cross-sections is not narrow enough to exclude that possible resonances may be lost due to interpolation. The data also contain oscillator strengths for about 4000 line components of hydrogen, compared to about 150 hitherto available.

As an example we have calculated the energy flux and polarization for a centered dipole model with a polar field strength of 320 MG — the value currently assumed for the prototype Grw+70°8247. Especially at wavelengths below about 5000 Å the predicted flux and polarization differ from the calculations in which simple approximations for the photoionization have been used. We have, however, not yet been able to achieve better agreement with the polarization observed in Grw+70°8247.

Models of 3 Magnetic White Dwarf Stars in Flux and Polarization

Angela Putney[1] and Stefan Jordan[2]

[1] Palomar Observatory, California Institute of Technology, Pasadena, CA 91125, USA
[2] Institut für Astronomie und Astrophysik der Universität, D-24098 Kiel, Germany

The isolated magnetic white dwarf stars G 99-47, KUV 813-14 (KUV 23162-1220), and G 227-35 were observed in linear and circular spectropolarimetry and then compared to calculated theoretical spectra to find a model for the magnetic field strength and structure. The comparisons were to Stokes' V/I (circular polarization) spectra in addition to total flux F_λ, and these add many constraints to the possible solutions. An off-centered dipole or a dipole+quadrupole configuration best fits the observations and this is consistent with the idea that the magnetic Ap and Bp stars — also showing deviations from a centered dipolar field — are the predecessors of the highly magnetic white dwarfs.

The magnetic field of a white dwarf star will vary in strength and direction over the surface of the star. The flux spectrum of such a star reveals the variations of field strength over the observable hemisphere, but gives no direct information about either the direction of the field or the orientation of the magnetic axis with respect to the observed hemisphere. Because of this, a good fit to the flux spectrum alone is possible in more than one region of parameter space. The circularly polarized spectrum (V/I) will give, in addition to field strength variations, the average direction of the visible field, which, in turn, tells us the inclination of the magnetic axis to the line of sight. The addition of circular polarization data to our comparisons caused the parameters we found for G 99-47 and KUV 813-14 to be somewhat different from previous solutions, which did not consider circular polarization. Thus we have demonstrated the usefulness and necessity of including circular spectropolarimetric data in the process of fitting a magnetic field model. The attempts at fitting G 227-35 have made it clear that work needs to be done on developing theories on continuum polarization (the models all showed strong polarization in the blue with a fairly constant decrease to the red, whereas the data showed strong polarization in the red, weakening to $\sim$ 5000 Å, and then increasing again at shorter wavelengths. No combination of parameters could force the models to resemble the polarization continuum data). This work has been submitted to the ApJ, contact authors for preprints/reprints.

Cyclotron Absorption Coefficients in the Stokes Formalism

H. Väth

Institut für Astronomie und Astrophysik der Universität, D-24098 Kiel, Germany
(supas097@astrophysik.uni-kiel.d400.de)

We extend the discussion of the calculation of the cyclotron opacities $\alpha_\pm$ of the ordinary and extraordinary mode (Chanmugam et al.) to the opacities κ, q, v in the Stokes formalism. We derive formulae with which $\alpha_\pm$ can be calculated from κ, q, v. We are hence able to compare our calculations of the opacities, which are based on the single-particle method, with results obtained with the dielectric tensor method of Tamor. Excellent agreement is achieved.

We continue by calculating the cyclotron radiation emitted in accretion shocks that form extended polar caps on magnetic white dwarfs. In previous calculations the three-dimensional extension of the shock was not considered, and instead only a two-dimensional approximation of the geometry of the emission region was made. Therefore rays that pass through the shock without originating from the surface of the white dwarf were neglected. We find that this results in a significant underestimate of the cyclotron flux. Furthermore, it underestimates the degree of linear polarization and overestimates the degree of circular polarization of the radiation emitted by the system.

Part IV
Atmospheres II

IUE Echelle Observations of the Photospheres and Circumstellar Environments of Hot White Dwarfs

J.B. Holberg

Lunar and Planetary Laboratory, University of Arizona, Tucson AZ, USA

Abstract: For well over a decade, narrow, interstellar-like features have been observed in the high resolution IUE spectra of many hot white dwarfs. It is now clear that these features originate in at least two different environments. In many of the hottest white dwarfs these features are formed in the stellar photosphere by trace amounts of key ions of C, N, O, Si, Fe and Ni. In these stars it is possible to establish photospheric abundances which can be directly compared with the theoretical predictions of radiative leviation. Such abundances are also critical to the interpretation the strong EUV and soft X-ray opacities observed in many white dwarfs. In other, primarily cooler, white dwarfs the narrow absorption features appear to be formed in circumstellar shells of unknown origin. Current IUE observational programs together with archive spectra have yielded important new insights into the occurrence of heavy element absorption features in white dwarfs spectra.

1 Introduction

The original motivation for observing hot DA white dwarfs with IUE at high dispersion was that these relatively nearby stars, with their virtually featureless continua, are ideal for probing the local interstellar medium. Although the goal of studying ISM absorption lines using hot white dwarfs has been realized, it was the unexpected discovery of narrow interstellar-like lines (Bruhweiler & Kondo 1981) from highly ionized species associated with the white dwarfs themselves which has produced the most interest. Subsequent observations of nearly every white dwarf bright enough to be observed with IUE in echelle mode has led to a growing of IUE echelle spectra of hot degenerate stars which is still being exploited. These observations have yielded our clearest insights into the nature of the heavy element content of the hottest white dwarfs and provided the abundance data to test the theoretical predictions of radiative levitation. In addition, observations of some of the cooler DAs and hotter He- rich stars provide the best evidence that these high gravity stars manage to produce circumstellar shells.

Listed below are some of the varied phenomena which have been investigated using the IUE echelle mode and the types of object in which these phenomena have been observed.

Photospheric Abundances — hot DA and DO white dwarfs
Constraints on EUV Opacity — hot DA and DO white dwarfs
Circumstellar Shells — cooler DA and a few hot DO stars
Active Mass Loss, Winds — some DO stars
Gravitational Redshifts — binary systems containing a hot WD
Intrasystem Gas — hot DA WDs in binary systems
Interstellar Lines of Sight — all WDs

This paper briefly reviews IUE echelle observations of white dwarfs. The primary topics are the determination of photospheric abundances, and the nature of the circumstellar gas. IUE observations of H-rich DA and DAO stars and He-Rich DO and DB stars are discussed. Largely neglected in this review is an equally large body of observations of planetary nebulae central stars and related objects. These stars exhibit similarly varied and interesting phenomena (see Feibelman & Bruhweiler 1990 and Tweedy 1995). The discussion here primarily deals with SWP spectra (1150 to 1950 Å), however, some results of recent Hubble Space Telescope (HST) spectroscopy are included.

2 A Brief Review of IUE Echelle Observations of Hot White Dwarfs

The first published IUE echelle observations of white dwarfs were those of Bruhweiler & Kondo (1981) who detected narrow interstellar-like lines due to C IV, N V and Si IV in the spectra of the hot DA G191−B2B. Because the observed velocity of these lines conflicted with the existing published radial velocity of the star, the authors hypothesized that these features arose from ionized gas in the vicinity of G191−B2B. Shortly there after Dupree, & Raymond (1982) observed similar features in the DA + dMe system, Feige 24. In this case, however, two distinct Doppler velocities were seen, one identified as originating in the photosphere of the star itself. This was later verified by Vennes et al.(1991) and Vennes & Thorstensen (1994) who demonstrated that the stellar features indeed exhibit the expected orbital motion of the DA.

Other important observations and analyses include those of Bruhweiler & Kondo (1983) and Vennes, Thejll, & Shipman (1991). In table 1 are listed these ions and wavelengths which have been reported. The source of these features, the stellar photosphere, a circumstellar environment, or the interstellar medium is also given together with an example of a star and reference in which such features can be found.

2.1 C, N, O, and Si Abundances in White Dwarfs

The theory of radiative levitation provides a natural explanation for the presence of selected heavy ions in the photospheres of hot DA and DO white dwarfs. As developed by Vauclair, Vauclair, & Greenstein (1979) and others, the outward

Table 1. Features Identified in the Spectra of White Dwarfs

Ion	λ(Å)	Source	Example	Ref(s)
C II	1334,1335⋆	CSM	CD-38°10980	H94
C III	$\lambda\lambda$1175 sextuplet	PHOT	G191-B2B	S92, H95
C IV	1548,1552	PHOT	G191-B2B	BK81
Si II	1190,1193,1260,			
	1304,1526	CSM	CD-38°10980	H94
Si II⋆	1264,1265,			
	1309,1533	CSM	CD-38°10980	H94
Si III	1206	CSM	CD-38°10980	H94
Si III⋆	$\lambda\lambda$1300 sextuplet	CSM	GD394,CD-38°10980	BK83
Si IV	1393,1402	PHOT,CSM	G191-B2B,GD 394	BK81,H94
N IV	1718	PHOT	RE2214-492	H93
N V	1238,1242	PHOT	G191-B2B	BK81
O IV	1338,1343	PHOT	RE2214-492	H93
O V	1371	PHOT	RE2214-492	H93
Al III	1660	PHOT	RE2214-492	H94
Fe V	numerous	PHOT	G191-B2B	H93
Fe VI	numerous	PHOT	RE2214-492	H93
Fe VII	1208,1226,1239,1332	PHOT	PG1034+001	FB90
Ni V	numerous	PHOT	RE2214-492	H94
N I	$\lambda\lambda$1200	ISM		
O I	1302	ISM		
C II	1334,1335⋆	ISM		
Si II	1190,1193,1260			
	1304,1526	ISM		
Si III	1206	ISM		
S II	$\lambda\lambda$1250	ISM		

⋆ Excited States
CSM - Circumstellar gas
PHOT - Photosphere
ISM - Interstellar Medium
BK81 - Bruhweiler & Kondo (1981)
BK83 - Bruhweiler & Kondo (1983)
H93 - Holberg et al. (1993)
H94 - Holberg et al. (1994)
H95 - Holberg et al. (1995)
S92 - Sion et al. (1992)

radiative forces on certain ions are large enough to overcome the high gravitational field of the white dwarf. Indeed, this explanation was put forward by Bruhweiler & Kondo (1983) to explain the presence of such ions as N V, C IV, and Si IV observed in several hot white dwarfs. To quantitatively test this hypothesis, it is necessary to estimate the photospheric abundances implied by the observed equivalent widths of the features seen in IUE spectra. The first such estimates were those of Wesemael, Henry, & Shipman (1984) for Feige 24 and Wolf 1346. These authors used the results of Henry, Shipman, & Wesemael (1985) who computed equivalent widths for a selection of astrophysically important lines as a function of effective temperature and abundance. These results are still widely used to estimate abundances from observed equivalent widths. The only published addition to these tables is that of Vennes, Thejll, & Shipman (1991) for Si III. All published work assumes a pure H, LTE stellar atmosphere.

The theory of radiative levitation is used to establish the predicted equilibrium abundances in photospheric layers. Calculations of this nature have been conducted by several investigators beginning with Morvan, Vauclair, & Vauclair (1986), Chayer et al. (1987), Vauclair (1989) and Chayer, Fontaine, & Wesemael (1989). At present the most comprehensive treatment of this problem is the work of Chayer, Fontaine, & Wesemael (1995) who consider the radiative forces and equilibrium abundances for 12 ions, up to and including Fe, over the relevant range of temperature and gravity for both H-rich and He-rich stars. This is complemented by Chayer et al. (1995) who provide a similar treatment of nickel ions.

The fundamental conclusion of Chayer, Fontaine, & Wesemael (1995) is that the quantitative predictions of radiative levitation are in disagreement with the observed abundances for many elements. Many features of radiative leviation appear to be qualitatively born out by observation. For example, the elements observed are in general those predicted to be present by radiative levitation. Many important aspects observed in the stars, however, are not reproduced by the theory. This is manifest in terms of individual stars with similar temperatures and gravity but significantly different abundances. For example HZ 43, which shows no heavy elements in its IUE spectrum, can be contrasted with other stars such as GD 246 in which features due to C, N, and Si are be seen. The inadequacies of present theory are also apparent in terms of the details of the observed abundance patterns. For example, Chayer, Fontaine, & Wesemael find predicted C abundances to be in excess of those observed while Si abundances are under predicted. The authors stress that the failure of radiative levitation are likely due the neglect of important physical processes in white dwarf atmospheres which compete with radiative forces. Among those suggested, is a mechanism for the effective expulsion of these ions from the stellar atmospheres. It is interesting to note that the circumstellar material discussed in Section 3 may be related to such processes.

It is not only the predictions of radiative levitation theory which are deficient. The abundances derived from LTE calculations and utilizing models containing only one element at a time may also be inadequate. The former is evident in G191−B2B, where strong features of both the C III and C IV ions are present. In figure 1 is shown the region of the coadded G191−B2B spectrum of Holberg et al. (1994) in which the C III λ 1176 multiplet can be seen. Using the LTE calculations of Henry, Wesemael, & Shipman (1985) to separately estimate the C abundance for each ion, yields log (C/H) of -6.4 and -5.7 from the observed C III and C IV lines, respectively. This is similar to what was done by Vennes, Thejll, & Shipman (1991) to suggest a non-photospheric origin for certain lines seen in other white dwarfs. Such an explanation is not tenable for G191−B2B, where all available evidence points to a photospheric origin for C as well as other lines. It has been demonstrated (Lanz & Hubeny 1994 and Hubeny & Lanz 1995) that non-LTE calculations may help to resolve this discrepancy. Lanz & Hubeny and Hubeny and Lanz also demonstrate the importance of considering possible correlations between the trace abundances of different ions such as C and Fe in some white dwarf atmospheres.

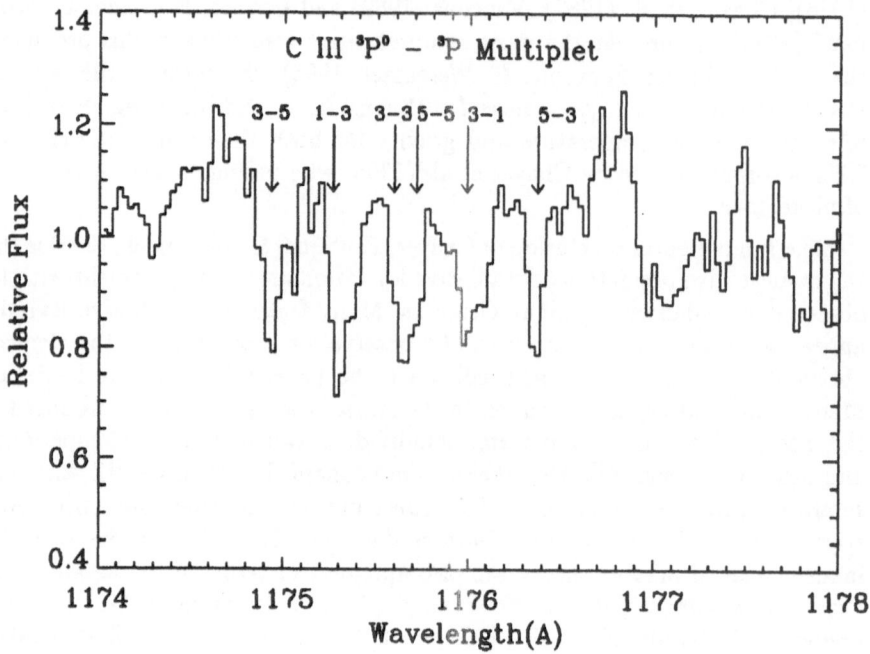

Fig. 1. Coadded IUE spectrum of G191−B2B showing C III features. The arrows indicate location and J values of each transition.

Another primary use of photospheric abundances derived from IUE observations is the interpretation of the white dwarf EUV spectra. In general the ions seen in the UV are also those which posses a number of strong transitions and ionization edges at EUV and soft X-ray wavelengths. Even at the resolution of the Extreme Ultraviolet Explorer (EUVE), these numerous features produce a heavy blanketing of the stellar continuiuum. From the review of white dwarfs observations at EUV wavelengths in this volume by Barstow et al. (1995), it is evident that individual components of this short wavelength opacity are often difficult to identify. The abundances derived by IUE therefore offer a good starting point for the theoretical modeling of such stellar atmospheres. It is also clear, however, from the examples shown in Barstow et al. (1995) that, even accounting for IUE derived abundances, observed short wavelength opacities are only partially accounted for.

2.2 Iron-Group Elements

Perhaps the most unexpected constituents of the photospheres of hot DA white dwarfs have been iron-group elements. Ions having atomic masses of 26, would seem too heavy to remain in the photosphere of a white dwarf. However, the higher ionization stages of these elements possess a large number of transitions in the UV as well as many strong EUV features. Calculations (Chayer, Fontaine, & Wesemael 1991 and Chayer et al. 1995) indicate that the radiative forces on these ions are indeed sufficient to levitate them in the photospheres of very hot DA stars.

Features due to Fe VII were initially reported by Feibelman & Bruhweiler (1990) in a number of planetary nebula nuclei and in several hot DOZ white dwarfs including PG 1159−035, PG 1034+001, and KPD0005+5106. In DA white dwarfs, features due to Fe V were first reported in HST Faint Object Spectrograph spectra of G191−B2B by Sion et al. (1992) and in coadded IUE spectra of this star by Tweedy (1991) and Vennes et al. (1992). The latter authors also observed Fe V and Fe VI features in Feige 24 and estimated Fe abundances from observed equivalent widths in both stars. Vidal-Madjar et al. (1994) have also reported Fe V lines in high dispersion HST spectra of G191−B2B. Strong Fe V lines are also reported in archive IUE spectra of the hot DAO star Feige 55, by Lamontagne et al. (1993) and Bergeron et al. (1993). Holberg et al. (1993) obtained IUE echelle observations of two bright DA stars, RE2214−492 and RE0623−377. In both these stars, a large number of lines due to Fe V and Fe VI were clearly evident. Subsequently, Holberg et al. (1994) reported on results from the coaddition of five SWP spectra of RE2214−492 and ten spectra of G191−B2B. Present in these higher S/N spectra are many features due to Ni V. Werner & Dreizler (1994) similarly report the detection of Ni V features in G191−B2B, Feige 24, RE2214−492 and RE0623−377.

Many of the above papers also determine photospheric abundances of Fe and Ni relative to H from observed equivalent widths. These abundance estimates are compared in table 2. Although the sources of many of these determinations are similar, the models, assumptions, and methods on which these estimates are based often differ. For example, Vennes et al. (1992) and Vidal-Madjar et al. (1994) both use LTE models to estimate Fe abundances in G191−B2B, while Holberg et al. (1994) and Werner & Dreizler (1994) use non-LTE models. In addition, Vennes et al. (1992), Holberg et al. (1994), and Werner & Dreizler (1994) estimate abundances by visually matching observed Fe features with those in synthetic spectra, while Holberg et al. (1994) quantitatively compare observed and measured Fe equivalent widths. In addition to the above differences, Lanz & Hubeny (1994) have pointed out the potential importance of other elements such as C on Fe abundances. Ideally the most reliable abundance estimates for Fe and Ni should be obtained from self-consistent analyses in which the abundances of all major ions are estimated simultaneously.

Table 2. Observed Fe and Ni Abundances in DA White Dwarfs

Star	V92	H93	H94	W94	VM94
		Log(Fe/H) Abundance			
G191-B2B	-5.17±0.35	....	-4.5	-5.15±0.15	-5.59±0.23
Feige 24	-5.15±0.15	....	....	-5.0±0.08	....
RE2214-492	....	-4.25±0.25	-4.0	4.25±0.25	....
RE0623-377	....	-4.25±0.25	-4.25±0.25	....	....
		Log(Ni/H) Abundance			
G191-B2B	....	....	-6.0±0.3	-6.0	....
Feige 24	....	....	....	-5.65±0.35	....
RE2214-492	....	....	-5.50.3	-4.4±0.3	....
RE0623-377	....	....	....	-4.65±0.35	....

V92 - Vennes et al. (1992)
H93 - Holberg et al. (1993)
H94 - Holberg et al. (1994)
W94 - Werner & Dreizler (1994)
VM94- Vidal-Madjar et al. (1994)

Werner & Dreizler (1994) have used their Fe and Ni abundances to suggest that Ni/Fe abundance ratios in hot white dwarfs are in excess of the solar value of 20. They suggest that this discrepancy is evidence that Fe and Ni are levitated by radiation. However, Chayer et al. (1995) have presented radiative levitation calculations which show, that while the observed levels of Fe and Ni are compatible with models, the predicted Fe/Ni ratios are not in accord with observation. They go on to suggest that this is further evidence for additional processes which compete with levitation.

In summary, the observed abundances derived for many of the heavier elements detected in hot white dwarfs can be used to test the predictions the theory of radiative levitation and to complement the analysis of EUVE spectra of these stars. In general, however, such comparisons of theory and observation show radiative levitation alone cannot account for the heavy element abundance patterns observed in hot white dwarfs. It is also clear from the limited number of analyses which have been done, that the EUV spectra of many white dwarfs contain significant opacity in addition to that which can be accounted for with observed UV lines alone.

3 Circumstellar Gas

As mentioned previously, Dupree & Raymond (1982) observed two Doppler components in the highly ionized lines in Feige 24, one of which was associated with the stellar photosphere. The second, stationary, component was attributed to interstellar gas photoionized by the hot white dwarf. Because Feige 24 is a pre- cataclysmic binary, it has been argued (Sion & Starrfield, 1984) that the photospheric features are the result of a wind from the red dwarf accreting on to

the white dwarf. Such a wind might also account for the presence of the circumstellar component. Indeed, Vennes & Thorstensen (1994) have shown that He
II λ 1640 features appear when the white dwarf is near inferior conjunction and
that there also exist stationary low velocity features due to the C IV resonance
doublet.

The existence of non-photospheric features in the IUE spectra of isolated
white dwarfs, however, is less easily explained. The discovery (Bruhweiler &
Kondo 1983) in the DA star GD 394 of quasi-metastable Si III transitions, whose
lower levels are 6.5 eV above ground was a clear indication that relatively dense
shells of ionized gas exist in the vicinity of at least some white dwarfs. In the
case of this star, a large velocity difference between the stellar photosphere and
the lines appear to rule out a photospheric origin. Similar Si III features were
subsequently observed in the cooler DA, CD -38^0 10980 by Holberg et al. (1985)
but, at that time, were assigned to the stellar photosphere. Bruhweiler (1984)
also found similar blue shifted features due to C IV, N V and Si IV in the DA
white GD 71.

The most systematic examination of circumstellar features in DA white
dwarfs is that of Vennes, Thejll, & Shipman (1991) who reviewed the IUE
echelle spectra of some 25 DA stars. Their work resulted in several new DA
stars in which differences between IUE and existing photospheric velocities indicate the existence of circumstellar gas. They further demonstrated that, in
some DA stars, abundances derived for different ionization states of the same
element lead to inconsistencies, if the ions are assumed to resided in the stellar
photosphere. This provides independent evidence of the non-photospheric origin
for the lines seen in several stars, in particular for CD -38^0 10980 and GD 394.
Finally, they made first-time estimates of lower limits on the electron density
required to collisionally excite Si II features seen in some stars.

Recent HST GHRS observations (Shipman et al. 1995) of GD 394, while
largely confirming earlier IUE results, have also revealed presence of the Si III
λ 1417 $^1P^0 - ^1S$ feature, in which the lower level is 10.3 eV above ground. This
implies a very strong source of collisional excitation. In addition to the Si III
and Si II features, GD 394 also posses some of the strongest Si IV resonance
lines (Bruhweiler & Kondo, 1983) found in any white dwarf (see figure 2).
Two interesting aspects of these lines are the large width, 100 km s^{-1}, and the
unsaturated line cores. This suggests that these lines are formed in a region with
a substantial velocity gradient.

Holberg, Bruhweiler & Andersen (1995) used an 11 spectrum coaddition of
SWP images of CD -38^0 10980 to conduct a detailed investigation of circumstellar features seen in this star. Using accurate ground based and IUE velocities
they showed that the relative velocity between stellar photosphere and the circumstellar features is -12.1 $\pm$2.0 km s^{-1}. They also arrived at first time column
densities for the excited and ground state populations of the Si III transitions.
This together with a measured Si III/Si II ionization ratio, allowed estimates of
the electron densities in the circumstellar gas. It was found that if only collisional
excitation is involved then the required electron densities must be in excess of

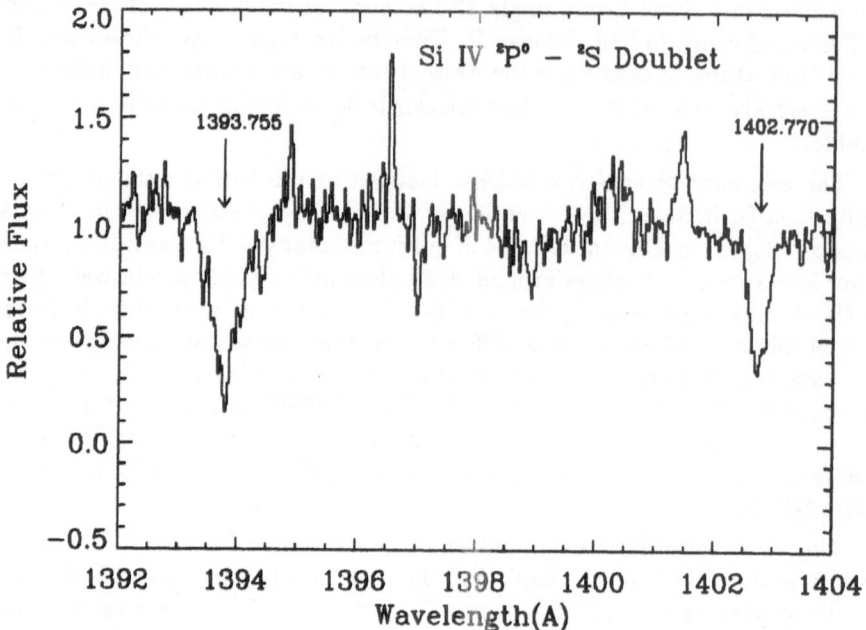

Fig. 2. Coadded IUE spectrum of GD 394 showing the extraordinarily broad components of the C IV doublet.

10^9 cm^{-3} . However, if the gas is near the star and photoexcitation is dominant then much lower electron densities are possible. Using this fact and the relative velocity between the circumstellar features and the stellar photosphere, the authors were able to locate the circumstellar gas deep within the gravitational well of CD -38^0 10980.

In summary, there is now compelling evidence for shells of ionized gas in the vicinity of some white dwarfs. How such circumstellar shells come in to being and how they are maintained remains unknown. In at least two stars, CD -38^0 10980 and GD 394, the electron densities in this gas must be substantial. It is equally unclear where these shells reside in relation to the stellar surface. The determination of precise photospheric velocities for the stars would help to clarify this situation considerably. Only in the case of CD -38^0 10980, with the help of precise velocities and circumstantial arguments concerning the electron density, has it been possible to be specific about the location of the gas. The apparent inability of the theory of radiative levitation to correctly predict the observed abundances of key ions in white dwarf photospheres has led to suggestions that mass loss may be responsible for the abundance discrepancies. The existence of circumstellar shells could well represent evidence of such a process.

4 IUE Archives and IUE Data

The IUE archives currently (1 Nov. 1994) contain the SWP echelle spectra of some 40 DA and DAO white dwarfs and 7 DO and DB stars. In all this represents a total of 161 spectra of H-rich stars and 16 for He-rich stars. The distribution of spectra is somewhat uneven with about 20 of the stars having only a single image while others have from 11 to 16 images. The brightest white dwarfs, such as Sirius B, can be fully exposed in as little 600 s, while the fainter stars at 15th magnitude require 45,000 s to 60,000 s; the average exposure is approximately 20,000 s.

The IUE SWP camera has a spectral resolution of approximately R = 10,000 over the range 1150 to 1950 ÅThis leads to an effective resolution of 0.08 Å or 30 km s^{-1} with data sampled at intervals of 0.035 ÅThus, interstellar and most of the weaker stellar features are not resolved and the available information consists of the centroid and the equivalent width of the feature. On the other hand, the IUE telescope and cameras have proved to be exceptionally stable over the years and well centered spectra have wavelength scales internally accurate to ±3 km s^{-1}.

It is well known that IUE echelle spectra are plagued by inherently low signal-to-noise (S/N). Although individual spectra of very bright stars may reach S/N 20, the spectra of most white dwarfs seldom achieve better than S/N 10 and are often much worse. This makes the identification of weak and narrow- interstellar like features more of an art than a science.

There are several techniques which can increase the effective S/N of IUE echelle data. A common method used to search for multiple lines from a single ion is to coadd the lines in velocity space. In this way, weak features sharing a common velocity can often be found. A second method is to coadd independent spectra of the same object, either over the entire 800 Å range of the SWP camera or in smaller wavelength windows. This method obviously requires multiple observations of the same object. Coaddition is often difficult to implement for white dwarfs, especially if there are few well defined stellar or interstellar features which can be used as fiducial points to register the spectra. The wavelengths of features can differ by small amounts due to a) the placement of the star within the SWP large aperture, b) satellite and earth Doppler velocities, and c) the empirical dispersion relations used by the IUE data extraction procedures. Items b) and c) are, in general, well corrected for by standard IUE spectral extraction procedures. However, it is best to check, especially for older data, that these corrections have been applied. One major problem that has recently come to light is the use of erroneous dispersion relations for images obtained between 1984 and 1990. However, a simple algorithm (Garhart 1992) can be used to correct assigned wavelengths. The placement of the stellar image within the 10 by 20 arcsecond IUE large aperture is the most significant factor in causing wavelength shifts.

The method this observer has found most effective with white dwarfs having weak features is to use the common interstellar features and the stellar features in each image to define a wavelength off set for that image. The spectra are then shifted, resampled and coadded. In general this procedure works quite well and yields a real improvement in S/N.

This work has been supported by NASA grant NAGW5–2738.

References

Barstow, M. A., Holberg, J. B., Koester, D., Nousek, J. A., & Werner, K., 1995, in this volume

Bergeron, P., Wesemael, F., Lamontage, R., & Chayer, P., 1993, *ApJ*, **407**, L85.

Bergeron, P., Wesemael, F., Beauchamp, A., Wood, M.A., Lamontagne ,R., Fontaine, G., & Liebert, J., 1994, *ApJ*, in press

Bruhweiler, F. C., 1984, in Future of Ultraviolet Astronomy Based on Six Years of IUE Research, ed. J. M. Mead, Y. Kondo & R. Chapman (NASA CP2349), p. 269.

Bruhweiler, F. C., & Kondo, Y., 1981, *ApJ*, **248**, L123.

Bruhweiler, F. C., & Kondo, Y., 1983, *ApJ*, **269**, 657.

Chayer, P., Fontaine, G., Wesemael, F., & Michaud, G. 1987, in IAU Colloq. 95, The Second conference on Faint Blue Stars, ed. A.G.D. Philip, D.S. Hayes & J. Liebert (Schenecttady: Davis), p653.

Chayer, P., Fontaine, G., & Wesemael, F., 1989, in IAU Colloq. 114, White Dwarfs, ed G. Wegner, (New York: Springer), 253.

Chayer, P., Fontaine, G., & Wesemael, F., 1991, in White Dwarfs, ed G. Vauclair & E.M. Sion (Dordrecht: Kluwer), 249.

Chayer, P., LeBlanc, F., Fontaine, G., Wesemael, F., Michaud, G., & Vennes, S., 1995, preprint

Chayer, P., Fontaine, G., & Wesemael, F., 1995, preprint

Dupree, A. K., & Raymond, J. C., 1982, *ApJ*, **263**, L63.

Feibelman, W. A., & Bruhweiler, F. C. 1990, *ApJ*, **357**, 548.

Garhart, M. P. 1993, IUE NASA Newsletter No. 51, 1

Henry, R. B. C., Shipman, H. L., & Wesemael, F., 1985, *ApJS*, **57**, 145.

Holberg, J. B., Wesemael, F., Wegner, G., & Bruhweiler, F. C., 1985, *ApJ*, **293**, 294.

Holberg, J. B., et al., 1993, *ApJ*, **416**, 806.

Holberg, J. B., Hubeny, I., Barstow, M. A., Lanz, T., Sion, E. M., & Tweedy, R. W., 1994, *ApJ*, **425**, L105.

Holberg, J. B., Bruhweiler, F. C., & Andersen, J., 1995, *ApJ*, in press

Hubeny, I, & Lanz, T., 1995 in this volume

Lamontagne, R., Wesemael, F., Bergeron, P., Liebert, J., Fullbright, M. S., & Green, R. F., 1993, in White Dwarfs: Advances in Observation and Theory, ed. M. A. Barstow (NATO, ASI Series), p.347.

Lanz, T. & Hubeny, I., 1994, *ApJ*, in press

Morvan, E., Vauclair, G., & Vauclair, S., 1986, *A& A*, **163**, 145.

Shipman, H. L. et al., 1995, preprint

Sion, E. M., & Starrfield, S. G., 1984 *ApJ*, **286**, 760.

Sion, E. M., Bohlin, R. C., Tweedy, R. W., & Vauclair, G. P., 1992, *ApJ*, **391**, L32.

Tweedy, R. W. 1991 PhD Thesis, Univ. of Leicester

Tweedy, R. W. 1995 preprint

Vauclair, G., Vauclair, S., & Greenstein, J. L., 1979, *A& A*, **80**, 79.

Vauclair, G. 1989, in IAU Colloq. 114, White Dwarfs, ed G. Wegner, (New York: Springer), 253.

Vennes, S., Chayer, P., Thorstensen, J. R., Bowyer, S., & Shipman, H. L., 1992, *ApJ*, **392**, L27.

Vennes, S., Thejll, P., & Shipman, H. L., 1991, in 7th European Workshop on White Dwarfs: White Dwarfs, eds., G. Vauclair & E. Sion, Kluwer, p.235.

Vennes, S., Thorstensen, J. R., Thejll, P., & Shipman, H. L., 1991, *ApJ*, **372**, L37.

Vidal-Madjar, A., Allard, N. F., Koester, D., Lemoine, M., Ferlet, R., Bertin, P., Lallement, R. & Vauclair, G., 1994, *A&A*, 287, 175.

Werner, K., & Dreizler, S., 1994, *A&A*, **286**, L31.

Wesemael, F., Henry, R., B., C., & Shipman, H. L., 1984, *ApJ*, **287**, 868.

The Mass Distribution of DA White Dwarfs

David S. Finley

Center for EUV Astrophysics, University of California, Berkeley,
2150 Kittredge St., Berkeley, CA 94720
david@cea.berkeley.edu

1 Introduction

Measurements of the masses of white dwarfs have played a central role in white dwarf studies since the initial astrometric mass determination for Sirius B (Boss 1910). Now, as the knowledge of stellar evolution and the quality of the available data on white dwarfs masses has improved, the factors that determine the white dwarf mass distribution can be better evaluated, including: the history of star formation in the Galaxy; stellar evolutionary timescales; the mass loss processes that determine the initial mass to final mass relation; white dwarf cooling rates as a function of mass and temperature; and the relative contributions of different channels for white dwarf formation.

Considerable insight may be gained by comparing the observed mass distribution with distributions calculated using galactic evolution models. First, the observed distribution is very narrow and sharply peaked. White dwarfs near the peak mass (the mode of the distribution) are formed from stars just above the galactic main sequence turnoff mass; hence the location of the mass peak establishes the initial-to-final mass relation at the low-mass end. The width of the observed mass peak then determines the slope of the initial-to-final mass relation. The location of the mass peak also determines the fraction of white dwarfs above and below the 0.55 $M_\odot$ lower threshold for planetary nebula (PN) formation, and thus sets the expected ratio of white dwarf to planetary nebula birth rates.

Given the finite age of the galactic disk, the intrinsic white dwarf mass distribution must have a sharp cutoff at the low-mass end. Hence the relative numbers of white dwarfs found below that cutoff reflects the fraction of white dwarfs formed in interacting binaries in which mass transfer has truncated core growth. On the high mass end, the steepness of the main sequence mass distribution is such that only very small numbers of massive white dwarfs can be formed from single star evolution. Hence a significant excess of massive white dwarfs would imply another channel for white dwarf formation, such as the merger of interacting binaries as suggested by Iben (1990).

The aim of this paper will be to give an overview of our current knowledge of the mass distribution, suggest a new tool for evaluating different mass distributions, briefly discuss how selection effects modify observed distributions, and present some implications of comparisons of selection-corrected theoretical mass distributions with observations.

2 Mass Determination Methods

Determination of the mass distribution for white dwarfs requires two essential factors: a large enough ensemble for good statistics when the data are binned into suitably small mass intervals (at least 50 objects, preferably in excess of 100), and accurate determinations of the masses (internal errors of $< 0.05\ M_\odot$). It is also desirable to have a homogeneous data set in order to avoid systematic effects due to different measurement, reduction, or analysis techniques. Too few gravitational redshift measurements or astrometric mass determinations are available for this purpose. Methods by which masses are obtainable for significant numbers of stars consist of either determining the temperatures and gravities, and then applying a mass-radius relation (the gravity method), or else using temperatures with distances from geometric parallaxes to infer the stellar radii, from which masses can then be directly derived using a mass-radius relation (the radius method). However, the accuracy of parallaxes is such that individual mass determinations can have errors that are, on average, an order of magnitude larger than the most accurate masses obtained from gravities.

Photometric data have been used to obtain masses for white dwarfs in the 8,000 K to 16,000 K range by Koester, Schulz, and Weidemann (1979, hereafter KSW), Shipman (1979), Wegner (1979), Shipman & Sass (1980), and Weidemann & Koester (1984, hereafter WK). Photometry is sensitive to the gravity (and hence mass) in this range because of the pressure-dependence of the H^- opacity, as discussed by KSW and by Shipman & Sass. However, the parameter determinations are of limited accuracy and are subject to systematic errors in the models (mostly due to uncertainties in parameterizing convection) as well as errors in the absolute flux calibrations.

In recent years, the new generation of CCD detectors has made it possible to obtain high-quality spectra of significant numbers of white dwarfs. With such spectra, detailed fits to the Balmer line profiles permit temperature and gravity determinations of significantly greater precision than was possible using photometry. Two spectroscopically determined DA mass distributions have been obtained to date. The first, covering the cool end (mostly 13,000 K − 30,000 K), was published in Bergeron, Saffer, and Liebert (1992, hereafter BSL). The second, from a spectroscopic survey of DA hotter than 24,000 K, will be reported here and is soon to be published (Finley, Koester, and Basri 1995, hereafter FKB).

3 Mass Distribution Results

Reviews of the previously published photometric results have been presented in
Weidemann (1989), Weidemann (1990), and in BSL, and the reader is referred to
those sources and the original articles for the details of those investigations. Here,
we will only present the results of Weidemann & Koester (WK), that are representative of the best that can be done with photometric mass determinations.
WK analyzed 70 stars between 8,000 K and 16,000 K, using specially constructed
broad-band multichannel colors for optimum signal-to-noise. Their masses were
derived from log g using the Hamada-Salpeter (Hamada & Salpeter 1961) zero-
temperature mass-radius relation. Therefore, we have used their temperatures
and gravities and Wood's evolutionary models (1990, 1992) with carbon cores
and helium layer masses of $M_{He} = 10^{-4} M_*$ to redetermine the masses.

The resulting mass distribution is plotted in Fig. 1, along with a Gaussian
fit to the data. Fitting a Gaussian represents a relatively unbiased means of
estimating the mode (most probable value) of the distribution, and the Gaussian
width also gives a measure of its breadth. On the other hand, the sample mean
and dispersion $\sigma(M)$ are strongly biased by the presence of outliers and are thus
less suited for determining the properties of the peak of the mass distribution.
The Gaussian fit results and other derived quantities that characterize the mass
distribution are listed in Table 1. The first column lists the mode, taken as the
centroid of the Gaussian fit. The second column gives the 1 σ uncertainty in the
mode, σ_{mode}, where $\sigma_{mode} \simeq \sigma_G \times (2/N)^{1/2}$. The third column contains the
Gaussian sigma (σ_G, the 1/e half width). The other columns list the median,
which is less affected by outliers than the mean; the straight (unweighted) mean
mass; $\sigma(M)$, the dispersion of the masses about the mean; and N, the total
number of objects in each sample.

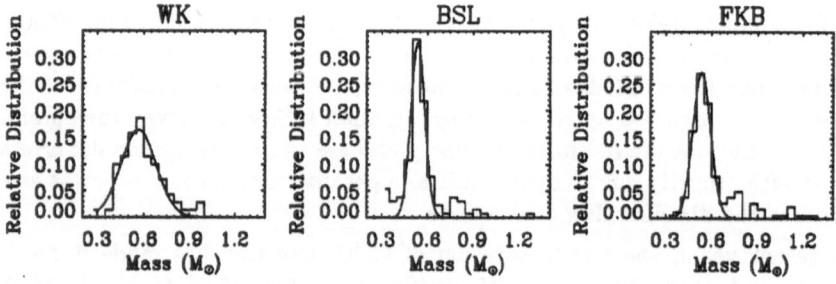

Fig. 1. Mass distributions, binned into 0.05 $M_\odot$ intervals, from Weidemann & Koester
(1984, WK), Bergeron, Saffer & Liebert (1992, BSL), and Finley, Koester, and Basri
(1995, FKB). Histograms are observed mass distributions, smooth curves are Gaussian
fits to the mass distributions.

Purely spectroscopic mass determinations rely on detailed fits to the Balmer
line profiles in order to derive temperatures and gravities. The Balmer lines

Table 1. Derived quantities for the different mass distributions.

Sample	Mode σ_{mode}		σ_G	Median	Mean	$\sigma(M)$	N
Weidemann & Koester	0.567	0.020	0.116	0.583	0.598	0.136	70
BSL	0.538	0.006	0.046	0.542	0.562	0.137	129
FKB	0.535	0.006	0.059	0.550	0.601	0.166	177
FKB < 30	0.522	0.007	0.035	0.536	0.573	0.121	56
FKB > 30	0.546	0.009	0.067	0.560	0.614	0.182	121

Note: Properties of two subsamples of FKB sample, with $T_{eff} < 30,000$ K and $> 30,000$ K, are listed as FKB < 30 and FKB > 30.

are sensitive to temperature and gravity beyond the observed DA T_{eff} limit of about 90,000 K. Therefore samples can be taken from stars in excess of the 15,000 K upper limit for photometric mass determinations, in the temperature range where convection is negligible. Flux calibration errors have little, if any, effect on parameter determinations obtained from line fitting. At S/N $\sim$ 100, masses may be obtained with an internal precision of $0.02 - 0.03 \, M_\odot$.

The two spectroscopic samples cover different temperature ranges. The BSL sample included 129 DA between 13,000 K and 40,000 K, with 90% of their sample having $T_{eff} < 30,000$ K. The FKB sample, intended to be a complete spectroscopic survey of known DA hotter than 25,000 K, currently includes 177 DA between 17,000 K and 95,000 K, with 90% being between 23,000 K and 75,000 K.

Except for the different T_{eff} ranges, the two data sets (BSL/FKB) are nearly identical in terms of the S/N (typically > 80), resolution (6/5 Å) and wavelength coverage ($H\beta$ through the Balmer limit). The white dwarf model atmosphere codes (P. Bergeron's (see BSL) and D. Koester's (see KSW)) incorporate the Hummer-Mihalas (Hummer & Mihalas 1988) occupation probability formalism, with the same parameterization (see Bergeron 1993). Both treat Stark broadening using recently calculated profiles for the first 16 Balmer lines, based on the Vidal, Cooper, and Smith (1970) unified theory, and provided by T. Schöning and K. Butler to both Bergeron and Koester. In both cases, pure H models were used. The analyses both involved simultaneous fitting of the Balmer lines, out to at most 150 Å from line center. BSL evidently fit $H\beta$ through H9 for their cooler sample, whereas in the case of the hotter stars in the FKB sample, lines beyond $H\epsilon$ were generally very weak and thus were usually not used. Both analyses used the carbon-core evolutionary models of Wood (1990) with C cores and $M_{He} = 10^{-4} \, M_*$ to calculate masses from T_{eff} and log g.

The mass distributions for the full BSL and FKB samples are plotted in Fig. 1, and the characteristics of each are listed in Table 1. The slightly greater width of the core of the FKB mass distribution was due to the decreased precision of mass determinations at higher T_{eff}, given that σ_G was 0.067 for a subsample of FKB stars hotter than 30,000 K, but was only 0.035 for those cooler than 30,000 K. The significantly larger mean mass for the FKB sample is due to its greater proportion of high-mass stars. Further implications of the mass distributions will be discussed in subsequent sections. We also investigated possible systematic differences between FKB and BSL by comparing parameter determinations for stars in common between FKB on one hand, and BSL and additional DA recently published in Bergeron et al. (1994), and found differences in the masses that were of the order of $0.02 - 0.03\ M_\odot$. Given that the models have identical input physics, these differences appear to reflect systematic effects of the respective data reduction and analysis techniques.

Other differences are apparent between the BSL and FKB mass distributions. Although the locations of the mass peaks are nearly identical, the BSL sample includes 11 WD below $0.4\ M_\odot$, while FKB has only 3. On the other hand, BSL has only 1 WD more massive than $1.0\ M_\odot$, while FKB has 9. The reason for such a pronounced difference between the high and low-mass populations of the two samples is suggested when one plots mass vs. T_{eff} for the stars, as shown in Fig. 2. Eight of the nine WD in the FKB sample with masses larger than 1 $M_\odot$ lie in the range of 32,000 to 47,000 K. On the low-mass end, all 14 of the WD less massive than $0.4\ M_\odot$ are cooler than 35,000 K. The isochrones plotted in Fig. 2 show that the cooling rate for the massive WD suddenly decreases around 50,000 K, so that those stars cool much more slowly than WD near the mass peak. Similarly, the low-mass WD cool much faster than the WD near the mass peak until they reach about 30,000 K. To investigate these effects, I used Wood's models to calculate cooling rates and thus determine the relative space densities for white dwarfs as a function of mass and temperature, and found that the differential cooling and observational selection effects do largely account for the observed T_{eff}–mass distribution. The results will be presented in more detail in FKB.

4 Comparisons with Theoretical Mass Distributions

The Weidemann & Koester (1984) mass distribution had a width that was dominated by the observational errors, given the much narrower distributions obtained from the spectroscopic measurements. The question naturally arises whether we have now resolved the mass distribution, or do observational errors still dominate? To answer that question, I compared the observed mass distribution for the T_{eff} range of 20,000 K to 28,000 K with the predicted mass distribution for the same T_{eff} range.

The observed mass distribution was obtained by combining the 38 FKB and 55 BSL measurements of DA in that range. For the 18 objects in common, the average mass difference was $M(BSL) - M(FKB) = 0.03\ M_\odot$. Therefore, I added half that difference to the masses of the 20 objects observed only by FKB, and subtracted half for the 37 that were unique to BSL. The FKB and BSL masses for the 18 stars in common were averaged together, giving a total of 75 stars in the sample. The resulting mass distribution, binned in $0.03\ M_\odot$ intervals, is plotted in Fig. 3. The peak mass per the Gaussian fit was $0.524\ M_\odot$. The distribution was very narrow, with the width of the best fit Gaussian being only $0.034\ M_\odot$. The average 1 σ error in the FKB mass determinations for this sample

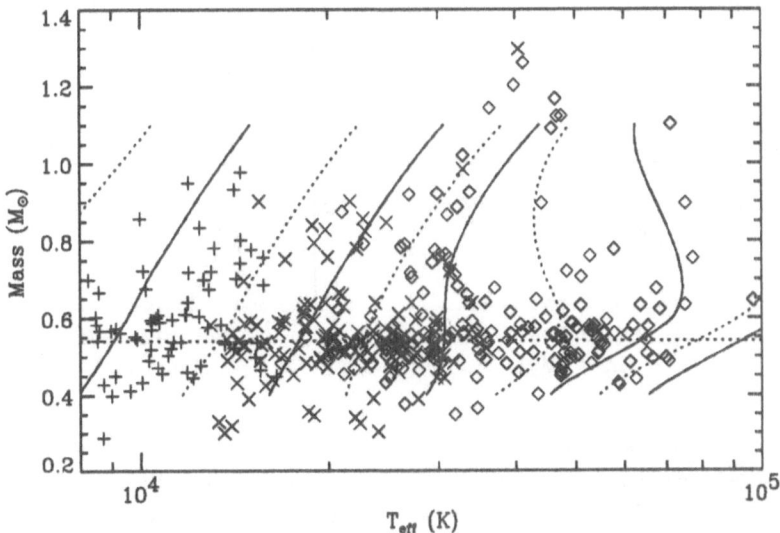

Fig. 2. Temperatures and masses of DA white dwarfs: crosses – Weidemann & Koester (1984); ×'s – Bergeron, Saffer & Liebert (1992); diamonds – Finley, Koester & Basri (1995). Horizontal dotted line marks 0.54 $M_\odot$ peak of BSL and FKB mass distributions. Curves are isochrones (in years) from Wood's models, for log τ_c = 5 to 9.5, from right to left, by steps of 0.5.

was 0.026 $M_\odot$ (the BSL errors should be similar), suggesting that the width of the observed mass distribution was primarily due to observational errors.

The theoretical mass distribution was calculated following Wood (1992). I used a constant star formation rate, a Salpeter IMF ($\phi(M) = (M/M_\odot)^{-2.35}$), assumed a pre-WD lifetime $t_{MS} = 10(M/M_\odot)^{-2.5}$ Gyr, and ignored any changes in scale height with age, which are small for the white dwarfs hotter than about 10,000 K per the study by Yuan (1992). The age of the disk was taken to be 10 Gyr. White dwarf cooling times were again taken from Wood's models.

I found that the critical factor in matching the observed and theoretical mass distributions was the parameterization of the initial-to-final mass relation. I chose a parametrization $M_f = A \exp((BM_i)^C)$. The power in the exponent was required to match the narrowness of the observed mass distribution. Informal fits were performed that showed that good matches between the predicted and observed distributions could be obtained by setting $A \simeq 0.50$, $B \simeq 0.13$, and $C \sim 1.5 - 2.0$. (Setting C as small as 1 steepened the $M_i - M_f$ relation and resulted in far too broad a peak for the mass distribution). The value for B was chosen such that an 8 $M_\odot$ MS star produced a 1.4 $M_\odot$ WD as per Weidemann & Koester (1983). Matching the location of the peak of the observed distribution required that a 1 $M_\odot$ progenitor at the disk turnoff mass produced the minimum

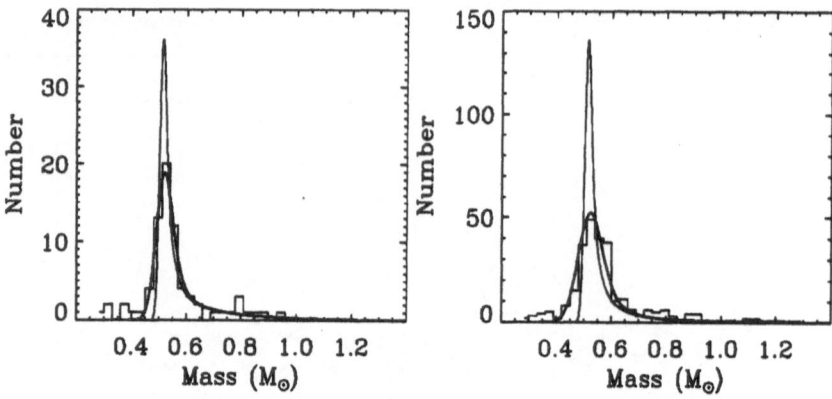

Fig. 3. Mass distributions, binned into 0.03 $M_\odot$ intervals, for combined BSL and FKB objects between 20,000 K and 28,000 K (left), and full combined samples (right). Histograms are observed mass distributions. Smooth light curves are predicted mass distributions with differential mass loss effects included. Smooth heavy curves are theoretical mass distributions with differential mass loss and observational errors included. Predicted distributions are scaled so that the total numbers (observed and theoretical) between 0.4 and 0.7 $M_\odot$ are equal.

WD mass of $\simeq 0.5\ M_\odot$. Matching the "knee" of the observed distribution (0.55 – 0.7 $M_\odot$) necessitated relations that were quite flat at the low-mass end, with 2 and 3 $M_\odot$ progenitors resulting in ~ 0.54 and $\sim 0.60\ M_\odot$ WD, respectively. The relations were thus qualitatively similar to curve "B" from Weidemann & Koester (1983).

The predicted mass distribution was calculated explicitly for the 20,000 K to 28,000 K temperature range using $A = 0.49$, $B = 0.129$, and $C = 1.7$ after weighting the relative space densities to match the observational selection effects that apply to a magnitude-limited sample. Differential mass loss and observational errors were then included by successively convolving the intrinsic mass distribution with Gaussians with widths σ_G of 0.01 and 0.026 $M_\odot$. The distribution including only differential mass loss is plotted as a thin solid curve in Fig. 3. The mass distribution with differential mass loss and observational errors included is plotted as a heavy solid curve. The resulting fit is seen to be excellent.

This comparison shows that the observed DA with masses < 0.4 $M_\odot$ represent a population completely distinct from the bulk of the DA, given a lower cutoff of the predicted mass distribution of about 0.51 $M_\odot$. Although the numbers are small, the predicted number of high-mass WD appears to agree with the observations in this T_{eff} range. While good fits are obtainable without differential mass loss, a dispersion in M_f of up to perhaps 0.02 $M_\odot$ is permissible.

In order to evaluate the implications for the WD birth rate vs. the planetary nebula birth rate, I used the above parameters to calculate the mass distribution for WD younger than 10^6 years. The fraction of WD below the 0.55 $M_\odot$ lower limit for producing visible PN was 70%, which would imply a 3.3:1 ratio of WD to PN birth rates. As discussed by Weidemann (1990), the observed ratio is 1:1 or smaller. If the WD to PN birth rate ratio were as high as 2:1, such a narrow mass distribution would still require a peak mass of about 0.55 $M_\odot$. Thus a systematic shift in masses of 0.025–0.035 $M_\odot$ (slightly more for FKB, slightly less for BSL) would be required to obtain a WD to PN birth rate ratio of 2:1–1:1. A slight mass increase may be obtainable by including finite thickness hydrogen layers in the evolutionary models from which masses are calculated (see following section). Otherwise, a systematic shift in log g determinations of about 0.05 dex will be required to match the observed birth rate ratio.

A similar comparison of observed and predicted mass distributions was made for the full combined FKB and BSL samples, comprising 273 DA between 13,000 K and 80,000 K, using the same parameters that were used for the fit to the cooler sample. The results are plotted in the right panel of Fig. 3. The breadth of the mass distribution is largely accounted for by the average observational error of 0.04 $M_\odot$, although the peak is shifted and skewed slightly due to systematic shifts in masses as a function of T_{eff}. Also, the full sample shows a significant discrepancy between the observed and predicted numbers of massive white dwarfs. While the observed mass distribution included 43 stars with masses greater than 0.7 $M_\odot$. the predicted distribution contained only 19 objects in that mass range. This result implies that about 24 out of 273, or about 9% of the stars in the combined sample are candidates for formation through mergers of interacting binary stars.

5 Do DA Have Thick or Thin Hydrogen Envelopes?

The FKB sample covers a wide range in T_{eff}. Finite-temperature effects increase in magnitude at higher temperatures, and white dwarfs with thick envelopes are more affected than those with thin envelopes. Given that selection effects cannot change the value of the mode of the mass distribution as a function of T_{eff}, consideration of the consistency of derived masses vs. T_{eff} should provide some constraint on the allowed envelope masses for DA.

The masses presented so far were calculated with Wood's models for C core WD with helium layers of $M_{He} = 10^{-4} M_*$ and no hydrogen. As a test of the thick-envelope hypothesis, we also calculated masses using another sequence of Wood's models, again with C cores, but with $M_{He} = 10^{-2} M_*$ and $M_H = 10^{-4} M_*$. We determined the modes of the mass distributions for the 56 stars in the FKB sample cooler than 30,000 K and the 77 stars hotter than 40,000 K using both model sequences. For thin envelopes, the hot sample had a mode of 0.542 with $\sigma_{mode} = 0.011$, and the cool sample mode was 0.524 with $\sigma_{mode} = 0.007$. Thus the difference was 0.019 ± 0.013, which is not statistically significant. However, for thick envelopes, the hot sample mode was 0.596 with $\sigma_{mode} = 0.009$, and the cool sample mode was 0.557 with $\sigma_{mode} = 0.008$, for a difference of 0.039 ± 0.012. Thus for thin envelopes the high and low temperature modes are discrepant at only a $1.5\ \sigma$ level, but are discrepant at a $3\ \sigma$ level if thick envelopes are assumed.

Unless there are systematic effects in $\log g$ determinations vs. T_{eff} that cause the depression of the masses derived for the cooler stars relative to the hotter stars by about $0.04\ M_\odot$, these results imply that the general presence of envelopes as massive as $M_{He} = 10^{-2}\ M_*$ and $M_H = 10^{-4}\ M_*$ might be ruled out for the majority of DA. However, Koester & Schönberner (1986) showed that radii of white dwarfs with $M_H = 10^{-6}\ M_*$ are virtually indistinguishable from stars with no H layers. Therefore if cooling sequences with intermediate envelope masses were to be calculated it would then be possible to determine which envelope mass within those limits is consistent with the observations. Still, proper evaluation of those results requires that further work be done to quantify the systematic uncertainties in spectroscopic mass determinations.

6 Summary

Successive improvements in observational and analytic techniques and the availability of sophisticated evolutionary models have resulted in a significant improvement of our knowledge of the white dwarf mass distribution and its implications. Spectroscopic mass determinations have shown that the mass distribution for the majority of DA is strongly peaked and very narrow, with the observed mass distribution having a FWHM $< 0.1\ M_\odot$. The narrowness is such that the $M_i - M_f$ relation is required to be very flat at the low-mass end, with 1-3 $M_\odot$ progenitors producing WD in the 0.5–0.6 $M_\odot$ range. DA below 0.4 $M_\odot$ are confirmed to comprise a distinct population, given a lower mass of about 0.5 $M_\odot$ for the theoretical mass distribution that matches the observations. It appears that only about half of the observed DA more massive than 0.7 $M_\odot$ can be accounted for via single star evolution. That conclusion is tentative, though, and requires further study of the parameterization of the $M_i - M_f$ relation and evaluation of selection effects in the observed samples. Differential cooling has a very significant effect on the space densities of WD as a function of T_{eff} and mass. Consequently, the mass distribution for a sample defined by some temperature range can differ strongly from the WD equivalent of the initial mass function (IMF). Theoretical mass functions being used for comparisons must therefore be calculated for the same range of parameters as the observed sample in order to obtain valid results. The nominal location of the observed mass peak is discrepant with the observed WD and PN birth rate ratio. Removal of the discrepancy requires that the derived masses be increased by about 0.03 $M_\odot$. Finally, at present it is not possible to conclusively rule out thick ($10^{-4}\ M_*$) hydrogen envelopes for the majority DA white dwarfs based on the location of the peak mass vs. T_{eff}, although the evidence appears to indicate that envelopes that thick are not allowed at the $3\ \sigma$ level.

References

Bergeron, P., Saffer, R.A. & Liebert, J. (1992): ApJ **394**, 228 (BSL)

Bergeron, P. (1993), in White Dwarfs: Advances in Observation and Theory, NATO ASI series, ed. M.A. Barstow (Dordrecht: Kluwer Academic Publishers), 267

Bergeron, P., Wesemael, F., Beauchamp, A., Wood, M.A., Lamontagne, R., Fontaine, G. & Liebert, J. (1994): ApJ **432**, 305

Boss, L. (1910), Preliminary General Catalogue (Washington, Carnegie Institution)

Finley, D.S., Koester, D. & Basri, G. in preparation (1995) (FKB)

Hamada, T. & Salpeter, E.E. (1961): ApJ **134**, 683

Hummer, D.G. & Mihalas, D. (1988): ApJ **331**, 794

Iben, I., Jr. (1990): ApJ **353**, 215

Koester, D., Schulz, H., Weidemann, V. (1979): A&A **76**, 262 (KSW)

Koester,D. & Schönberner, D. (1986): A&A **154**, 125

Shipman, H.L. (1979): ApJ **228**, 240

Shipman, H.L. & Sass, C.A. (1980): ApJ **235**, 177

Vidal, C.R., Cooper,J. & Smith, E.W. (1973): ApJS **25**, 37

Wegner, G. (1979): A&A **84**, 1384

Weidemann, V. (1989): IAU Coll. 114, White Dwarfs, ed. G.Wegner (Springer: Berlin), p. 1

Weidemann, V. (1990): ARA&A **28**, 103

Weidemann, V. & Koester, D. (1983): A&A **121**, 77

Weidemann, V. & Koester, D. (1984): A&A **132**, 195 (WK)

Wood, M.A. PhD thesis, University of Texas at Austin (1990)

Wood, M.A. (1992): ApJ **386**, 539

Yuan, J.W. (1992): A&A **261**, 105

PG 1159 Stars and Their Evolutionary Link to DO White Dwarfs*

Stefan Dreizler[1], Klaus Werner[2] and Ulrich Heber[1]

[1] Dr.Remeis–Sternwarte Bamberg, Universität Erlangen–Nürnberg, Sternwartstraße 7, D-96049 Bamberg, Germany
[2] Institut für Astronomie und Astrophysik der Universität, D-24098 Kiel, Germany

1 Introduction

After its detection by McGraw et al. (1979) the star PG 1159-035 has become the prototype of a new spectroscopic class of hydrogen deficient [pre-] white dwarfs. The spectra of these stars are characterized by the absence of hydrogen as well as a broad absorption trough due to He II 4686 Å and neighbouring C IV lines. Today 26 stars have been assigned to this class of stars. Major spectroscopic and photometric differences between the members required a more precise classification scheme. According to Werner (1992, W92 henceforth) three spectroscopic subtypes can be identified: Those stars displaying pure absorption spectra are called type **A**, those exhibiting emission in the cores of He II 4686 Å and in C IV 4659 Å and 5801/5812 Å are called type **E** PG 1159 stars. PG 1159 stars with relatively low gravities and strong central emissions, represent the subtype **lgE**. After the detection of a nebula associated with PG 1520+525 (Jacoby & van De Steene 1993) fourteen PG 1159 stars were confirmed to be central stars of planetary nebulae (CSPN), however, a search for PNe around several other PG 1159 stars was unsuccessful (Kwitter et al. 1989, Méndez et al. 1988, Tweedy et al. 1994). Except for the lgE stars there is no sign of ongoing mass loss (in particular of PG 1159-035, Fritz et al. 1990). Low amplitude non-radial g-mode pulsations were detected in eight PG 1159 stars which define a new class of very hot variables, the DOV or GW Vir (=PG 1159-035) stars whereas twelve are definitely stable against pulsations (see also Vauclair, these proceedings). In Table 2 we list all PG 1159 stars summarizing their known spectral and pulsational characteristics.

First analyses of seven objects with line blanketed LTE models of optical and UV data by Wesemael et al. (1985, WGL85) showed that the effective temperatures exceed 80 000 K and that the gravity is around $\log g \sim 7$. A He/H ratio exceeding ten was concluded from the absence of the Balmer lines. Other analyses using unblanketed LTE or NLTE models have been performed (Barstow &

---

* Based on observations obtained at the European Southern Observatory, La Silla, Chile and at the German–Spanish Astronomical Center, Calar Alto, operated by the Max-Planck-Institut für Astronomie Heidelberg jointly with the Spanish National Commission for Astronomy

Tweedy 1990, Barstow & Holberg 1990, Husfeld 1987). The high effective temperature and the peculiar chemical composition of these stars, however, require highly sophisticated NLTE atmospheres in order to derive reliable atmospheric parameters. These models have become available after the development of the so-called Accelerated Lambda Iteration method (Werner 1986, Dreizler & Werner 1993). Line blanketed model atmospheres including all elements from H–Ni can be computed with this method.

In their pioneering work Werner, Heber & Hunger (1991, henceforth WHH91) analysed each one pulsating and non-pulsating representative of type A and E followed by the analysis of more unique objects (Werner & Heber 1991; Werner 1991; Motch, Werner & Pakull 1993) Their analyses confirmed that these stars are extremely hot, the type A stars have effective temperatures of $\sim 100\,000$ K and the type E stars between 140 000 and 170 000 K. The surface gravity ranges from $\log g = 5.5$ to $\log g = 8.0$. The typical composition of PG 1159 stars (He=33%, C=50%, and O=17% by mass) indicates that C- and O-rich 3α processed material mixed with He-rich intershell matter is displayed at the surface. Therefore these stars have lost their entire H-rich and part of the He-rich layers. The most extreme case, H 1504+65, has even lost the complete He-rich layer and displays a pure CO core. The atypically presence of nitrogen (1.5% by mass) and the lower oxygen abundance in PG 1144+005 indicate a less drastic mass loss event in this star. The position in HR diagram reveals that these stars are post–Asymptotic–Giant–Branch (post–AGB) stars populating the area of highest effective temperatures in low–mass star evolution and are immediate progenitors of the white dwarfs.

2 Observations

Several new PG 1159 stars were detected in the recent years in follow-up spectroscopy of the stellar component in the Hamburg–Schmidt survey (HSS, Hagen et al. 1994). The HSS was originally designed as a quasar survey, however, it is also a very rich source of hot evolved stars. Up to now spectra of 328 objects have been obtained (Table 1), the vast majority with the 3.5m telescope at the Calar Alto observatory using the TWIN spectrograph which covers the optical range completely with a resolution between 3 and 6 Å. Among the more ordinary white dwarfs and subdwarfs we also found several very interesting objects, most of them will be discussed in more detail below: Four PG 1159 stars, five hot He-rich white dwarfs (DO), and two DOs showing signature of a very hot, compact wind (see Werner et al., these proceedings). It is worthwhile to note that the number of DOs could be significantly increased, only about a dozen has been known previously. More details about the survey, recent results as well as a list of the most interesting objects can be found in Dreizler et al. 1994b (D94b)[1].

---

[1] Please note that one magnetic WD in Tab. 2 of D94b was misidentified: HS1440+7518 ($\alpha = 14^h40^m14\overset{s}{.}0$, $\delta = 75\overset{\circ}{}18'19''$, B= $14\overset{m}{.}9 \pm 0.5$) rather than HS1412+6115 is a magnetic WD.

Table 1. Summary of spectral types of the stellar component in the HSS.

type	number	type	number	type	number	type	number
DA	70	DB	7	DO	5	DAO	1
DO(Z)+wind	2	DAB/DBA	2	DA+DA	3	DA+M	3
mag. WD	3	sdB	70	sdOB	16	sdO	26
He sdO	30	He sdB	2	PG 1159	4	HBA	21
HBB	13	sdF/G	20	CV	3	QSO	2
other	25					Sum	328

Napiwotzki & Schönberner (1991) detected a new spectroscopic class of CSPNe showing characteristic spectral features of PG 1159 stars e.g. the C IV/He II absorption trough. However, in these stars hydrogen can clearly be detected, in contrast to all previously known PG 1159 stars. They therefore coined the term "hybrid PG 1159 star" or in the notation scheme of Werner (1992), lgEH for those resembling lgE PG 1159 stars (NGC 7094, Abell 43) or type AH for the one resembling a type A PG 1159 star (Sh 2-68). D94b presented the first lgEH PG 1159 stars where no associated nebula could be detected up to now.

3 New spectroscopic analyses

A large number of cooler PG 1159 as well as hybrid PG 1159 stars were analysed during the last year. Preliminary results on hot PG 1159 stars are presented by Rauch & Werner in these proceedings. Previous results are shortly summarized here and can be found in the review of W92.

3.1 "Cool" PG 1159 stars

All previously analysed PG 1159 stars have effective temperatures at or above 100 000 K. The analysis of HS 0704+6153, however, showed, that PG 1159 stars are also found way below that temperature. Dreizler et al. 1994a (D94a) found T_{eff}=65 000 and $\log g = 7.0$ for this object. The model grid computed for that analysis was extended and used for the analysis of other "cool" PG 1159 stars: PG 0122+200, PG 2131+066, MCT 0130-1937, and HS 1517+7403. The analysis of the remaining type A PG 1159 stars will follow in near future. Model atoms and fitting procedures are identical to those in D94a. The detectability of a (very weak) He I 5876 Å line gives quite tight constraints to the effective temperature. The surface gravity and carbon abundance is determined from several C IV and He II lines. An oxygen abundance can not be determined from the optical spectrum since the effective temperature is too low to exhibit O VI lines (O V lines in the optical are too weak for analyses). From existing low-resolution IUE spectra the oxygen abundance is derived from the O V 1371 Å line. Due to the poor S/N ratio of these spectra a large error range has to be accepted. Fig. 2 displays the spectra of the "cool" PG 1159 stars overlayed with theoretical spectra. Atmospheric parameters derived from these analyses are summarized in Table 2. All these stars are cooler than $100\,000$ K and have $\log g = 7.5$. They lie within the parameter range of hot He-rich DO white dwarfs (Fig. 1, see also below).

3.2 Hybrid PG 1159 stars

These stars are intermediate between the PG 1159 stars and the H-rich white dwarfs progenitors. First analyses of the CSPNe with H-He NLTE models were performed by Napiwotzki (1993). W92 estimated C/H=0.17 for NGC 7094 from exploratory model atmosphere calculations. These model atmosphere calculations have now been extended to the degree of sophistication known from the analyses of other PG 1159 stars i.e. Stark broadening of all lines contributing to the C IV/He II trough is accounted for in the NLTE iterations (see D94a for details). Due to the higher H abundance the hydrogen model atom needs to be more detailed compared to WHH91. It is extended to 10 NLTE levels and 36 lines. The C atom is identical to that employed for the cool PG 1159 stars. It is not necessary to include O in the model atmosphere calculations since the O abundance is significantly lower than in normal PG 1159 stars. It is instead iterated together with N in subsequent line formation calculations. Fits are presented in Fig. 2, the derived parameters are listed in Table 2, and the positions in the HR diagram are shown in Fig. 1.

3.3 DO white dwarfs

It is quite natural to propose the hot He-rich DO white dwarfs to be descendants of the PG 1159 stars. When the PG 1159 stars contract and cool down the gravitational settling will remove all metals from their atmosphere. Small traces of metals might be supported against the gravity due to radiative levitation for a while.

First LTE analyses of DOs from the Palomar Green survey (PGS, Green et al. 1986) have been presented by WGL85. They introduced a separation into two subclasses depending on the detection of the He I 4471 Å line. Those exhibiting this line are termed "cool" DOs, those without "hot" DOs. C, N, and O abundances of the hot DO PG 1034+001 were determined by Sion et al. (1985). The EUVE spectrum of RE 0503-289 was analysed by Barstow et al. (1994). Recently, a detailed NLTE analyses of all hot DOs was started in our group. Results on the hottest two, KPD 0005+5106 and PG 1034+001, are presented by Werner et al. (1994, 1995). In KPD 0005+5106 the evidence for a hot corona could be found from the analysis of ROSAT data (Fleming et al. 1993). Preliminary results of seven further DOs will be presented here. Effective temperatures and gravities are very similar to those of the cool PG 1159 stars, the NLTE model atmosphere calculations are therefore extended to lower metal abundances. Model atoms and fitting procedures are identical to those for cool PG 1159 stars. Fits are presented in Fig. 2, the derived parameters are summarized in Table 2. While PG 0046+0746 and HS 0111+0012 clearly exhibit the He I 4471 Å line and are cool DOs according to WGL85, all other DOs analysed here are hot DOs. A weak He I 5876 Å line can, however, still be detected in two other DOs classified as "hot" by WGL85. The C abundances cover a wide range even among DOs with similar T_{eff} and gravity. Even more surprising is the appearance of DOs and (mainly cool) PG 1159 stars at the same location in the HR diagram, extending the variety of metal abundances for stars with comparable parameters. As in the analysis of KPD 0005+5106 and PG 1034+001 the He II 4686 Å line in HS 0727+6003 and HS 0742+6520 can not be reproduced in its strength by any model atmosphere. A similar problem was encountered in the analysis of the DOs probably surrounded by a wind and in K 1-27, a hot He dominated CSPN (Rauch et al. 1994). Test calculations show, however, that the problem might be overcome with a much more detailed He II model atom accounting for Stark broadening and level dissolution in the model atmosphere calculations (Hummer & Mihalas 1988, Werner, Dreizler & Wolff these proceedings for more details).

4 Evolutionary status

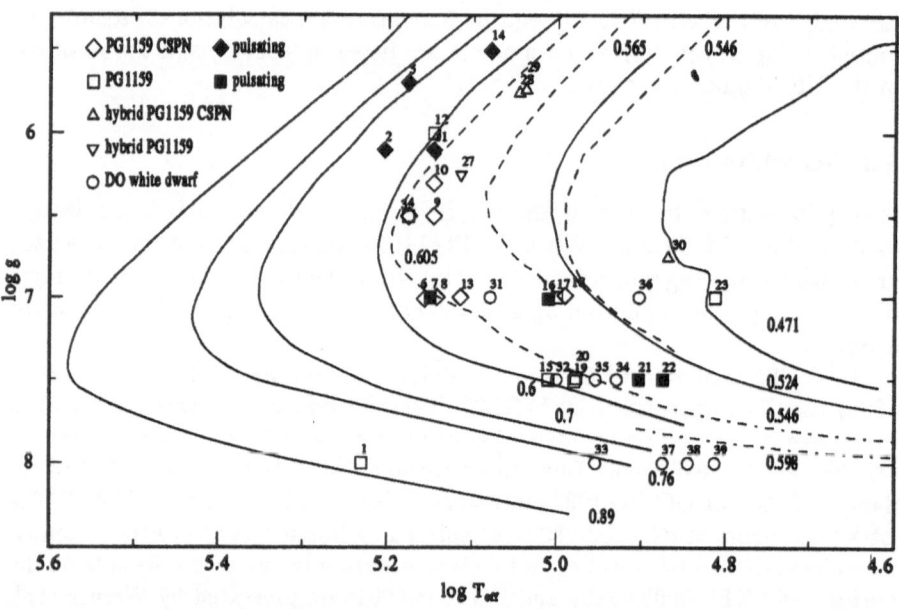

Fig. 1. Position of [pre-] white dwarfs in the ($\log T_{eff}$-$\log g$) plane. Theoretical evolutionary tracks are labeled with the respective stellar mass (Wood & Faulkner 1986 solid lines, Schönberner 1983 dashed lines, Koester & Schönberner 1986 dash-dotted lines, Blöcker 1994 solid line 0.524M$_\odot$, and Dorman et al. 1993 solid line 0.471M$_\odot$, a redward loop was omitted for clarity). Numbers refer to the stars in Table 2

Table 2. The known PG 1159 stars, hybrid stars, and DO white dwarfs, their spectroscopic subclassification and atmospheric parameters (: =uncertain). Columns 3 and 4 denote if the star is variable or has a planetary nebula. T_{eff} is given in kK. Abundances are given in % mass fraction except for the DOs where they are given in log (number ratios). References can be found in Table 3. The consecutive numbers in the last column refer to the labels in Fig.2.

star	type	Var.	PN	T_{eff}	log g	H	He	C	N	O	ref.	Nr.
H 1504+65	Ep	no	no	170	8.0			50		50	5,12	1
RXJ 2117.1+3412	lgE	yes	yes	160	6.1		38	56		6	‡,9	2
PG 1144+005	Ep	no	no	150	6.5		39	58	1.5	1.6	5,13	3
Jn 1	E	no	yes	150:	6.5		30	46		24	‡,2,11	4
NGC 246	lgE	yes	yes	150:	5.7		38	56		6	‡,2,5,8	5
PG 1520+525	E	no	yes	140	7.0	<8	30	46	<0.3	16	5,14	6
PG 1159-035	E	yes	no	140	7.0	<8	30	46	<0.3	16	14,18	7
NGC 650	E		yes	140:	7.0						11	8
Abell 21=Ym29	E		yes	140:	6.5		35	51		14	‡,11	9
Longmore 3	lgE	no	yes	140:	6.3		38	56		6	‡,1	10
K 1-16	lgE	yes	yes	140:	6.1		38	56		6	‡,5	11
PG 1151-029	lgE	no	no	140:	6.0		35	51		14	‡,5	12
VV 47	E	no	yes	130:	7.0		35	51		14	‡,2	13
Longmore 4	lgEp	yes	yes	120	5.5		46	43		11	1,15	14
HS 0444+0453	A		no	100:	7.5						†	15
PG 1707+427	A	yes	no	100	7.0	<8	30	46	<0.3	16	6,14	16
PG 1424+535	A	no	no	100	7.0	<8	30	46	<0.3	16	5,14	17
IW 1	A	no	yes	100:	7.0						2,11	18
MCT 0130-1937	A	no	no	95	7.5		50:	30		20	†	19
HS 1517+7403	A		no	95	7.5		61	37		2	†	20
PG 2131+066	A	yes	no	80	7.5		50:	30		20	†,3	21
PG 0122+200	A	yes	no	75	7.5		50	30		20	†,7,*	22
HS 0704+6153	A		no	65	7.0	<10	44	26		20	4	23
NGC 6852	lgE:		yes								11	24
NGC 6765	lgE:		yes								11	25
Sh 2-78	A		yes								11	26
HS 2324+3944	lgEH		no	130	6.2	18:	37	44	<1	<1	†	27
Abell 43	lgEH		yes	110	5.7	42:	51	5	<1	<1	†	28
NGC 7094	lgEH		yes	110	5.7	42:	51	5	<1	<1	†	29
Sh 2-68	AH		yes	74	6.8						10	30

star	type			T_{eff}	log g	H/He	C/He	N/He	O/He	ref.	Nr.
KPD 0005+5106	DOZ			120	7.0	<-1.0	-3.5	-4.0	<-4.0	16	31
PG 1034+001	DOZ			100	7.5	<-1.0	-5.0	-3.2	-4.1	17	32
PG 0108+101	DOZ			90	8.0		-2.2:			†	33
HS 0727+6003	DO			85	7.5		<-4.0			†	34
HS 1830+7209	DO			>80	7.5		<-4.0			†	35
HS 0742+6520	DOZ			80:	7.0					†	36
PG 0109+101	DOZ			75	8.0		-3.2			†	37
PG 0046+0746	DO			70	8.0		<-4.0			†	38
HS 0111+0012	DOZ			65	8.0		-2.5:			†	39

Table 3. References for Table 2.

1	Bond & Meakes 1990	11	Napiwotzki & Schönberner 1995
2	Bond & Ciardullo 1993	12	Werner 1991
3	Bond et al. 1984	13	Werner & Heber 1991
4	D94a	14	WHH91
5	Grauer et al. 1987	15	Werner et al. 1992
6	Grauer et al. 1992	16	Werner et al. 1994
7	Hill et al. 1987	17	Werner et al. 1995
8	Husfeld 1987	18	Winget et al. 1991
9	Motch et al. 1993	†	Dreizler et al. these proceedings
10	Napiwotzki 1993	‡	Rauch & Werner these proceedings
		*	Vauclair et al. these proceedings

Although the number of reliable NLTE analyses of PG 1159 stars and of their probable progenitors has grown enormously in the recent years, the latest stages of low–mass star evolution are by far not yet understood. All stars shown in Fig. 1 are post–AGB stars except for HS 0704+6153 and Sh 2-68 which can be explained either by a post–Horizontal Branch or binary evolution. The highest temperature range in the post–AGB evolution is populated by a variety of stars with similar T_{eff} and gravity but very different (exotic) chemical compositions. We find He-C-O rich PG 1159 stars, H-He-C rich hybrid PG 1159 stars, He dominated DOs with a variety of carbon abundances as well as He dominated CSPN (K 1-27 and LoTr 4, Rauch et al. 1994, Rauch & Werner these proceedings).

The interpretation that PG 1159 stars exhibit deep intershell matter is corroborated by the detection of Ne in three lgE PG 1159 stars (Werner & Rauch 1994). Neon is strongly enhanced to 2% by mass in the helium burning shell by the $^{14}N(\alpha, \gamma)^{18}F(\beta, \nu)^{18}O(\alpha, \gamma)^{22}Ne$ reaction (e.g. Iben & Tutukov 1985). The atmospheric Ne abundance therefore is a sensitive indicator to what extent the stellar envelope was eroded. The derived Ne abundances are in very good agreement with the evolutionary calculations and therefore strongly support the idea that these stars have lost their entire H-rich and most of the He-rich layers.

Such a drastic mass loss is not predicted by standard post–AGB evolution calculations and more exotic scenarios like the late He–shell flash (Iben et al. 1983, Iben & MacDonald, these proceedings) have to be employed. We recall that a spectacular mass loss event could be directly observed in the lgE PG 1159 star Longmore 4 (Werner et al. 1992). This star turned its spectral type into [WC2-3] and within a few days back again. On one hand this corroborates the evolutionary link between WC–CSPNe suggested by Werner & Heber (1991) and Werner & Koesterke (1992). On the other hand, provided that such mass loss events occur frequently enough, they could explain the strong erosion of the stellar surface in PG 1159 stars. Werner et al. (1992) speculated about a connection between GW Vir pulsations and such events.

In HS 2324+3944 the carbon abundance is comparable to the high values in the PG 1159 stars while Abell 43 and NGC 7094 are less enriched in He and C. O and N are trace elements ($< 1\%$ by mass) in all hybrid PG 1159 stars.

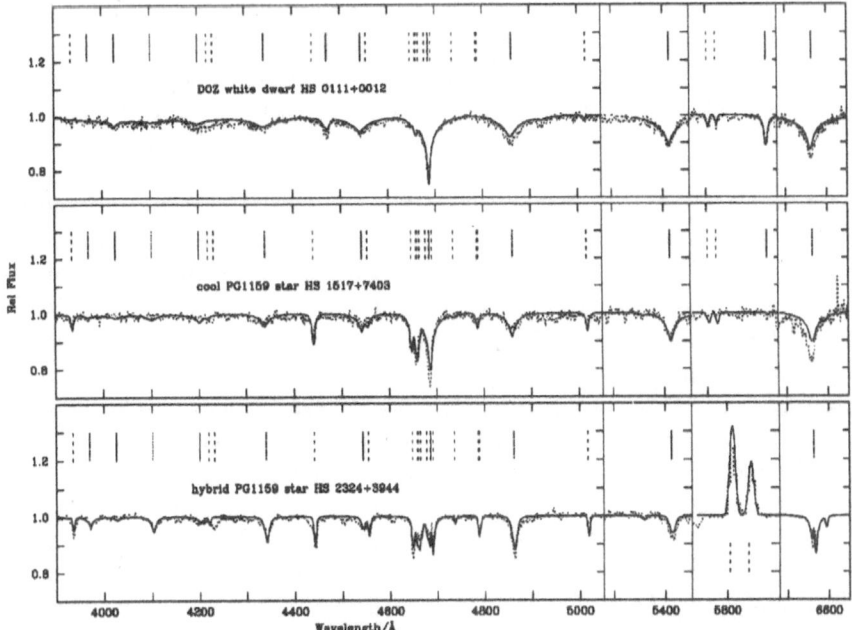

Fig. 2. A DOZ white dwarf, a cool and a hybrid PG 1159 star overlayed with synthetic spectra of the final parameters (see Table 2). He lines are marked with solid lines, blends of He and H lines with narrow dashed, C IV lines with wide dashed lines.

Detailed analyses of the metal abundances in hybrid PG 1159 stars will presented elsewhere (Dreizler et al. in prep.). Neon was searched for in HS 2324+3944 but is below the detection limit. Together this abundance pattern can be explained by a less drastic mass loss compared to H-deficient PG 1159 stars. The surface is composed of 3α (C,He-rich)– and CNO (N,He-rich) processed material mixed with parts of the primordial surface. The reason for the less drastic mass loss is not clear. It may be speculated that the slightly lower luminosities of the hybrid PG 1159 stars are the reason for weaker stellar winds. The difference in the surface gravity between normal and hybrid PG 1159 stars is, however, not significant to proof that hypothesis. An alternative explanation is connected with the GW Vir instability. Due to the high H abundance the hybrid PG 1159 stars should be stable according to the prediction of Starrfield (1987, see also below). Mass loss events like in Longmore 4 are then impossible, provided that there is a connection between pulsations and such events.

5 GW Vir instability strip

The question what makes a PG 1159 star a GW Vir pulsator is not yet decided. Within the spectroscopic twins (PG 1707+427/PG 1424+535 and PG 1159-035/PG 1520+525) the first ones are pulsating while the second ones are stable.

Within the error limits of the optical analysis these stars are indistinguishable. These findings left the exact location of the GW Vir instability strip undetermined. Some new insights in the location of the blue edge of this instability strip came from the analysis of HST-FOS and EUVE spectra. The analysis of the HST-FOS spectrum of PG 1159-035 particularly the evaluation of the ionization balance of O v/vi made a temperature determination of unprecedented precision possible (Werner & Heber 1993). The effective temperature of 140 000 K (WHH91) was confirmed within an error of 3%. On the other side the analysis of the EUVE spectrum of PG 1520+525 indicated an effective temperature slightly higher than given by WHH91. Werner et al. (in prep.) suggest that the blue edge of the GW Vir instability strip lies between 140 000 and 150 000 K.

Since we find pulsating PG 1159 stars with T_{eff} below that of the non-pulsating star PG 1424+535 (100kK) the red edge of the instability strip has become more fuzzy. Together with the results from above it is quite obvious that the location of the GW Vir instability is not only dependent on the effective temperature but probably also on the chemical composition. Starrfield et al. (1984) showed that cyclic ionization ($\kappa - \gamma$ effect) of carbon and oxygen can drive non-radial g-mode pulsations. The driving mechanism requires a low helium abundance ($< 25\%$ Cox 1986, Starrfield 1987), the complete absence of hydrogen, and a high oxygen abundance. In general, the hydrogen abundance in hot H-deficient atmospheres is very difficult to determine. All hydrogen lines are blended with He ii lines, so that a H abundance as high as H/He≤ 0.1 remains undetected. Apart from the hottest PG 1159 stars, showing O vi lines in the optical, the O abundance can only be determined from UV spectra. Due to the low S/N ratio of the accessible (low-res.) IUE spectra uncertainties in the determination of the oxygen abundance particularly in the cooler PG 1159 stars and the high lower limit for the hydrogen abundance makes a spectral discrimination between the pulsating and non-pulsating PG 1159 stars very difficult. Both aspects will be further investigated with high resolution optical spectroscopy to derive tighter limits for the H abundance and with HST to obtain more precise O abundances.

The amount of H remaining in the atmospheres of PG 1159-035 is also important for the identification of the DOs as descendants of the PG 1159 stars. The hybrid PG 1159 stars show that even stars with a significant fraction of H left in the atmosphere can look very similar to normal PG 1159 stars. Due to the large upper limit enough H can be left in the atmosphere to turn a PG 1159 star into a DA white dwarf after gravitational settling has removed all heavier elements (including He).

The transition from the PG 1159 stars to the DOs is also very likely not only dependent on T_{eff} and gravity but also on the chemical abundances, since the ($\kappa - \gamma$) mechanism which might drive the pulsations as well as convection, thus preventing the gravitational settling. Slightly different metal abundances in the precursor stars may determine the onset of diffusion and therefore explain the co-existence of DOs and cool PG 1159 stars in the same parameter region. In general, the metal abundances can however not be explained by existing diffusion calculations (see Werner et al. 1995 for a more detailed discussion; Unglaub & Bues, these proceedings) although they have been further improved recently.

Acknowledgements

We would like to thank our colleagues in Hamburg providing us with candidate lists and finding charts from the HSS. This research was supported by grants from the DFG (He 1356/16-2, We 1312/6-1) and the BMFT (50 OR 9409 1). Several travel grants by the DFG are acknowledged. Computations were performed on CRAY computers in Erlangen, Munich and Kiel.

References

Barstow M.A., Tweedy R.W. 1990, MNRAS 242, 484

Barstow M.A., Holberg J.B. 1990, MNRAS 245, 370

Barstow M.A., Werner K., Sion E.M., Holberg J.B. Hubeny I. 1994, CCP7 Newsletter 21, 18

Blöcker T 1994, Acta Astron., 43, 305

Bond H.E., Meakes M.G. 1990, AJ 100, 788

Bond H.E., Ciardullo R. 1993 IAU Symp 155, 489

Bond H.E., Grauer A.D, Green R.F., Liebert J. 1984, ApJ 279, 751

Cox A.N. 1986, in Highlights in Astronomy, ed. J.P. Swings, 7, 229

Dorman B., Rood R.T., O'Connell R.W. 1993, ApJ 419, 596

Dreizler S., Werner K. 1993, A&A 278, 199

Dreizler S., Werner K., Jordan S., Hagen H.J. 1994a, A&A 286, 463 (D94a)

Dreizler S., Heber U., Jordan S., Engels D. 1994b, in Hot Stars in the Galactic Halo, ed S.J. Adelman, Cambridge University Press, p. 228 (D94b)

Fleming T.A., Werner K., Barstow M.A. 1993, ApJ 416, L79

Fritz M.L., Leckenby H.J., Sion E.M. 1990, AJ 99, 908

Grauer A.D, Bond H.E., Green R.F., Liebert, J. 1987, IAU Coll. 95, p. 231

Grauer A.D, Green R.F., Liebert J. 1992, ApJ 399, 686

Green R.F. Schmidt M., Liebert J. 1986, ApJS 61, 305

Hagen H.J., Engels D., Groote D., Reimers D. 1994 A&A submitted

Hill J.A., Winget D.E., Nather R.E., 1987, IAU Coll. 95, p. 627

Hummer D.G., Mihalas D. 1988, ApJ 331, 794

Husfeld D., 1987 IAU Coll. 95, p. 237

Iben I. Jr., Tutukov A.V. 1985, ApJS 58, 661

Iben I. Jr., Kaler J.B., Truran J.W., Renzini A. 1983, ApJ 264, 605

Jacoby G.H., Van De Steene G. 1993, BAAS 25, 1369

Koester D., Schönberner D. 1986, A&A 154, 125

Kwitter K.B., Massey P., Congdon C.W., Pasachoff J.M. 1989, AJ 97, 1423

McGraw J.T., Starrfield S.G., Liebert J., Green R. 1979, IAU Coll. 53, p. 377

Méndez R.H., Gathier R., Simon R.P., Kwitter K.B. 1988, A&A 198, 287

Motch C., Werner K., Pakull M.W. 1993, A&A 268, 561

Napiwotzki R. 1993, PhD. thesis, University Kiel

Napiwotzki R., Schönberner D. 1991, A&A 249, L16

Napiwotzki R., Rauch T. 1994, A&A 285, 603

Napiwotzki R., Schönberner D. 1995, A&A submitted
Rauch T., Köppen J., Werner K. 1994, A&A 286, 543
Schönberner D. 1983, ApJ 272, 708
Sion E.M., Liebert J., Wesemael F. 1985, ApJ 292, 477
Starrfield S. 1987, IAU Coll. 95, 309
Starrfield S., Cox A.N., Kidman R.B., Pesnell W.D. 1984, Apj 281, 800
Tweedy R., Kwitter K.B. 1994, ApJ 433, L93
Werner K. 1986, A&A 161, 177
Werner K. 1991, A&A 251, 147
Werner K. 1992, Lecture Notes in Physics Vol. 401, Springer, p. 273 (W92)
Werner K., Heber U. 1991, A&A 247, 476
Werner K., Koesterke L. 1992, Lecture Notes in Physics Vol. 401, Springer, p. 288
Werner K., Heber U. 1993, NATO ASI Series C Vol. 403, p. 303
Werner K., Rauch T. 1994, A&A 284, L5
Werner K., Heber U., Hunger K. 1991, A&A 244, 437 (WHH91)
Werner K., Hamann W.R., Heber U., Napiwotzki R., Rauch T., Wessolowski U. 1992,
    A&A 259, L69
Werner K., Heber U., Fleming T. 1994, A&A 284, 907
Werner K., Dreizler S., Wolff B. 1995, A&A in press
Wesemael F., Green R.F., Liebert J. 1985, ApJS 58, 379
Winget D.E., Nather R.E., Clemens J.C., et al. 1991, ApJ 378, 326
Wood P.R., Faulkner D.J. 1986, ApJ 307, 659

Detection of Ultra-Hot Pre-White Dwarfs?*

Klaus Werner[1], Thomas Rauch[1], Stefan Dreizler[2] and Ulrich Heber[2]

[1] Institut für Astronomie und Astrophysik der Universität, D-24098 Kiel, Germany
[2] Dr.Remeis–Sternwarte Bamberg, Universität Erlangen–Nürnberg, Germany

1 Discovery and spectroscopy

Optical follow-up spectroscopy of faint blue stars from the Hamburg-Schmidt objective prism surveys of the northern sky (Hagen et al. 1994) and the southern hemisphere (Wisotzki 1994) has revealed two objects exhibiting unique absorption line spectra. Spectra of HE 0504−2408 were taken at ESO in 1992 (NTT) and 1993 (3.6m telescope with EFOSC 1, displayed in Fig.1). At first glance its spectrum resembles that of a hot DO white dwarf, i.e. it is dominated by broad He II absorptions. Close inspection, however, reveals the presence of unidentified absorption lines, the strongest of which are at 6060Å and 5670Å, which were never observed before in any hot star.

A first hint as to the origin of these absorption came from a direct comparison with the hot DO white dwarf KPD 0005+5106. Ultrahigh-excitation O VIII emission lines were detected in the latter, whose origin was attributed to shock fronts in a wind from the star (Werner et al. 1994). One of the strongest emissions was at 6068Å, close to the position of one of the unidentified absorption lines in HE 0504−2408. Since O VIII is a one-electron ion, we began to look for lines of the isoelectronic ions of carbon and nitrogen, C VI and N VII. To our surprise we found absorption features at all positions where there is a $\Delta n=1$ transition from these ions. These transitions can be expected to be the strongest. Even all expected $\Delta n=2$ transitions are detectable in HE 0504−2408 as weak absorptions. In addition we find an absorption line of He-like C V at 4945Å from which we conclude the presence of lines from the isoelectronic N VI and O VII ions which, however, cannot be resolved from the C VI and N VII lines, respectively. A more thorough discussion on the line identification can be found in Werner et al. (1995, W95).

Subsequently, a second object (HS 0713+3958) with a similar spectrum was discovered which appears to be even more extreme since we identified absorption lines of Ne IX and Ne X (for details see W95).

---

* Based on observations obtained at ESO (La Silla, Chile) and at DSAZ (Calar Alto, Spain)

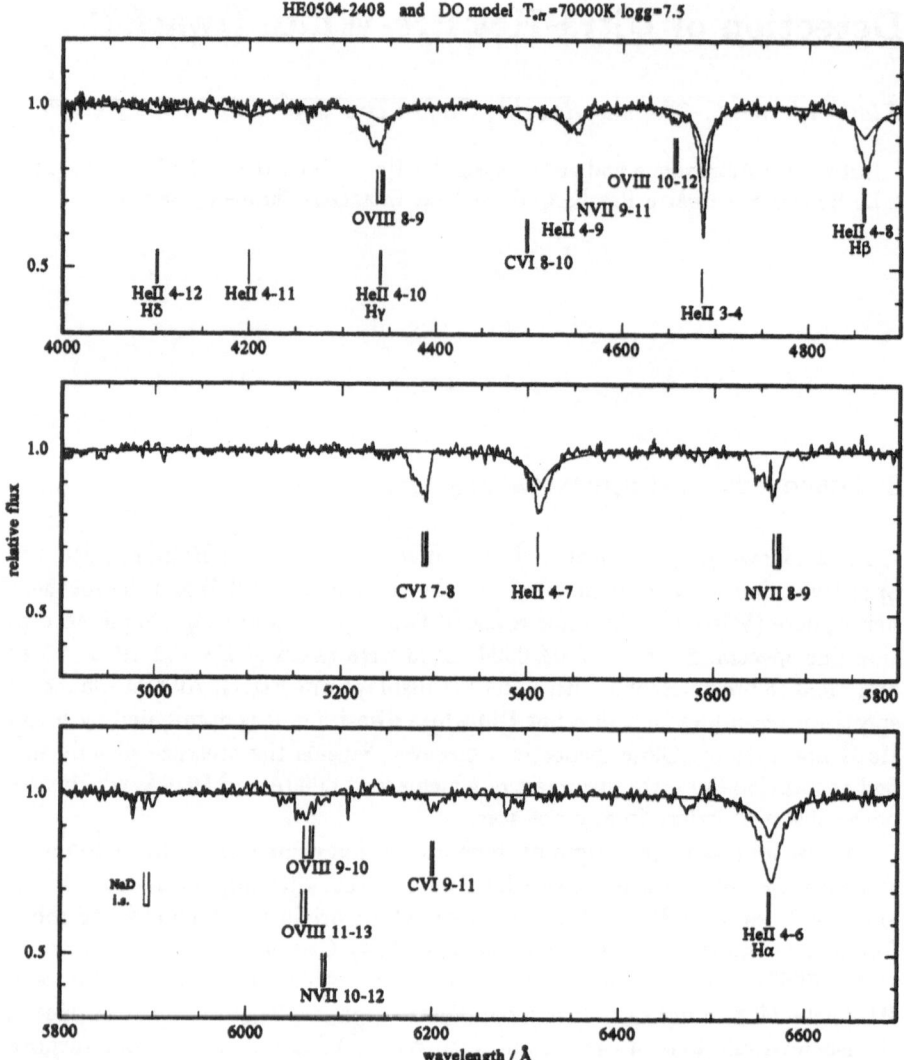

Fig. 1. Optical spectrum of HE 0504−2408. Note the unique absorption lines by ultra-highly ionized C, N, O. Overplotted is a DO model flux. The observed He II lines are unusually strong.

2 NLTE models for super-hot atmospheres

The presence of such ultrahigh-excitation lines in absorption is difficult to understand. We considered the possibility that they are pressure broadened lines of photospheric origin, from a super-hot compact atmosphere, although several reasons argue against such an explanation. Nevertheless, such very hot stars may exist, as they are predicted from post-AGB evolutionary theory. Figure 2 shows

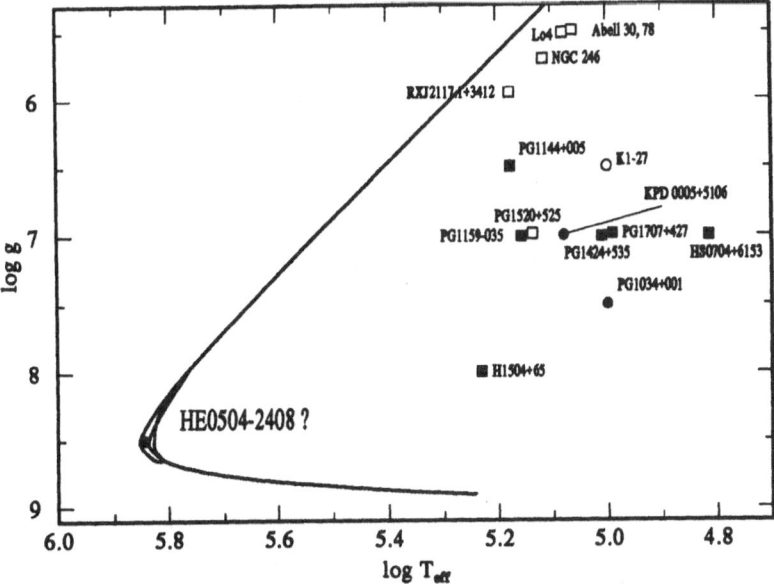

Fig. 2. Paczynski (1970) $1.2M_\odot$ post-AGB track and positions of H-deficient stars (open symbols: associated planetary nebula). The temperature along the track may reach a maximum value near $T_{\text{eff}}=700$kK.

a $1.2M_\odot$ track which reaches a maximum temperature of about $T_{\text{eff}}=700$kK, well in excess of even the hottest PG 1159 stars. But the corresponding evolutionary time scale is extremely short, making the probability to detect a star in this short hot phase very small. And then the question arises if we can expect at all the co-existence of these extreme absorption features together with quite strong He II lines. Wouldn't helium be completely ionized in such a hot atmosphere?

To answer this question we began to compute NLTE models in excess of $500\,000$ K. In a very first step we calculated crude models including H, He, C, and O and mimicked the missing X-ray opacities which would be provided by other metals by introducing an artificial opacity. Figure 3 shows the oxygen ionisation stratification in a model with $T_{\text{eff}}=710$kK and $\log g =8.5$. O VIII is the dominant ion in the line forming regions ($\log \text{m}\approx0$) and O VII is also present so that we can expect the lines of these ions to come out qualitatively as observed. In fact, two O VIII lines can be seen in the synthetic spectrum in Fig. 4. The fact that the He II 4686Å line is still present in the model, although much too weak, encouraged us to construct much more sophisticated atmospheres hoping that He II 4686Å would become stronger. Similarly, other He II lines like 4860Å were almost absent in the crudely calculated spectrum, in contradiction to the observations.

In the next step 45 ions of 13 species were included (H I-II, He I-III, C V-VII, O VI-IX, Ne VIII-IX, Na IX-X, Mg IX-XI, Al X-XII, Si X-XIII, P XI-XIV, S XII-XV, Ar XII-XVI, Ca XI-XVI) in the calculations. Numerous tests pre-

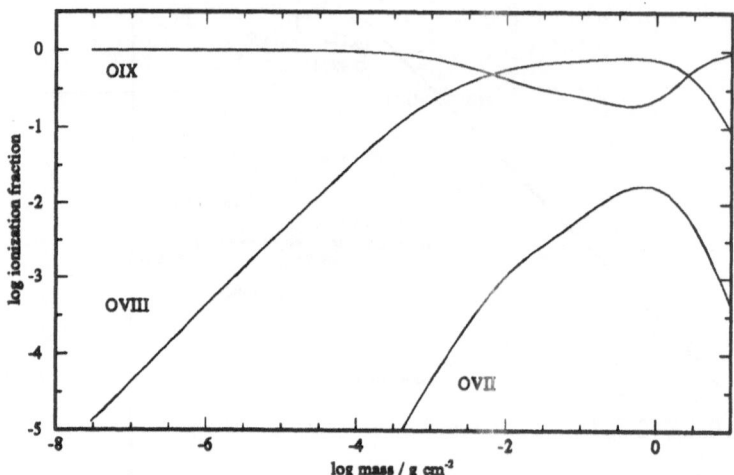

Fig. 3. Oxygen ionization stratification in a $T_{\text{eff}}=710$kK, $\log g =8.5$ model atmosphere.

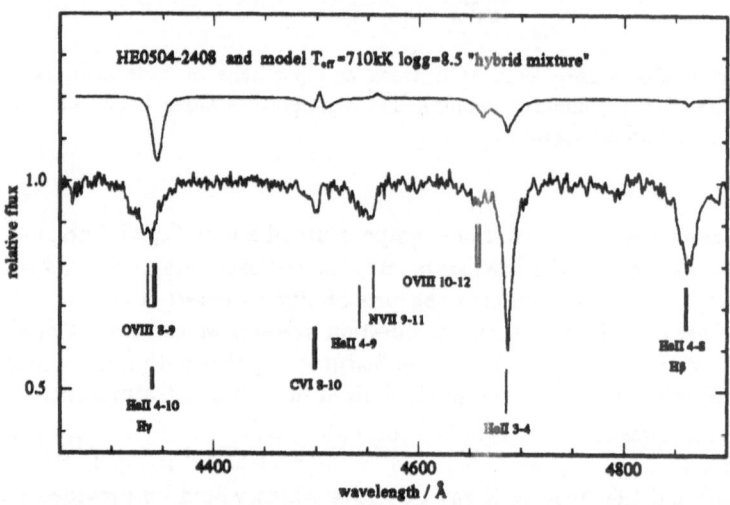

Fig. 4. Computed and observed line profiles of HE 0504−2408. The element abundances match those of the H-He-C rich "hybrid" central stars.

dicted these ions to be dominant. Much effort has been put into the atomic model design of the metals. The latest atomic data from the Opacity Project were considered. Figure 5 shows the model fluxes over a wide range. The X-ray flux is mainly blocked by strong O and C edges and it is redistributed into the (E)UV and optical regions. However, the optical line profiles of these models confirmed that the ultrahigh-excitation features cannot be formed in a photosphere together with the He II lines.

But what is the nature of HE 0504−2408 and HS 0713+3958? We have spec-

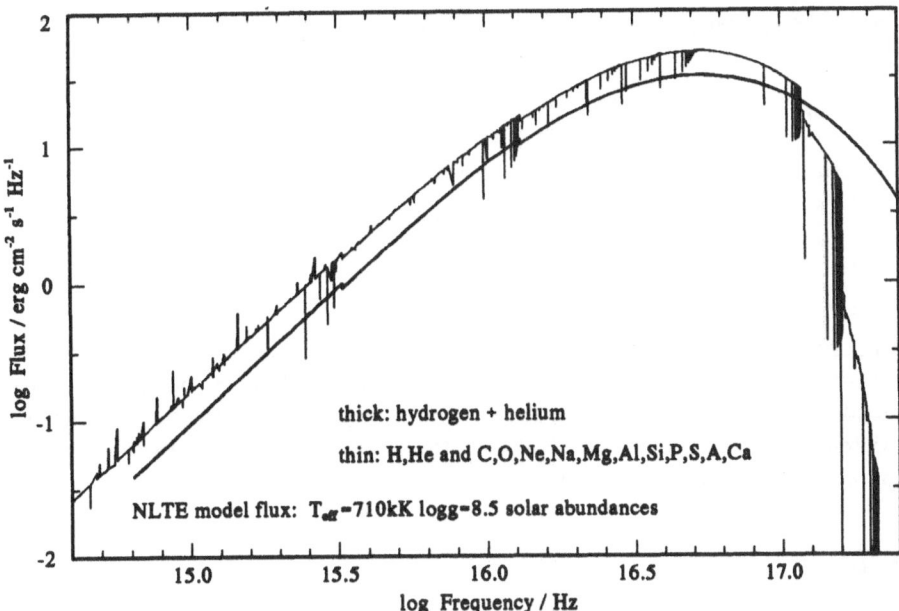

Fig. 5. Emergent fluxes of $T_{eff}=710$kK, $\log g=8.5$ model atmospheres from the optical to the X-ray region. The flux maxima are located near 50Å. The strongest absorption edges are due to the ground states of C VI, O VII, O VIII. Numerous edges due to light metals are less conspicuous.

ulated that the broad metal absorption lines are formed in a hot, compact wind flowing out from a DO white dwarf (W95). This would explain the detectable blueshift and asymmetry of the lines. The He II lines are unshifted and symmetric but a simple photospheric explanation for their shape is not sufficient. The He II lines, particularly 4686Å, are markedly deeper than in other hot DOs and we completely failed to fit a model (see Fig. 1). We are confident that further observations (in the ultraviolet with HST) will help to explain these bizarre spectra.

Acknowledgements. We thank our colleagues L. Wisotzki and H.-J. Hagen from the Hamburger Sternwarte for target selection. This research was supported by grants from the DFG (We 1312/6-1, He 1356/16-2) and the BMFT (50 OR 9409 1).

References

Hagen H.-J., Engels D., Groote D., Reimers D. 1994, A&A in press
Paczynski B. 1970, Acta Astr. 20, 47
Werner K., Heber U., Fleming T.A. 1994, A&A 284, 907
Werner K., Dreizler S., Heber U., Rauch T., Wisotzki L., Hagen H.-J. 1995, A&A in press (W95)
Wisotzki L. 1994, in: IAU Symp. 161, 723

White Dwarfs in Old Planetary Nebulae

Ralf Napiwotzki

Dr. Remeis-Sternwarte, Sternwartstr. 7, 96049 Bamberg, Germany

Abstract: A review is given on recent work on the central star – white dwarf transition zone. Classifications and analyses of central stars of old planetary nebulae are discussed and it is shown that present day observations imply the existence of two seperated spectral evolutionary sequences for hydrogen-rich and -poor central stars/white dwarfs.

1 Introduction

Central stars of planetary nebulae (PN) are the immediate precursors of white dwarfs. Most stars enter the white dwarf cooling sequence through this evolutionary channel (Drilling & Schönberner 1985). After the maximum temperature is reached (100,000 K and more) the nuclear burning ceases, the surrounding nebula disperses, and the central star stage ends. Thus the nuclei of old PNe mark the transition to the white dwarfs. During this stage the onset of gravitational settling, which causes the chemically very pure atmospheres of many white dwarfs, can be observed.

However, not long ago this region of the HR diagram was terra incognita. Much work was done by Weinberger and collaborators (see e.g. Hartl & Weinberger 1987 and references therein) and Kwitter et al. (1988) by discovering new faint and evolved PNe and identifying their nuclei. However, only for a small number of central stars of old PNe reliable spectral classifications were known (see the compilations of Méndez 1991, and Acker et al. 1992).

In the following we will start with a short review on previous investigations in related fields and the emerging picture of (pre-)white dwarf spectral evolution. Then we turn to the original topic of this review the classification and analysis of white dwarf central stars. The results of a recent systematic survey of the nuclei of old PNe are summarized. Finally we will discuss the new picture of white dwarf spectral evolution drawn from this investigation and recent results on white dwarfs.

2 Related Fields

Central stars of PNe and white dwarfs can be divided into two major groups: hydrogen-rich and hydrogen-poor ones. Méndez (1991) reported that about 1/3 of the CPNe of his compilation are H-deficient. An extensive analysis of high resolution spectra of H-rich PNNi was presented by Méndez et al. (1988a). However, most of these central stars possess young PN and have still rather low surface gravities.

A rich source of hot white dwarfs was provided by the Palomar-Green (PG) survey (Green et al. 1986; Wesemael et al. 1985), which for the first time allowed statistically meaningful investigations of stars at the hot end of the white dwarf cooling sequence. Four subtypes are found in this region: DA (pure hydrogen), DAO (hydrogen with traces of helium), DO (helium dominated), PG 1159 stars (helium- and carbon-rich atmospheres). A review on the analysis of DO and PG 1159 stars can be found in Dreizler et al. (1995).

Liebert (1986) reported a ratio of H-rich to H-poor stars of 7($\pm$3):1 for the hot white dwarfs in the PG survey, but noted a lack of very hot DA white dwarfs. Holberg (1987) selected a sample of the apparently hottest DA/DAO white dwarfs from the PG survey and concluded that none of them appear to have a temperature in excess of 80000 K, with T_{eff} typically in the range 60000 K to 70000 K. However, detailed analyses of these stars were not presented. Fontaine & Wesemael (1987) discussed this seemingly lack of very hot H-rich white dwarfs in the context of "evidences" for very thin hydrogen layers of DA white dwarfs (EUV/X-ray observations, the DB gap, pulsational properties of ZZ Ceti stars) and proposed that all white dwarfs evolve through one evolutionary channel: the PG 1159/DO stars. Most of these evidences are seen in new light today (cf. many contributions in these proceedings): the EUV opacity is now believed to be caused by trace metals, the ZZ Ceti pulsations can be modeled with thick H layers, and new results on hot DA/DAO stars were published recently by Bergeron et al. (1994). These results together with new data on central stars of old PNe, presented in the next section, will be used for a rediscussion of the spectral evolution of hot (pre-)white dwarfs.

3 Central Stars of Old PNe

3.1 Classification

As already noted in the introduction the number of central stars of old PNe with a spectral classification remained very small (cf. Méndez 1991; Acker et al. 1992). The first systematic investigation of old PNNi (Napiwotzki & Schönberner 1991, 1995; Napiwotzki 1992) now yielded a total of 38 classifications. Most of the observed objects have high surface gravities, many are white dwarfs. About three quarters (28) of the PNNi display H-rich atmospheres. Seven new members of the rare class of helium and carbon dominated PG 1159 stars were discovered. Three H-rich stars belong to the previously unknown class of hybrid central stars. They display a He II/C IV absorption trough typical for PG 1159 stars, but strong Balmer lines, too. Sample spectra of stars of these three classes are shown in Fig. 1. The remaining three stars are close binary PNNi. The ratio of H-rich to -poor objects (4:1) is in reasonable agreement with the findings of Liebert (1986) and Méndez (1991). This is not in accordance with the predictions of the

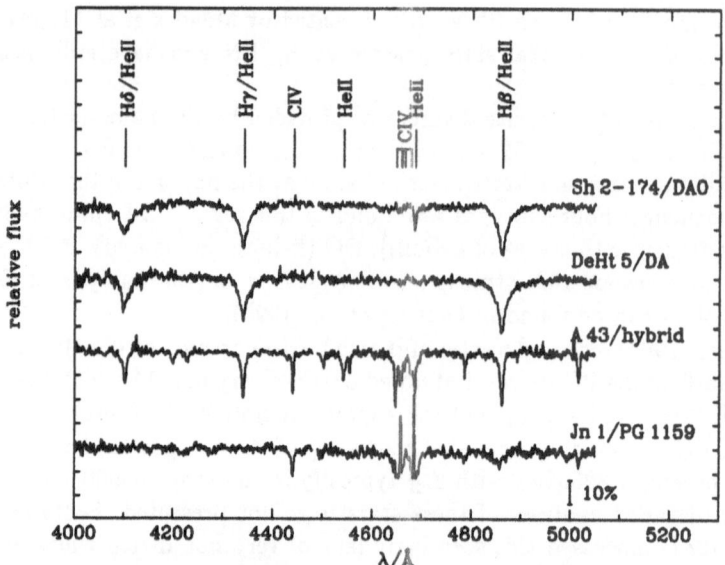

Fig. 1. Sample spectra of the central stars of Sh 2-174 (DAO), DeHt 5 (DA), A 43 (hybrid-PG 1159), and Jn 1 (PG1159). Important spectral lines are marked

one channel scenario of Fontaine & Wesemael (1987), but supports the idea of two distinct spectral sequences. A further discussion of this point will be given below.

3.2 Analysis

In this review we will only discuss the analysis of the hydrogen-rich central stars. A review on the properties and the analysis of PG 1159 stars is given in Dreizler et al. (1995). Let us note that for most PG 1159 central stars of the Napiwotzki & Schönberner (1995) sample reliable parameters can be estimated by a simple comparison with the PG 1159 stars analyzed in Werner et al. (1991; see Napiwotzki 1993b and Napiwotzki & Schönberner 1995). Before the start of the survey of Napiwotzki & Schönberner (1991, 1995) only three analyses of high gravity CPN were published: A 7 (Méndez et al. 1981), NGC 7293 (Méndez et al. 1988), EGB 6 (Liebert et al. 1989). The much enlarged number of analyses presented in Napiwotzki (1993a,b) now allows a more meaningful investigation of the stellar evolution in this region of the HR diagram.

The central stars of the Napiwotzki & Schönberner (1995) sample were analyzed by Napiwotzki (1993a,b) with a grid of hydrogen and helium composed NLTE model atmospheres, calculated with the ALI code developed by Werner (1986). Details on the models are given in Napiwotzki et al. (1993). Due to the lack of other temperature indicators both, effective temperature and gravity, must be determined from the Balmer lines. For DA white dwarfs this was successfully done by simultaneous line profile fitting of all available Balmer lines

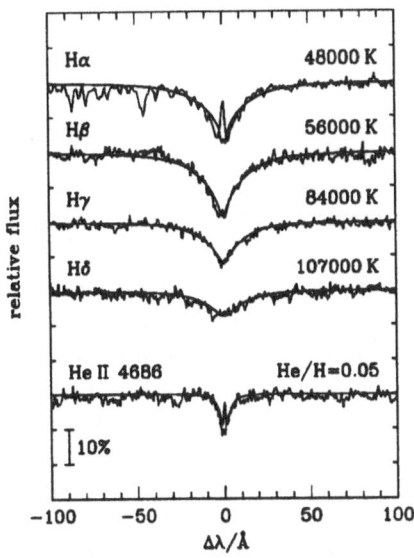
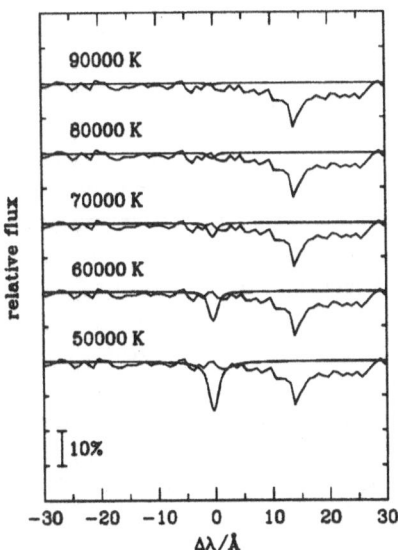

Fig. 2. Balmer line fits to NGC 7293. Individual temperatures for the Balmer lines are indicated. Additional parameters are $\log g = 7.0$ and $n_{\text{He}}/n_{\text{H}} = 0.05$

Fig. 3. Comparison of the observed spectrum to synthetic He I 5876 Å profiles for different temperatures

(e.g. Bergeron et al. 1992; Napiwotzki et al. 1993). Generally the observed line profiles are well reproduced by model spectra with the optimum parameters.

However, this method failed for nearly all hydrogen-rich central stars of old PNe (Napiwotzki 1992; Napiwotzki & Rauch 1994): no consistent fit to the Balmer lines (Hα to Hδ were used) was possible. In Fig. 2 we show the result for the central star of NGC 7293 as a typical example. A strong trend is obvious: fitting of higher Balmer lines yields higher temperatures. Méndez et al. (1988) analyzed this central star before. They fitted the Hγ line only and derived $T_{\text{eff}} = 90000\,\text{K}$ and $\log g = 6.9$. This is within the error limits in agreement with our result for Hγ: $T_{\text{eff}} = 84000\,\text{K}$, $\log g = 7.0$.

Napiwotzki & Schönberner (1993) and Napiwotzki & Rauch (1994) discussed possible reasons of the Balmer line problem: wind effects, magnetic fields, pressure ionization and line quenching, deficits in the line broadening theory, and modifications of the atmospheric structure due to line blanketing of heavy elements. The authors concluded that none of these effects is sufficient to explain the Balmer line problem. Two more points were investigated more recently: K. Werner (priv. comm.) investigated the non-coherent electron scattering, discussed in Hummer & Mihalas (1967) and found no influence on the emergent spectrum. Kubat (1995a, b) found a moderate sphericity effect on Hα and, to an lesser extent, on Hβ, too, for pre-white dwarfs with $\log g \approx 6$. These effects are considerably weaker for larger gravity and too small to account for the Balmer line problem.

The line blanketing by heavy elements deserves some further comments. Bergeron et al. (1993) were able to remove the Balmer line problem in a "cool" DAO ($T_{\text{eff}} \approx 60000\,\text{K}$) with a LTE model atmosphere containing an arbitrary amount of iron. However, Werner & Dreizler (1993) calculated NLTE model atmospheres including the line blanketing of iron and C, N, O and found only weak influence on the line profiles for metal abundance up to ten times the solar value. However, we must be aware that our knowledge of EUV opacity is only sketchy, as was shown by recent EUVE spectroscopic observations of hot DA stars. Thus this point needs further investigation.

Because of the Balmer line problem, independent temperature estimates are highly desirable. For elements with two or more observable ionization stages (He, C, N, O, and Fe in hot stars) the ionization equilibria can be utilized for very accurate temperature determinations. Unfortunately most white dwarf central stars are too faint to allow the detection of these weak features. For some stars of the old PNe sample an useful lower temperature limit can be derived from the He I 5876 Å line. The non-detection of He I in NGC 7293 indicates $T_{\text{eff}} \geq$ 70000 K (Fig. 3). This is consistent with the temperatures derived from Hγ and Hδ (84000 K and 107000 K respectively), but excludes the lower temperatures from Hα and Hβ (48000 K and 56000 K). Similar results were obtained for 4 additional CPNe. This implies that the (highest) temperature from the highest (observed) Balmer line (Hδ) gives the most reliable T_{eff} estimate.

This hypothesis is supported by the analysis of the hot, hydrogen-rich sdO star BD+28°4211 presented in Napiwotzki (1993a,b). The spectral appearance and parameters of BD+28°4211 are very similar to the nuclei of the old PNe sample and it suffers from the Balmer line problem, too: T_{eff} determinations range from 50000 K for Hα up to 85000 K for Hϵ. From the analysis of the weak He I 5876 Å line detected in a high resolution spectrum $T_{\text{eff}} = 82000\,\text{K}$ is derived in good agreement with the Hϵ result. A preliminary analysis of an IUE high resolution spectrum yielded consistency of the ionization equilibria of C, N, O, and Fe with this temperature (for iron see Dreizler & Werner 1993).

An additional tool for the T_{eff} determination of PN nuclei is provided by photoionization modeling of the nebula. Clegg & Walsh (1989) determined $T_{\text{eff}} =$ 120000 K for the central star of NGC 7293 from the nebula lines. This is in good agreement with the temperature determined from Hδ ($T_{\text{eff}} = 107000\,\text{K}$).

We can conclude that from the different temperatures derived from different Balmer lines the temperatures derived from the highest Balmer lines are the most reliable, close to the "real" temperature of the central stars.

3.3 Results and discussion

The results of the analyses of Napiwotzki (1993a,b) are summarized in Table 1. A T_{eff}-g diagram with the results is given in Fig. 4. The mean mass amounts to 0.58 $M_\odot$ in good agreement with mass determinations of white dwarfs (cf. the review of Weidemann 1995). The hottest and most massive central star is WeDe 1 with $T_{\text{eff}} \approx 200000\,\text{K}$ and 0.77 $M_\odot$. A recent LTE analysis carried out by Liebert et al. (1994) yielded $T_{\text{eff}} = 165000\,\text{K}$ and $\log g = 7.65$. The central stars of old PNe fit nicely into the reported "gap" of the hydrogen-rich sequence.

Table 1. Parameters of the hydrogen-rich central stars from Napiwotzki (1993a,b). The PNNi are denoted by their PN G designation taken from the catalogue of Acker et al. (1992) and by their common names. The Hδ temperature is given. Masses were derived from the comparison with the evolutionary tracks in Fig. 4

PN G	Name	$T_{\mathrm{eff}}/\mathrm{K}$	$\log g$	$n_{\mathrm{He}}/n_{\mathrm{H}}$	$M/M_\odot$
030.6+06.2	Sh 2-68	74000	6.75	0.10	
034.1−10.5	HDW 11	61000	6.40	0.10	
036.0+17.6	A 43	110000	6.00	0.10	0.56
036.1−57.1	NGC 7293	107000	7.00	0.05	0.58
047.0+42.4	A 39	118000	6.50	0.10	0.56
060.8−03.6	NGC 6853	99000	6.75	0.10	0.55
063.1+13.9	NGC 6720	137000	7.00		0.60
066.7−28.2	NGC 7094	140000	6.25	0.10	0.59
111.0+11.6	DeHt 5	82000	6.75	<0.004	0.53
120.3+18.3	Sh 2-174	65000	6.80	0.003	
124.0+10.7	EGB 1	133000	7.50	<0.03	0.69
156.3+12.5	HDW 4	50000	7.55	<0.001	0.51
156.9−13.3	HaWe 5	43000	7.40	<0.01	
158.5+00.7	Sh 2-216	85000	6.90	0.01	0.54
197.4−06.4	WeDe 1	217000	7.35	<0.10	0.77
215.5−30.8	A 7	109000	6.75	0.10	0.56
219.1+31.2	A 31	92000	6.30	0.10	0.54

Worth mentioning in this context is the recent article of Bergeron et al. (1994) on the analysis of hot DA/DAO stars, which yielded a number of very hot white dwarfs without a nebula. Now a continuous hydrogen-rich sequence from the central star to the white dwarf region is revealed and therefore evolutionary scenarios claiming a hydrogen-poor stage of all pre-white dwarfs (Fontaine & Wesemael 1987) can be ruled out.

Most stars are well explained by the canonical post-AGB evolution (Fig. 4). However, the position of three stars (HDW 11, Sh 2-174, Sh 2-68) lie in a region of the HR diagram not crossed by post-AGB tracks (even with very conservative error bars). Two similar objects (EGB 5, PHL 932) were discovered by Méndez et al. (1988a,b). The HR position of these stars can be explained by evolution from the extreme horizontal branch (EHB). However, post-EHB evolution is too slow, to account for the presence of a PN (Dorman et al. 1993). The position of two stars (HDW 4 and HaWe 5) is (within the error bars) in agreement with post-AGB evolution. However, the time since the AGB was left would amount to more than $3 \cdot 10^6$ years, if standard evolution is assumed. This is at variance with the kinimatical ages (< 10000 years) of their nebulae.

All these objects are most likely the outcome of a close binary evolution. During the giant stage (on the AGB or RGB) of the central star precursor, it can fill its Roche lobe or even enclose the companion in its atmosphere (common envelope). Especially the latter causes heavy mass loss and a modification of the

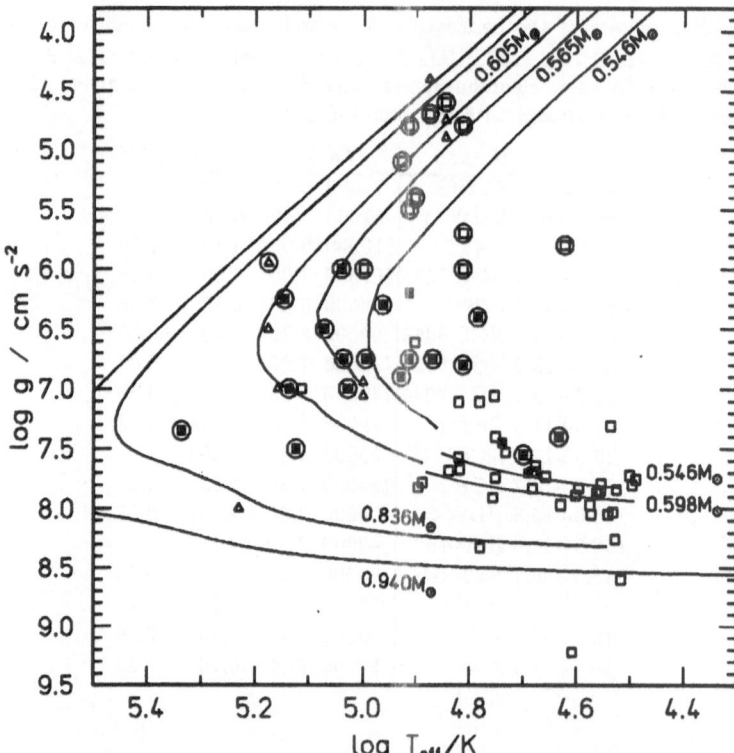

Fig. 4. T_{eff}-g diagram with the results of the analysis of Napiwotzki (1993a; filled symbols) and of Méndez et al. (1981, 1985, 1988a), Bergeron et al. (1992, 1994), Napiwotzki et al. (1993), Husfeld et al. (1989), Werner et al. (1991), Werner & Heber (1991), Werner (1991), Motch et al. (1993). Hydrogen-rich objects are marked by squares, hydrogen-poor by triangles. Central stars are encircled. Evolutionary tracks were taken from Schönberner (1983), Koester & Schönberner, and Blöcker & Schönberner (1990).

stellar evolution. The results is a hot star inside an expanding shell, which can mimic a normal PN (see e.g. the review of Livio 1993). The case of Sh 2-174 is discussed in detail in Tweedy & Napiwotzki (1994).

A photometric search for close binary CPNe has been carried out by Bond (1989). The author estimated a close binary fraction of $10\ldots15\%$. Méndez (1989) looked for radial velocity variations indicating binarity. However, both methods are only sensitive to rather close binaries. Recently Napiwotzki & Werner started a photometric and spectroscopic search in the infrared J, H, K bands for late type companions of white dwarf CPNe and PG 1159 stars. For these relatively low luminosity stars the detection of very cool main sequence companions (down to M6 or even later) is possible. Thus a high detection rate and a meaningful result of this survey can be expected. A first interesting result was derived for GD 561 the central star of Sh 2-174: it has a cool companion separated by $\approx 4''$. It will be used for the determination of a spectroscopic parallax. Furthermore the

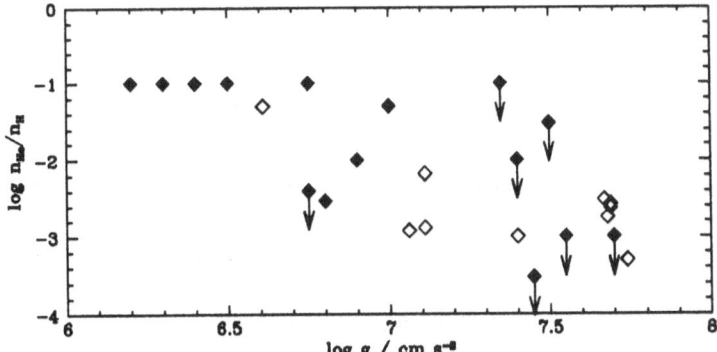

Fig. 5. Photospheric helium abundance vs. gravity. Data were taken from Napiwotzki (1993a; filled symbols) and from Bergeron et al. (1994; open symbols). Upper limits are marked by arrows

preliminary analysis revealed that the central star itself has an infrared excess of $0\overset{m}{.}5$ in K indicating binarity. However, more observations of more CPNe are necessary before we can draw firm conclusions.

A wide spread of helium abundances is observed in the atmospheres of hydrogen-rich central stars. From theoretical considerations (no observable helium traces can be supported in atmospheres of white dwarfs by radiative levitation; Vennes et al. 1988) it was expected that the atmospheres of DAO stars are chemically layered with a very thin H layer floating on top of the helium envelope. However, Napiwotzki & Schönberner (1993) and Bergeron et al. (1994) have shown, that the spectra are best fitted by homogeneous and not by layered atmospheres.

Napiwotzki (1993a) and Bergeron et al. (1994) noted a correlation between helium abundance and gravity. However, Bergeron et al. (1994) did not consider it significant and stressed a stronger correlation with T_{eff}. A diagram of $n_{\mathrm{He}}/n_{\mathrm{H}}$ vs. g for both samples is shown in Fig. 5. Now there is a strong trend visible: for all stars with $\log g \leq 6.7$ the He abundance is close to the "solar" value $n_{\mathrm{He}}/n_{\mathrm{H}} = 0.1$. For stars with higher gravity the helium abundance decreases with increasing gravity. The scatter seems to be larger than expected from the observational errors alone. However, the sample of Napiwotzki (1993a) consists of hot central stars, while the Bergeron et al. (1994) sample is dominated by cooler DAO stars. Thus the stars in Fig. 5 cover a wide range in temperature. Therefore we conclude that the helium abundance is mainly correlated with g. An interesting point in this context is that all DA stars in the sample of Bergeron et al. (1994) have $\log g$ in excess of 7.5. Fig. 5 shows that gravitational settling is prevented until the surface gravity exceeds $\log g \approx 6.7$ and an equilibrium with the helium abundance still higher than predicted by radiative levitation alone is reached for higher gravity. A likely candidate for the process counteracting gravitational settling is mass loss. It strongly decreases with increasing gravity and can therefore easily explain the observed behavior.

4 Conclusions

Our knowledge of the central star – white dwarf transition region has dramatically increased during the last years. The survey of central stars of old PNe by Napiwotzki & Schönberner (1995) yielded a total of 38 classifications, multiplying the number of previously known stars. The ratio of hydrogen-rich to -poor star amounts to 4:1. This is consistent with the findings of Liebert (1986) and Méndez (1991).

The analysis of the H-rich CPNe is hampered by the Balmer line problem: for most stars a consistent fit to all Balmer lines was not possible. Generally the highest Balmer line yields the highest temperature. Possible reasons were discussed, but this problem is yet unsolved. However, it was outlined that the temperature derived from Hδ (in most cases the highest useable Balmer line) is in good agreement with other temperature indicators and can therefore be used.

The investigation revealed that the reported gap in the hydrogen-rich evolutionary is not real. Taking into account the hot white dwarfs analyzed in Bergeron et al. (1994) we have now a continuous H-rich sequence from the CPNe to the white dwarfs. The observational findings are well explained by a two channel scenario with a H-rich and a H-poor sequence. It is no longer necessary to invoke a one channel scenario (Fontaine & Wesemael 1987) to explain the spectral evolution of pre-white dwarfs. Five central stars are not in agreement with standard post-AGB evolution. They are probably the result of a close binary evolution.

Napiwotzki & Schönberner (1993) and Bergeron et al. (1994) have shown that the atmospheres of DAO stars are chemically homogeneous and not layered. The observed correlation of helium abundance and gravity can be explained by the presence of a weak stellar wind counteracting the gravitational settling.

References

Acker, A., Ochsenbein, F., Stenholm, B., et al. 1992, Strasbourg-ESO catalogue of galactic planetary nebulae, ESO, Garching bei München
Bergeron, P., Saffer, R.A., Liebert, J. 1992, ApJ 394, 228
Bergeron, P., Wesemael, F., Lamontagne, R., Chayer, P. 1993, ApJ 407, L85
Bergeron, P., Wesemael, F., Beauchamp, A., et al. 1994, ApJ 432, 305
Blöcker, T., Schönberner, D. 1990, A&A 240, L11
Bond, H.E. 1989, in: IAU Symp. No. 131, Planetary nebulae, ed. S. Torres-Peimbert, Kluwer, p. 251
Clegg, R.E.S., Walsh, J.R. 1989, in: IAU Symp. No. 131, Planetary nebulae, ed. S. Torres-Peimbert, Kluwer, p. 223
Dorman, B., Rood, R.T., O'Connell, W.O. 1993, ApJ 419, 596
Dreizler, S., Werner, K. 1993, A&A 278, 199
Dreizler, S., Werner, K., Heber, U. 1995, these proceedings
Drilling, J.S., Schönberner, D. 1985, A&A 278, 199
Fontaine, G., Wesemael, F. 1987, in: IAU Coll. 95, The second conference on faint blue stars, eds. A.G.D. Phillip, D.S. Hayes & J. Liebert, L. Davis Press, p. 285
Green, R.F., Schmidt, M., Liebert, J. 1986, ApJS 61, 305
Hartl, H., Weinberger, R. 1987, A&AS 69, 519

Holberg, J.B. 1987, in: IAU Coll. 95, The second conference on faint blue stars, eds. A.G.D. Phillip, D.S. Hayes & J. Liebert, L. Davis Press, p. 285

Hummer, D.G., Mihalas, D. 1967, ApJ 150, L57

Husfeld, D., Butler, K., Heber, U., Drilling, D.S. 1989, A&A 222, 150

Koester, D., Schönberner, D. 1986, A&A 154, 125

Kubat, J. 1995a, these proceedings

Kubat, J. 1995b, A&A, submitted

Kwitter, K.B., Jacoby, G.H., Lydon, T.J. 1988, AJ 96, 997

Liebert, J. 1986, in: Hydrogen deficient stars and related objects, eds. K. Hunger, D. Schönberner & N. Kameswara, Reidel, Dordrecht, p. 367

Liebert, J., Green, R., Bond, H.E., et al. 1989, ApJ 346, 251

Liebert, J., Bergeron, P., Tweedy, R.W. 1994, ApJ 424, 817

Livio, M. 1993, IAU Symp. No. 155, Planetary nebulae, eds. R. Weinberger & A. Acker, Kluwer, p. 279

Méndez, R.H. 1989, IAU Symp. No. 131, Planetary nebulae, ed. S. Torres-Peimbert, Kluwer, p. 261

Méndez, R.H. 1991, in: IAU Symp. 145, Evolution of stars: the atmospheric abundance connection, eds. G. Michaud & A. Tutukov, Kluwer, Dordrecht, p. 375

Méndez, R.H., Kudritzki, R.P., Gruschinske, J., Simon, K.P. 1981, A&A 101, 323

Méndez, R.H., Kudritzki, R.P., Simon, K.P. 1985, A&A 142, 289

Méndez, R.H., Kudritzki, R.P. Herrero, A., Husfeld, D., Groth, H.G. 1988a, A&A 190, 113

Méndez, R.H., Groth, H.G., Husfeld, D., Kudritzki, R.P., Herrero, A. 1988b, A&A 197, L25

Motch, C., Werner, K, Pakull, M.W. 1993, A&A 268, 561

Napiwotzki, R., 1992, in: The Atmospheres of early-type stars, Lecture Notes in Physics, eds. U. Heber & C.S. Jeffery, Springer Verlag, Berlin, p. 310

Napiwotzki, R., 1993a, PhD thesis, Universität Kiel

Napiwotzki, R., 1993b, Acta Astr. 43, 343

Napiwotzki, R., Rauch, T. 1994, A&A 285, 603

Napiwotzki, R., Schönberner, D. 1991, in: White dwarfs, eds. G. Vauclair & E. Sion, NATO ASI Series C, No. 336, Kluwer, Dordrecht, p. 39

Napiwotzki, R., Schönberner, D. 1993, in: White dwarfs: advances in observation and theory, ed. M.A. Barstow, NATO ASI Series C, No. 403, Kluwer, Dordrecht, p. 99

Napiwotzki, R., Schönberner, D. 1995, A&A, submitted

Napiwotzki, R., Barstow, M.A., Fleming, T. et al. 1993, A&A 278, 478

Schönberner, D. 1983, ApJ 272, 708

Tweedy, R.W., Napiwotzki, R. 1994, AJ 108, 978

Vennes, S., Pelletier, C., Fontaine, G., Wesemael, F. 1988, ApJ 331, 876

Weidemann, V. 1995, these proceedings

Werner, K. 1986, A&A 161, 177

Werner, K. 1991, A&A 251, 147

Werner, K., Dreizler, S. 1993, Acta Astr. 43, 321

Werner, K., Heber, U. 1991, A&A 247, 476

Werner, K., Heber, U., Hunger, K. 1991, A&A 244, 437

Wesemael, F., Green, R.F., Liebert, J. 1985, ApJS 58, 379

New Analyses of Helium-rich Pre-White Dwarfs*

*T. Rauch** and K. Werner***

Institut für Astronomie und Astrophysik der Universität, D-24098 Kiel, Germany

1 Introduction

One of the most interesting spectral classes of immediate white-dwarf progen-itors is constituted by the PG 1159 stars which are named after the prototype PG 1159-035 (GW Vir). They exhibit broad and shallow absorption lines of He II and C IV which form a characteristic absorption trough around He II λ 4686Å. 26 PG 1159 stars are known, with different spectral appearance, therefore we in-troduced a spectral subclassification (Werner 1992). Many of the PG 1159 stars were already analyzed by our group. The effective temperatures and surface gravities ($T_{\rm eff} = 65 - 170$kK, $\log g = 5.5 - 8.0$ in cgs units) cover a wide range in the $\log T_{\rm eff}$ - $\log g$ diagram. Observations have shown that about every other star is associated with a planetary nebula and/or shows variability due to non-radial g-mode pulsations. They are thought to be in an evolutionary phase between the [WCE] central stars and the DO white dwarfs. For a detailed review on the PG 1159 stars and their link to the DO stars see Dreizler et al.(1995).

In contrast to the PG 1159 stars two central stars of planetary nebulae (CSPN), LoTr 4 and K 1-27, are found in the same region of the $\log T_{\rm eff}$ - $\log g$ dia-gram like the PG 1159 stars but they do not show any carbon line in their spectra, i.e. they exhibit an almost pure He II absorption line spectrum. It is worthwhile to note that LoTr 4 and K 1-27 are the only objects of spectral subtype O(He) (Méndez 1991). According to evolutionary calculations of Iben et al.(1983) it is possible that they evolve into PG 1159 stars if mass loss is still in progress. The now exposed helium buffer layer of the former AGB star may subsequently be eroded to an extent that carbon-rich matter becomes visible at the surface.

We present spectral analyses of both known O(He) stars (LoTr 4, K 1-27) and first, still preliminary, results for seven PG 1159 stars (NGC 246, PG 1151-029, K 1-16, Longmore 3, A 21, Jn 1, and VV 47). In addition, a re-analysis of RX J2117+3412 has begun with new observational material. We determine ef-fective temperatures, surface gravities, and element abundances.

* Based on observations obtained at the European Southern Observatory, La Silla, Chile
** Visiting Astronomer, German-Spanish Astronomical Center, Calar Alto, operated by the Max-Planck-Institut für Astronomie Heidelberg jointly with the Spanish National Commission for Astronomy

2 Observations

Most of the spectra used for the present analysis were recently taken at the Calar Alto observatory and at ESO (Tab. 1). All images were flatfielded with suitable dome flats and wavelength calibrated with He-Ar comparison spectra taken directly before or after each exposure. A careful background subtraction was carried out especially in case of CSPN spectra in order to eliminate nebula emission. The spectra were co-added (two or three exposures, indicated by the single exposure times in Tab. 1) in order to improve the S/N ratio. The achieved spectral resolution is about 1.5Å for the TWIN spectra, $\lambda/\Delta\lambda \approx 1450$ for EMMI, and $\lambda/\Delta\lambda \approx 1750$ for EFOSC 1. The S/N ratio is about $30 - 50$.

Table 1. Observation log of the analyzed spectra taken at the Calar Alto (3.5m, Spain) and ESO (NTT and 3.6m, La Silla, Chile) observatories

object	telescope	spectrograph	dispersion Å mm^{-1}	exposure time / min	date	observer
LoTr 4	3.6m	EFOSC 1	32-63	40 + 75	Jan. 1993	Rauch
K 1-27	NTT	EMMI	39	60 + 90	Jan. 1992	Werner
NGC 246	3.5m	TWIN	36	60	Sep. 1992	Werner
PG 1151-029	3.6m	EFOSC 1	32-63	2 × 60	Jan. 1993	Rauch
K 1-16	3.5m	TWIN	36	60	May 1994	Rauch
RX J2117+3412	3.5m	TWIN	36	60	Sep. 1992	Werner
Longmore 3	3.6m	EFOSC 1	32-63	3 × 60	Jan. 1993	Rauch
A 21	3.6m	EFOSC 1	32-63	2 × 60	Jan. 1993	Rauch
Jn 1	3.6m	TWIN	36	2 × 60	Jul. 1992	Rauch
VV 47	3.6m	TWIN	36	60	Nov. 1989	Werner

3 Model atmospheres

The plane-parallel, hydrostatic NLTE model atmospheres used for the spectral analysis were calculated with our ALI code (Werner 1986). For details of atomic data and the construction of model atoms see Rauch & Werner 1988, Rauch 1993, and Rauch et al. 1994.

For the analysis of the O(He) stars K 1-27 and LoTr 4 a grid of H-He model atmospheres was calculated. These models were used to determine T_{eff}, $\log g$, and the H/He ratio. Another grid of models including nitrogen was used to determine the N/He ratio.

In previous analyses of PG 1159 stars a "typical" photospheric abundance pattern He:C:O = 33:50:17 (by mass) was found (Werner 1993). To begin with the present analysis we started with a slightly decreased oxygen fraction (38:56:6) in order to determine T_{eff} and g. H-He-C-O model atmospheres were calculated using standard model atoms (Werner et al. 1991). Subsequently, line formation calculations (fixed temperature structure) with more detailed C and O model atoms were carried out. For some parameters model atmospheres with different abundance ratios were calculated to improve the line profile fits (Tab. 2). These calculation are still in progress, thus we present preliminary results. The complete analysis (including fine tuning of the photospheric chemical composition) will be presented in a forthcoming paper.

4 Results

4.1 O(He) stars

In a recent analysis of K 1-27, Rauch et al.(1994) found $T_{\mathrm{eff}} = 100 \pm 15\mathrm{kK}$, $\log g = 6.5 \pm 0.5$, $n_{\mathrm{H}}/n_{\mathrm{He}} < 0.2$ (by number), and $n_{\mathrm{N}}/n_{\mathrm{He}} = 0.005 \pm 0.5\mathrm{dex}$. An upper limit $n_{\mathrm{C}}/n_{\mathrm{He}} < 0.005$ was determined. They concluded that K 1-27 is a late helium flash object according to the born-again post-AGB scenario of Iben et al.(1983).

We analyzed LoTr 4 analogously and determined $T_{\mathrm{eff}} = 130^{+10}_{-30}\mathrm{kK}$, $\log g = 5.5^{+0.4}_{-0.2}$, $n_{\mathrm{H}}/n_{\mathrm{He}} = 0.5 \pm 0.5\mathrm{dex}$, and $n_{\mathrm{N}}/n_{\mathrm{He}} = 2.5 \cdot 10^{-4} \pm 0.5\mathrm{dex}$ ($5 \times$ solar value). Upper limits of $n_{\mathrm{C}}/n_{\mathrm{He}} < 0.004$ and $n_{\mathrm{O}}/n_{\mathrm{He}} < 0.008$ (solar values) can be given. In Fig. 1 we compare synthetic spectra of our final models of K 1-27 and LoTr 4 with the observed spectra. The problem to fit He II λ 4686Å and N V$\lambda\lambda$ 4603, 4619 Å in K 1-27 is discussed in detail in Rauch et al.(1994).

Note that the presence of hydrogen in LoTr 4 is beyond doubt. Therefore K 1-27 (no hydrogen detected) could be in a more advanced stage of evolution than LoTr 4.

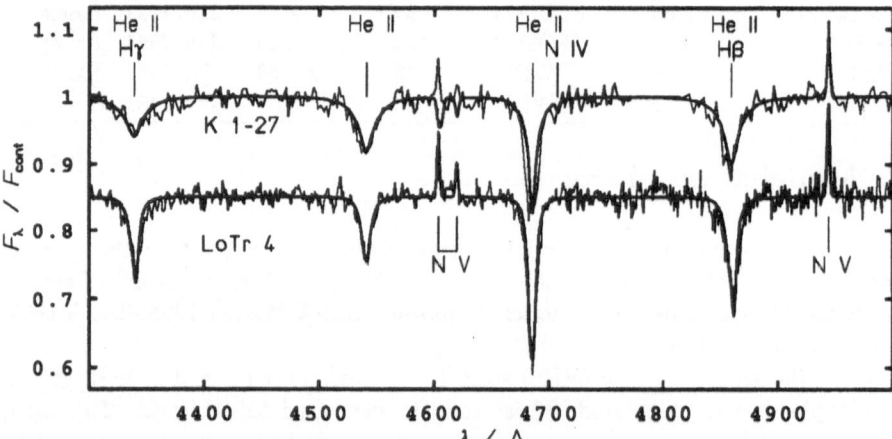

Fig. 1. Normalized spectra of the helium-rich central stars LoTr 4 and K 1-27 compared to synthetic spectra of our final models (for parameters see text)

4.2 PG 1159 stars

At the outset of the analysis we assumed element abundances of He:C:O = 38:56:6 (by mass) and tried to fit T_{eff} and g by reproducing the absorption trough around C IV λ 4658Å / He II λ 4686Å as well as He II $\lambda\lambda$ 4200, 4340, 4541, 4861, 6560 Å, C IV $\lambda\lambda$ 4440, 5801/5812Å, and O VI λ 5290Å simultaneously. The surface gravity is determined using the absorption trough and the He II lines.

The O VI $\lambda\,5290$Å and C IV $\lambda\,5801/5812$Å lines are sensitive indicators for T_{eff} at fixed abundance ratios. However, a precise abundance determination is still necessary for a final fit. We found the He II lines of the Pickering series to be good indicators of the $C + O$ content in the photosphere. Thus, some models with a different photospheric compositions are calculated in order to achieve good agreement of the theoretical He II lines and observation. Three examples are displayed in Fig. 2. The derived parameters for these stars and the other objects of our sample are summarized in Tab. 2. At the current stage of our analyses we estimate the error ranges to ± 20kK for T_{eff} and 0.5dex for $\log g$. The O abundances may be in error by 0.5dex. The C/He ratio should be correct within a factor of 2.

Table 2. Photospheric parameters of the two He-rich central star LoTr 4 and LoTr 4 (neither C nor O detected) and preliminary results for eight PG 1159 stars investigated in this work. The spectral subtypes of the PG 1159 stars are given according to Werner (1992)

object	spectral subtype	T_{eff} / kK	$\log g$	mass fractions				
				H	He	C	N	O
LoTr 4	O(He)	130	5.5	10	85	<2	0.2	
K 1-27	O(He)	100	6.5	<5	92	<1	2	<2
NGC 246	lgE	150	5.7		38	56		6
PG 1151-029	lgE	140	6.0		35	51		14
K 1-16	lgE	140	6.1		38	56		6
RX J2117+3412	lgE	160	6.1		38	56		6
Longmore 3	lgE	140	6.3		38	56		6
A 21	E	140	6.5		35	51		14
Jn 1	E	150	6.5		30	46		24
VV 47	E	130	7.0		35	51		14

RX J2117+3412 has been analysed before (Motch et al. 1993). Based on our new, better observations we obtain slightly different parameters which are, however, well within the error ranges. NGC 246 has been investigated before as well (Husfeld 1986), but more realistic model atmospheres accounting for the high carbon abundances are available now. We obtain a slightly different T_{eff} (higher by 20kK), a higher C/He ratio and an equal value for the surface gravity.

The analysis of our sample of very hot PG 1159 stars together with the new analysis of some cooler PG 1159 stars (Dreizler et al. 1995) almost completes the analysis of known PG 1159 stars. A summary of the parameters and a comparison with evolutionary tracks is shown by Dreizler et al. (1995).

Acknowledgements. This research was supported by the BMFT under grant 50 OR 9409 1 (TR), and by the DFG We 1312/6-1 (KW). Several travel grants by the DFG to the Calar Alto are acknowledged.

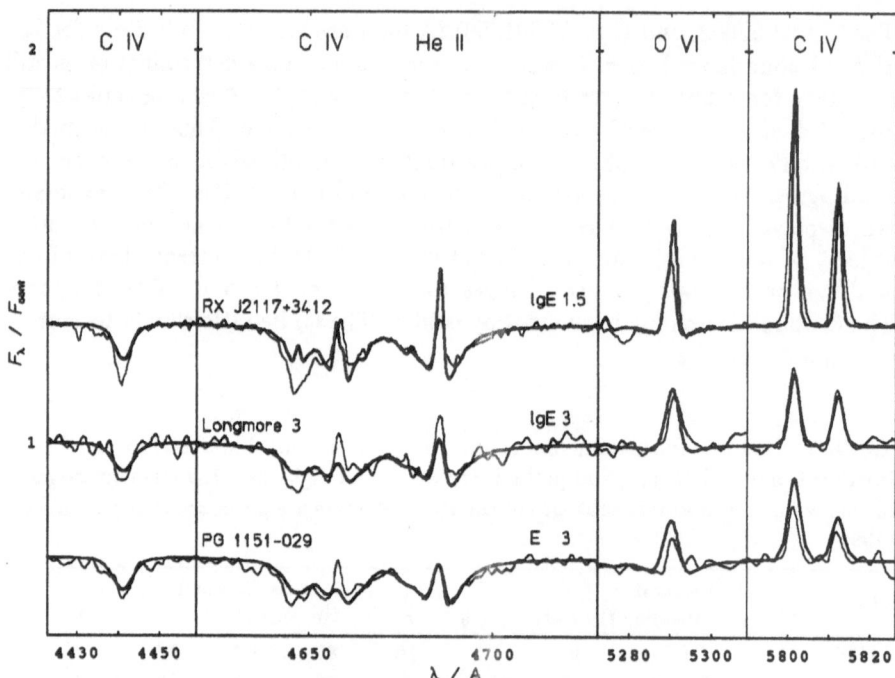

Fig. 2. Normalized spectra of three PG 1159 stars from our sample around
C IV λ 4440Å (5p-6d), C IV λ 4658Å (5g-6h etc.) / He II λ 4686Å, O VI λ 5290Å (7g-8h
etc.), and C IV λλ 5801/5812Å (3s-3p) compared with synthetic spectra of our final
models (parameters see Tab. 2). In the second panel the spectral subclasses and the
resolution of the spectra (in Å) are indicated. The theoretical spectra are convoluted
with Gaussians of respective FWHM

References

Dreizler S., Werner K. Heber U., 1995, these proceedings
Husfeld D. 1986 Ph. D. thesis, Universität München
Iben I.Jr., Kaler J.B., Truran J.W., Renzini A. 1983, ApJ 264, 605
Méndez R.H. 1991, IAU Symp. 145, Kluwer, Dordrecht, p. 375
Motch C., Werner K., Pakull M.W. 1993, A&A 268, 561
Rauch T. 1993, A&A 276, 171
Rauch T., Werner K. 1988, A&A 202, 159
Rauch T., Köppen J., Werner K. 1994, A&A 286, 543
Werner K. 1986, A&A 161, 177
Werner K. 1992, in: Heber U., Jeffery C.S. (eds.) Proc. Kiel/CCP7 Workshop, Atmo-
    spheres of Early-Type Stars. Lecture Notes in Physics, Springer, Berlin, p. 273
Werner K. 1993, in: Barstow M.A. (ed.) White Dwarfs: Advances in Observation and
    Theory. NATO ASI Series C, Vol. 403, p. 67
Werner K., Heber U., Hunger K. 1991, A&A 244, 437

The High Mass White Dwarf in HR 8210

Wayne Landsman[1], Theodore Simon[2] and P. Bergeron[3]

[1] Hughes STX, NASA/GSFC, Code 681, Greenbelt, MD, 20771
[2] Institute for Astronomy, University of Hawaii, 2680 Woodlawn Drive,
  Honolulu, HI, 96822
[3] Département de Physique, Université de Montréal, C.P. 6128, Succ. Centre Ville,
  Montréal, Québec, H3C 3J7 Canada

1 Introduction

HR 8210 (HD 204188, IK Peg) is a spectroscopic binary consisting of a $V = 6.05$, A8V star and a massive ($M > 1.15\ M_\odot$), hot ($T_{\rm eff} \sim$ 35,000 K) DA white dwarf. HR 8210 had been suspected to have a massive ($> 1\ M_\odot$) companion ever since Harper (1927) derived a circular orbit with a period of 21.7 days and a large mass function $f(m) = 0.161$. But not until HR 8210 was discovered to be a strong EUV source in the ROSAT Wide Field Camera survey (Pounds et al. 1994), and until subsequent followup observations with IUE, was the companion in fact shown to be a massive DA white dwarf. Since HR 8210 has a well-determined orbit, it may provide an opportunity to measure a gravitational redshift and a dynamical mass, and thus to test the white dwarf mass-radius relationship at large masses.

In this contribution we survey recent work on HR 8210 including an important revision of its orbit by Latham et al. (1994). We then summarize the results of our unsuccessful search for an ultraviolet eclipse in HR 8210. Finally, we outline the goals of our Cycle 4 GHRS program, and present a preliminary search for metal lines in the GHRS spectra.

2 Background

The IUE spectrum of HR 8210 is dominated by the A star at wavelengths longer than 1600 Å, and by the white dwarf at wavelengths shorter than 1400 Å (Wonnacott et al. 1993, Landsman et al. 1994). The IUE low-dispersion data is insufficient to constrain simultaneously both $T_{\rm eff}$ and $\log g$, and the addition of *Voyager* data (Barstow et al. 1994a) does not improve this situation. However, by using an estimate of the photometric distance to the HR 8210 primary of 38-53 pc, Landsman et al. could constrain the likely values of $T_{\rm eff}$ and $\log g$, and then use the evolutionary models of Wood (1992) to estimate a white dwarf mass of $1.15(^{+0.05}_{-0.15})\ M_\odot$. The orbital mass function and an assumed mass of 1.7 $M_\odot$ for the A star primary also allowed them to place a lower limit on the dynamical mass of $> 1.1\ M_\odot$.

How reliable is the mass determination and photometric distance of the primary in HR 8210? The separation of the components ($\sim 46\ R_\odot$) in HR 8210 is such that, although currently non-interacting, the system probably went through a common envelope phase. It is thus important to check the primary for any possible deviations from normal main-sequence evolution. One mildly puzzling aspect of HR 8210 has been its classification as an Am star. The Am stars are typically slowly rotating non-pulsators, and evolved off the main-sequence, but HR 8210 exhibits δ Scuti pulsations, a main-sequence gravity ($\log g = 4.25$), and a rather large rotational velocity (v sini = 83 km s^{-1}; quoted by Hoffleit and Warren 1991). Wonnacott et al. (1994) critically discuss the evidence for abundance anomalies in HR 8210 and conclude it is at most a mild Am star. They also show that the value of $v \sin i$ quoted in the Bright Star catalogue is too high, and that the correct value is < 50 km s^{-1}, a result subsequently confirmed by Latham et al. (1994). Finally, Wonnacott et al. show that Strömgren and Geneva photometry, as well as Hβ spectroscopy, all yield consistent stellar parameters, and they estimate a photometric distance to HR 8210 of 43.5 ± 1.0 pc.

Barstow et al. (1994b) observed HR 8210 with the *EUVE* spectrograph and found an excellent fit with a pure hydrogen model atmosphere. Due to the absence of metal features at short wavelengths, the EUVE spectrum of HR 8210 can provide a strong constraint on both T_{eff} *and* $\log g$. Barstow et al. find a best fit of $T_{\text{eff}} = 34{,}540$ K, and $\log g = 8.95$ with relatively small uncertainties, and use these parameters to derive a mass of 1.12 $M_\odot$. The white dwarf parameters then imply a distance of 44 pc to HR 8210, in excellent agreement with the photometric distance of the primary.

In support of our GHRS observations, Latham et al. (1994) have recently obtained a modern orbit for HR 8210. They derive a mass function $f(m) = 0.202 \pm .0025$ which is significantly *larger* than the value from Harper's orbit, primarily due to a larger value of the velocity semi-amplitude (44.7 km s^{-1} vs. 41.4 km s^{-1}). This difference is almost certainly because of the superior quantity and quality of the radial velocity data used in the Latham et al. orbit and not to evolution in the orbital parameters. With the new mass function, and assuming a primary mass of 1.65 $M_\odot$, the minimum white dwarf mass (for an inclination of 90°) is 1.17 $M_\odot$, already in marginal disagreement with the mass derived from EUVE spectroscopy.

3 Search for an Ultraviolet Eclipse

If the orbital inclination of HR 8210 is *exactly* 90°, then the flux from the white dwarf should be eclipsed for 5.7 hours as it passes behind the A star. However, because of the relatively wide component separation, the eclipse time drops rapidly with inclination angle, and *no* eclipse occurs for $i < 88°$. Evidence from the white dwarf mass that the inclination must be near 90° prompted our search for an ultraviolet eclipse.

Previously Wonnacott et al. (1993) had used the Harper ephemeris to search for an EUV eclipse in the ROSAT WFC data. However, at the current epoch the accumulated phase error in the Harper ephemeris amounts to several days, and so their negative ROSAT results are inconclusive. On 1994 May 26 we obtained 5 short-wavelength IUE images (SWP 50899 – 50903) during a shift centered around the mid-eclipse predicted by the Latham et al. ephemeris. In order to improve our time resolution, the star was placed in two or three positions in the large aperture of IUE, resulting in multiple spectra. The individual spectra in each image were extracted using a multiple Gaussian fit perpendicular to the dispersion direction. These individual spectra required a modest ($< 10\%$) correction due to the variation of the sensitivity of the IUE camera perpendicular to the dispersion.

Figure 1 shows that any variations of the IUE flux during the predicted eclipse time are well within the assumed 3% photometric reproducibility. From these data, it can be assumed that any eclipse must be shorter than 20 minutes, and that the orbital inclination must be less than 88°.

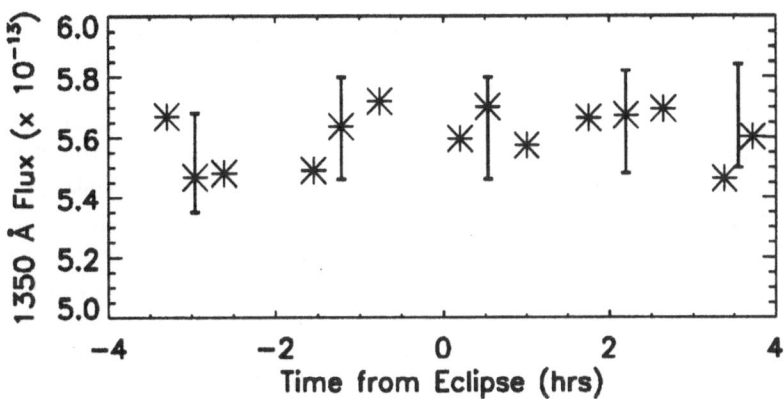

Fig. 1. The flux from HR 8210 in an 80 Å band around 1350 Å is shown as a function of time from the predicted white dwarf eclipse, according to the ephemeris of Latham et al. In each image the star was placed at two or three positions within the large aperture giving multiple spectra. The error bars give the total flux of each IUE image (assuming a 3% photometric uncertainty), while the asterisks show the flux measured in individual spectra within an image.

4 GHRS Observations

We have recently obtained medium-resolution GHRS Lyα spectra of HR 8210 at both radial velocity quadratures and at conjunction with the white dwarf star in front. The quadrature observations were successfully completed on 1994 Aug

16 and 1994 Oct 8, but 75% of the data obtained near conjunction on 1994 Aug 30 was lost due to problems with the on-board tape recorder.

The primary goal of our GHRS program is to measure the gravitational redshift of HR 8210 using the Lyα line. Our simulations suggest that our total integration time is sufficient to measure the velocity to within 10 km s^{-1}, with the important proviso that systematics can be identified and removed. For a mass of 1.2 $M_\odot$, this translates into an accuracy in the mass determination of 0.02 $M_\odot$.

Our GHRS program has four secondary goals. First, a measurement of the Lyα velocity at three orbital phases will provide a crude estimate of the white dwarf orbital amplitude, which combined with the orbit of the primary gives the component mass ratio, independent of inclination angle. Second, the GHRS spectrum includes the Si III λ1206 line, which is a prominent feature in the ultraviolet spectrum of white dwarfs that show metal lines (Vennes et al. 1991). The detection of any narrow metal lines, even if non-photospheric, could provide a much more accurate tracer of the white dwarf orbital velocity than the broad Lyα absorption. Third, unlike the case with IUE low-dispersion spectra, medium-resolution GHRS spectra of the Lyα core can be used to constrain simultaneously both T_{eff} and log g. Finally, the GHRS spectra will be used to estimate the interstellar column toward HR 8210, both directly from the Lyα absorption profile and from the N I λ1200 triplet.

Figure 2 compares the blue Lyα wing of our first GHRS spectrum of HR 8210 with an archival GHRS spectrum of GD 394, a somewhat hotter DA white dwarf with prominent (non-photospheric) metal lines in its ultraviolet spectrum (Vennes et al. 1991). The two stars are at a similar distance and direction in the sky, and the interstellar N I λ1200 features closely match. However, the strong Si III λ1206 absorption in GD 394 is entirely absent in HR 8210 (if anything, a slight *emission* is present). The absence of ultraviolet metal features in HR 8210 is not unexpected, both from theoretical considerations of diffusion in a high-gravity atmosphere, and from the absence of absorption features in the EUVE spectrum. However, their absence greatly complicates the determination of the white dwarf orbital velocity amplitude.

5 Conclusions

HR 8210 appears to provide an excellent laboratory for the study of the high mass white dwarf, in that the components appear to be non-interacting, the A star primary appears perfectly normal, and the white dwarf can be modeled with a pure hydrogen model atmosphere. However, it may not be possible to constrain precisely the inclination angle of the system, given both the absence of an eclipse, and the probable absence of narrow metal lines to serve as a velocity tracer. The lower limit on the dynamical mass of 1.17 $M_\odot$, derived by assuming $i = 90°$, suggests that, like Sirius, the radius derived from the mass via the Hamada-Salpeter relation, is larger than the photometrically derived radius (Thejll and Shipman 1987). The best hope for a more precise test of the mass-

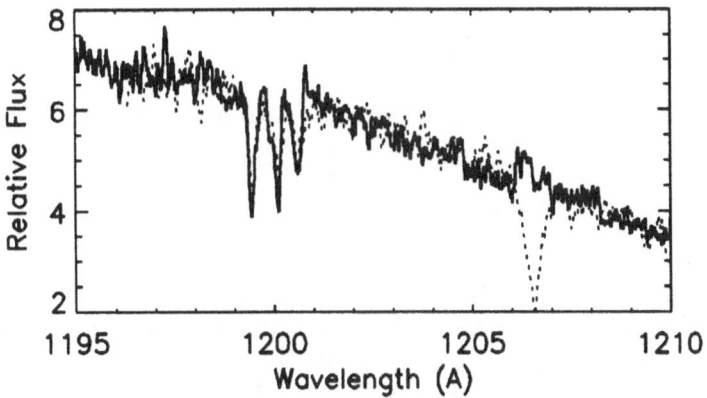

Fig. 2. The GHRS spectrum of HR 8210 (solid line) from our 1994 Aug 30 observation is compared with an archival GHRS spectrum of GD 394 (dotted line).

radius relation in HR 8210 is a measurement of the Lyα gravitational redshift, which will be the subject of a subsequent paper.

Acknowledgements

Support for this work was provided by NASA through grant number GO-5405.01-93A to Hughes STX from the Space Telescope Science Institute, which is operated by the Association of Universities for Research in Astronomy.

References

Barstow, M.A., Holberg, J.B., Fleming, T.A., Marsh, M.C., Koester, D., Wonnacott, D. 1994a, MNRAS, 270, 499
Barstow, M.A., Holberg, J.B., Koester, D. 1994b, MNRAS, 270, 516
Harper, W.E. 1927, Publ. Dom. Astrophys. Obs., 4, 167
Hoffleit, D., Warren, W. 1991, Bright Star Catalogue, Preliminary 5th edition, National Space Science Data Center CD-ROM
Landsman, W.B., Simon, T., Bergeron, P. 1993, PASP, 105, 841
Latham, D. et al. 1994, in preparation
Pounds et al. 1993, MNRAS, 260, 77
Thejll, P., and Shipman, H.L. 1987, PASP, 98, 922
Vennes, S., Thejll, P., Shipman, H.L. 1991, in White Dwarfs, ed. G. Vauclair & E. Sion (Dordrecht:Kluwer), 235
Wonnacott, D., Kellett, B.J., Stickland, D.J. 1993, MNRAS, 262, 277
Wonnacott, D., Kellett, B.J., Smalley, B., Lloyd, C. 1994, MNRAS, 267, 1045
Wood, M.A. 1992, ApJ, 386, 539

The Blue Edge of the ZZ Ceti Instability Strip

Detlev Koester[1], Nicole Allard[2,3] and Gerard Vauclair[4]

[1] Institut für Astronomie und Astrophysik der Universität, D-24098 Kiel, Germany
[2] Observatoire de Paris-Meudon, Meudon, France
[3] CNRS Institut d'Astrophysique, Paris, France
[4] Observatoire Midi-Pyrénées, Toulouse, France

Wesemael et al. (1991, Nato ASI Ser. 336, p.159) have summarized all past studies of the ZZ Ceti instability strip using many different observational techniques. Since then two more studies have been published, using the IUE archive of ultraviolet spectra (Kepler & Nelan 1993, AJ, 105, 608; Koester & Allard 1993, Nato ASI Ser. 403, p. 237). Taking all results at face value the blue edge — defined in most but not all papers by G117-B15A — varies between 12130 and 13640 K. Other candidates for the position of "hottest" ZZ Ceti that appear in the studies cited are GD165, L19-2, and G226-29.

We obtained a spectrum of G117-B15A with the FOS on the Hubble Space Telescope, and use for comparison IUE spectra of L19-2 and G226-29, concluding that all three stars must have very similar temperatures. The observed spectrum was analyzed with a χ^2 technique, using a part of our grid of white dwarf model atmospheres. A special feature of the synthetic spectra calculated is the inclusion of the latest version of the Lyman α satellite features according to the calculations of Allard & Koester (1992, A&A, 258, 464), Koester & Allard (1993), and Allard et al. (1994, A&AS, 108, 1). Convection is treated with the usual mixing-length approximation. Our version is almost identical to what is commonly called ML1; the standard grid uses a mixing-length parameter of 1.0 and additional grids were calculated for $\alpha = 2.0$, as well as ML2 and ML3.

A satisfactory fit with our standard models was not possible for any combination of T_{eff} and log g. After a careful study of other possibilities we traced this problem back to an incorrect temperature gradient in the outer layers, determined mostly by convection theory. Using the other model grids, ML3 could be clearly excluded, whereas satisfactory fits could be obtained for ML1 with $\alpha = 2.0$ and ML2. For the former case our final conclusion for the parameters of G117-B15A is $T_{eff} = 12250 \pm 125$ K, log g $= 8.10 \pm 0.15$; the results for L19-2 and G226-29 are very similar. For GD165 no UV spectrum is available; however, it seems very likely now that the blue edge of the ZZ Ceti instability strip is in the range 12250 to 12500 K, much lower than assumed up to now. Although from the UV spectra alone a clear preference emerges for intermediate convective efficiency, a similar analysis should be extended to include optical data, which are influenced in a different way by the parameters in the theory.

Acknowledgements: N. Allard acknowledges support of this work through grant 920167 from the NATO International Scientific Exchange programme. Work on HST observations in Kiel is supported by grant 50 OR 94091 from the Deutsche Forschungsagentur für Raumfahrtangelegenheiten (DARA).

L745-46A: Lyman Alpha Broadening and Metal Abundances

Detlev Koester[1] and Nicole Allard[2,3]

[1] Institut für Astronomie und Astrophysik der Universität, D-24098 Kiel, Germany
[2] Observatoire de Paris-Meudon, Meudon, France
[3] CNRS Institut d'Astrophysique, Paris, France

L745-46A was observed by HST/FOS in the UV; the spectrum shows Mg, Fe, and probably Si lines. Together with the optical scan by Stone and Baldwin (MN205, 347, 1983), it gives a complete, well defined energy distribution. This distribution is compared with He/metal models with T_{eff} from 7800 to 7200 K. The result is, that a large fraction of the flux predicted by models is missing below 3000 Å.

L745-46A is known to show Hα; the relative abundance was determined as $2\,10^{-4}$. Including Lα with Van der Waals broadening by neutral He in the calculations gives energy distributions with a Lα wing extending far into the visible region, which clearly cannot be realistic so far from the line center. Work on improved line profiles is in progress (N.A.), and will be reported in a future publication. In order to be able to interpret the energy distribution and the metal lines in the HST spectrum we have taken resort to a crude and unphysical method, weakening the line wing artificially with an exponential function until the continuum shape was roughly reproduced by the calculations.

We have encountered another difficulty: interpretation of the ionization balance requires more free electrons than are predicted from our standard procedure for pressure broadening, based on the Hummer-Mihalas formalism (metals, at the abundances determined, are only minor electron contributors). In order to increase the number of free electrons to fit the metal lines we have artificially increased the Ca abundance.

Work on this object continues, and we do not consider this to be a final analysis of L745-46A. However, with all caution, we give the element abundances obtained as current best estimates from the HST spectra, because they are of interest for calculations within the accretion/diffusion scenario, as follows: H/He $= 1.9\,10^{-4}$, Mg/He $= 4.7\,10^{-10}$, Fe/He $= 1.0\,10^{-10}$, Si/He $\leq 2.0\,10^{-10}$.

Acknowledgements: N. Allard acknowledges support of this work through grant 920167 from the NATO International Scientific Exchange programme. Work on HST observations in Kiel is supported by grant 50 OR 94091 from the Deutsche Forschungsagentur für Raumfahrtangelegenheiten (DARA).

A Model for the WC-type Central Star WR 72

Lars Koesterke and Wolf-Rainer Hamann

Institut für Astronomie und Astrophysik der Universität, D-24098 Kiel, Germany

Some central stars of planetary nebulae have very high mass-loss rates so that their spectra which are dominated by broad and strong emission lines are very similar to those of Wolf-Rayet (WR) stars. Hence the models for extended atmospheres, high mass-loss rates and high expansion velocities which we have developed for Pop I WR stars (Hamann et al. 1992: A&A 255, 200) can be applied to [WR]-type central stars as well (Hamann & Koesterke 1993: Proc. IAU Symp. No. 155, p87).

Here we present a He-C-N-O-model for WR 72 (Sand 3) based on the optical observation of Torres & Massey (1987: ApJ Suppl. 66, 459) and IUE spectra. We find a fair agreement in both line strength and shape for a model with $T_* = 130\,\mathrm{kK}$, $R_* = 0.14\,\mathrm{R_\odot}$, $\dot{M} = 10^{-6.31}\,\mathrm{M_\odot/yr}$, $v_\infty = 1200\,\mathrm{km/s}$ and $\beta_{\mathrm{CNO}} = (0.30, 0.02, 0.12)$ by mass. The He/C ratio is well determined, while β_{O} and β_{N} are estimates so far.

The rather high amount of nitrogen (cf. N V 1240Å, 4604-4620 Å) gives a hint concerning the evolutionary status of WR 72. Werner & Heber (1991: A&A 247, 476) have pointed out in their analysis of PG 1144+005 that the "born-again post-AGB" scenario of Iben et al. (1983: ApJ 264, 605) can explain the presence of a large amount of nitrogen, when hydrogen was convected downwards during the late thermal pulse and burned with carbon.

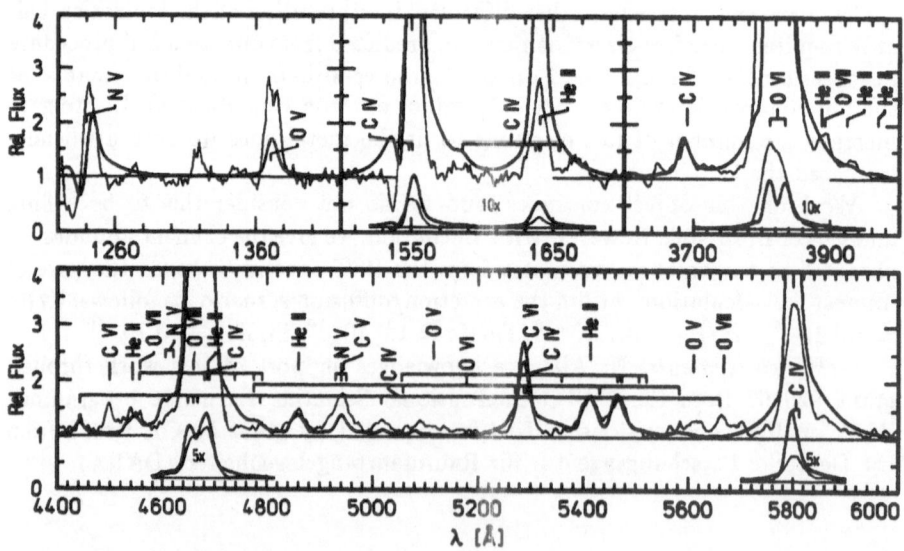

Fig. 1. Observed spectra of WR 72 compared with the model (thick lines).

White Dwarf Observations with the Ultraviolet Imaging Telescope

Wayne Landsman[1], Paul Hintzen[2,3] and Theodore Stecher[2]

[1] Hughes STX, LASP/NASA/GSFC, Code 681, Greenbelt, MD, 20771
[2] Laboratory for Astronomy and Solar Physics, NASA/GSFC, Code 681, Greenbelt, MD 20771
[3] Department of Physics and Astronomy, California State University, Long Beach, California 90840

Ultraviolet images of 66 astronomical targets were obtained with the Ultraviolet Imaging Telescope (UIT) during the Astro-1 mission in 1990 December. Astro-1 is an attached Shuttle payload that consists of UIT and the co-pointed Hopkins Ultraviolet Telescope (HUT) and the Wisconsin Ultraviolet Photopolarimeter Experiment (WUPPE). UIT contains two cameras with broadband sensitivities of 1250 – 1900 Å for the FUV camera, and 1800 Å – 3200 Å for the NUV camera. The UIT field of view is about 40′ and the images are recorded on film with about 3″ spatial resolution. A reflight of the Astro mission is currently scheduled for March 1995. UIT is especially useful for studying the following aspects of white dwarfs and their immediate progenitors.

1. **Hot, evolved stars in globular clusters:** The large field of view and solar-blind detectors of UIT make it ideal for observations of globular clusters. Recent discussions of UIT globular cluster observations are given by Landsman (1994) and Whitney et al. (1994).

2. **Serendipitous detections of white dwarfs in UIT fields:** The limiting sensitivity of UIT is about 17.5 at 1520 Å, and about 18.6 at 2600 Å. Since a hot unreddened white dwarf has $m_{1520} - V \sim -4.9$, the limiting sensitivity of UIT is about $V \sim 22$ for such stars. Unfortunately, due to target acquisition problems on Astro-1, only a few images achieved this limiting sensitivity. Spectroscopic followup of the white dwarf candidates is in progress.

3. **White dwarfs in open clusters:** No open clusters were observed during Astro-1 but several old and intermediate age clusters will be observed during Astro-2. UIT should a provide a nearly complete census of the hot white dwarfs, including those that are optically hidden in close binaries.

References

Landsman, W.B. 1994, in Hot Stars in the Galactic Halo, ed. S. Adelman, A. Upgren, & C. Adelman (Cambridge University Press, New York), 156
Whitney, J. et al. 1994, AJ, 108, 1350

LB 8827: A DB Star With Peculiar Line Profiles

F. Wesemael[1], A. Beauchamp[1], James Liebert[2], and P. Bergeron[1]

[1] Département de Physique, Université de Montréal, C.P. 6128, Succ. Centre Ville, Montréal, Québec, Canada H3C 3J7

[2] Steward Observatory, University of Arizona, Tucson, AZ 85721, USA

LB 8827 (PG 0853+163, Gr 904) is a ~23,000 K DBA white dwarf which was discovered to display optical He I lines with stunted cores when compared to other DB stars in that effective temperature range. Synthetic spectrum calculations suggest that these cores are not the signature of a high surface gravity ($\log g \sim 9$), as the 2^3P-6^3D $\lambda3819$ transition is predicted much too shallow, and the 2^1S-3^1P $\lambda5015$ transition too deep, in high-gravity models. The possibility that the line profiles are affected by a weak (~MG) magnetic field can also be excluded on the basis of red optical spectropolarimetry, secured by G. Schmidt, which allows us to place upper limits of 43 ± 29 kG at $\lambda5876$, and of 11 ± 42 kG at $\lambda4921$, on the effective field strength. The possibility that LB 8827 is a rapidly rotating DB star is also examined. A preliminary analysis of this possibility suggests that the optical spectrum of LB 8827 could be fit reasonably well by a rotating DBA star with $v \sin i \sim 600\,\mathrm{km\,s}^{-1}$ ($\Pi \sim 90$ s), although the match to the $\lambda3819$ and the 2^3S-3^3P $\lambda3889$ transitions remains somewhat problematic. Additional alternatives, including the possibility that LB 8827 is a binary system, need to be explored in some depth before a definitive picture of the nature of this object can be formed.

This work was supported in part by the NSERC Canada, by the Fund FCAR (Québec), and by the NSF grant AST 92-17961.

Magnetic or Pressure Shifts of Carbon Bands in Very Cool Helium-rich White Dwarfs?

Irmela Bues and Ludwig Karl-Dietze

Dr. Remeis-Sternwarte Bamberg, Sternwartstr. 7, D-96049 Bamberg

For helium-rich white dwarfs below 10000K the Deslandres-d'Azambuja band as well as the Swan band of C_2 have been observed and analyzed with or without magnetic fields, where the strengths of the features depend on the abundance ratios of He/C, C/H and C/O. Nonmagnetic objects like LP93-21 or G47-18 and magnetic ones like LP790-29 occur in the same range of T_{eff}. Our model atmosphere analysis of LP790-29 reproduced the magnetically shifted Swan bands with a polar field strength of 2×10^8G (Bues, I. 1993, NATO ASI **403**, 213). For cooler helium-rich objects, however, an increase of pressure by a factor of 100 may cause collision interaction, which might be the reason for pressure shifted bands, as discussed by Greenstein and Liebert (1990, ApJ **360**, 662). That is why we investigated the role of C_3 in the determination of opacity and pressure effects in the atmosphere of these very cool objects. From our dissociation equilibrium computation, a ratio of $C_2/C_3 \approx 10$ can be derived for conditions in model atmospheres with $T_{eff} \leq 6500$K, log g = 8, He/C $\geq$ 1000. For stars with strong bands of C_2 like G99-37, the 4050 Å band is present, as shown in Fig. 1. G99-37 has observed polarisation and a polar field of 2×10^7G. The structure of atmospheres without magnetic fields can be derived with normal assumptions for the equation of state and the collision cross sections down to temperatures of 4000K. Opacity by C_2 and C_3 reduces the gas pressure by a factor of 10-100, thus reducing the pressure shift for objects like LHS1126. From our computations a magnetic shift as for LP77-57 (5800K) is more likely than a pure pressure shift.

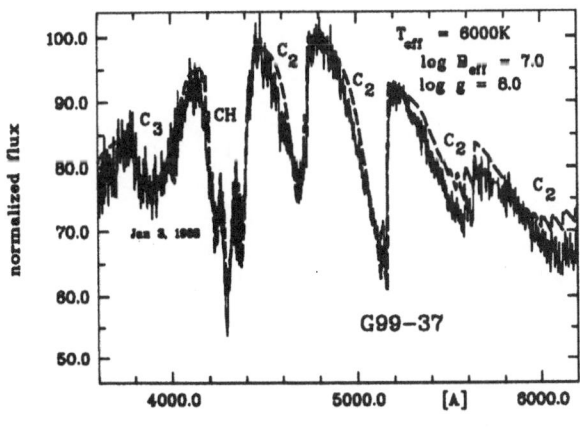

Fig. 1. G99-37 compared to a model atmosphere with parameters: 6000K, log g = 8, He/H = 1000, C/H = 35.5, B_{eff} = 10^7 Gauß.

IUE and Optical Observations of DAO White Dwarfs

J.B. Holberg[1], R.W. Tweedy[2], & J. Collins[1]

[1] Lunar and Planetary Laboratory, University of Arizona, Tucson AZ, USA
[2] Steward Observatory, University of Arizona, Tucson AZ, USA

We have obtained new IUE and optical observations of the metal-rich DAO white dwarf Feige 55. Four new IUE echelle spectra plus an archive spectrum indicate that Feige 55 exhibits radial velocity variations of at least 70 km s^{-1} Recent J and K band photometry indicates that the star possesses no main sequence companion down to 0.1 solar masses. If the observed velocity variations are orbital, then the companion may be degenerate object. In addition, the interstellar features seen in Feige 55 appear to be highly unusual and may be an indication of circumstellar gas associated with the star. Optical spectra of the H alpha profile of the Feige 55 show a narrow NLTE emission core.

CD -38° 10980 Revisited

J.B. Holberg[1], F.C. Bruhweiler[2], & J. Andersen[3]

[1] Lunar and Planetary Laboratory, University of Arizona, Tucson AZ, USA
[2] Department of Physics, Catholic University of America, Washington DC, USA
[3] Astronomical Observatory, Niels Bohr Institute of Astronomy, Physics, and Geophysics, University of Copenhagen, Denmark

A new coadded IUE echelle spectrum of the bright DA white dwarf CD -38° 10980, together with a newly determined radial velocity for the this star indicate that the UV sharp lined, Si and C, absorption features seen are clearly circumstellar in origin. Absorption in both excited and ground state transitions occurs at a velocity displaced by -12.1 ±2.0 km s^{-1} with respect to the photospheric velocity. Weak features due to the Si IV doublet are seen at a velocity intermediate between that of the circumstellar features and the photosphere. First time estimates of column densities for excited and ground states of C II, Si II and Si III are derived. These quantities are used with electron density estimates to determine the location and physical conditions of the circumstellar gas in the vicinity of CD -38° 10980. Electron densities in the circumstellar gas are found to exceed 10^9 cm^{-3}, unless strong photoexcitation occurs. Strict limits are placed on the photospheric abundance of Si and C in the star. The interstellar line-of-sight to CD -38° 10980 is also investigated.

Analysis of the DO White Dwarf PG 1034+001: Solution of the He II 4686Å Line Problem

Klaus Werner[1], Stefan Dreizler[2], and Burkhard Wolff[1]

[1] Institut für Astronomie und Astrophysik der Universität, D-24098 Kiel, Germany
[2] Dr.Remeis–Sternwarte Bamberg, Sternwartstr. 7, D 96049 Bamberg, Germany

We present a NLTE analysis of the hot DO white dwarf PG 1034+001 based on archival HST GHRS spectra, on new optical spctra and pointed ROSAT PSPC observations. The temperature is higher than previously thought (T_{eff}=100 000 K, $\log g$ =7.5). We determined abundances of C, N, O, Si (C= −5, N= −3.2, O= −4.1, Si= −5, log of number ratio relative to He) and, for the first time in a DO white dwarf, the Fe abundance on hand of newly identified Fe VI lines in the HST spectra (Fe= −5). An upper limit for Ni is derived (Ni< −5). Comparison with predictions of diffusion theory are partly contradictory. We find M=0.59M$_\odot$ and propose that PG 1034+001 is a descendant from the PG 1159 stars. Further details can be found elsewhere (A&A in press).

Moreover we have now solved a specific problem that was encountered when the line profile fit to He II 4686Å failed, although fits to the other He II lines were satisfactory. Problems with this line were reported earlier when NLTE analyses of sdO and DO stars ended up with similar discrepancies. We now found that at least in the case of the DOs this is due to a drawback in the NLTE models. Unlike in usual LTE models, level dissolution due to high particle densities was neglected up to now. We recently implemented the Hummer-Mihalas formalism into our code, following closely Hubeny et al. (1994, A&A 282, 151). Quite unexpectedly the model structure changes in a way that the 4686Å line is strongly affected whereas the other optical He II lines remain almost unchanged.

Fig. 1. Improving the He II 4686Å line fit to PG 1034+001: The fully drawn line is from a model that includes the Hummer-Mihalas occupation probability formalism.

Upper Limits for Mass–Loss Rates of PG 1159 Stars

Uwe Leuenhagen

Institut für Astronomie und Astrophysik der Universität, D-24098 Kiel, Germany

All hot PG 1159 stars ($T_{\mathrm{eff}}>100000\,\mathrm{K}$) show some narrow emission features, e.g. He II 4686 Å, 6560 Å, C IV 4650 Å, 5801,5811 Å and O VI 3811,3834 Å, 5290 Å. No P–Cygni profiles are present in the UV, e.g. C IV 1550 Å. What's the origin of this spectral appearance? Could it be caused by a weak stellar wind?

Possible progenitors of PG 1159 stars are [WC]–PG 1159 objects (Abell 30, Abell 78) which show broader optical emission features and very strong P–Cygni profiles at UV wavelengths. Obviously the mass–loss is still going on there. Analyses by means of spherically expanding model atmospheres have been worked out for [WC]–PG 1159 stars recently (Leuenhagen U., Koesterke L., Hamann W.-R., 1993, *Acta Astron.* **43**, 329 and references therein). Here we present a sequence of such models with gradually decreasing mass–loss rate. In Fig. 1 the synthetic profiles of some important lines are shown. Its a remarkable fact that the UV–P–Cygni features (which are strong for $\log \dot{M} = -7.3$, $[\dot{M}]=M_\odot\,\mathrm{yr}^{-1}$) change into absorption lines for $\log \dot{M} = -8.0$, whereas some weak optical emissions are still present for $\log \dot{M} = -9.0$. These synthetic spectra of the model atmospheres can be classified as [WC]–PG 1159 type for $\log \dot{M} = -7.3$ and as PG 1159 type for $\log \dot{M} = -9.0$. Assuming a decrease of the mass–loss rate during evolution this result confirms the supposition of an evolutionary link between [WC]–PG 1159 and PG 1159 stars.

The spectra of PG 1159−035 itself and RXJ 2117+3412 are compared to appropriate models in order to determine an upper limit for the mass–loss rate of these stars. The comparison yields $\log \dot{M} < -10.0$ and $\log \dot{M} < -9.5$ for PG 1159 and RXJ 2117, respectively.

We acknowledge DFG support under grant We 1312/2-3.

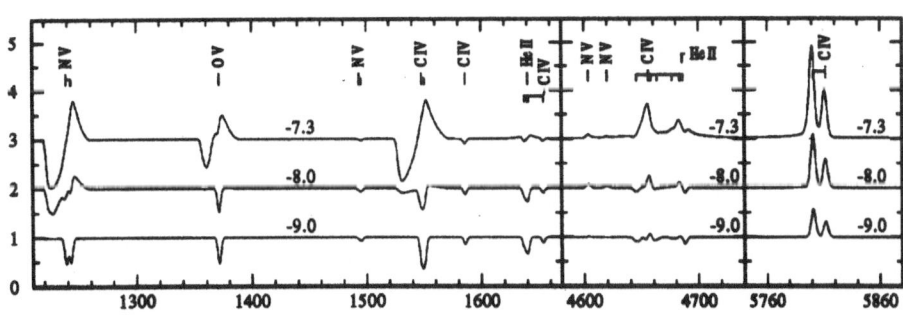

Fig. 1. Synthetic spectra for models with different mass–loss rates (labelled by $\log[\dot{M}/(M_\odot\,\mathrm{yr}^{-1})]$).

Part V

Binaries, Cataclysmic Variables, Subdwarfs

Exposed White Dwarfs in Dwarf Novae

Edward M. Sion

Department of Astronomy and Astrophysics, Villanova University, Villanova, PA 19085, e-mail: emsion@ucis.vill.edu

Abstract: Hubble Space Telescope far ultraviolet spectroscopic observations of cataclysmic variable white dwarfs, exposed during dwarf nova quiescence, have yielded a number of new insights on accretional heating, photospheric abundances of the accreted atmosphere and rotational velocities of the underlying degenerates. Recent results of synthetic spectral analyses of HST spectra (and, in some cases, *IUE* spectra) are highlighted together with comparisons between time-dependent theoretical simulations of the heating effect of the dwarf nova accretion event on the underlying degenerate and actual spectroscopic observations of white dwarf surface cooling.

1 Introduction

Two major developments have heralded a new era of investigations into the short term and long term effects of accretion processes on the underlying white dwarfs in cataclysmic variables (CVs): (1) the Hubble Space Telescope provides, for the first time, high quality spectra with fully-resolved line profiles and orbital phase resolution for these very faint degenerate stars (cf. Wood *et al.* 1994; Marsh *et al.* 1994; Horne *et al.* 1994, Sion *et al.* 1994), during dwarf novae quiescence and during the low brightness states of nova-like variables; (2) recent comprehensive *IUE* archival investigations (Deng, Zhang and Chen 1994; LaDous 1991) of the *International Ultraviolet Explorer (IUE)* low resolution SWP spectra of dwarf novae during quiescence have provided mounting evidence that the dominant source of light in the far ultraviolet (SWP) wavelength region (1200-2000Å) of many (if not all) of these cataclysmic systems appears to be the white dwarf photosphere. This second development, if confirmed with further studies, would provide a greatly widened sample of degenerate stars in non-magnetic cataclysmics with which to carry out synthetic spectral analyses and time-dependent theoretical simulations of the accretion/boundary layer physics.

HST FOS and GHRS studies of the white dwarf photospheres in non-magnetic cataclysmic variables promise to yield, and have already delivered: (1) the cooling response of the white dwarf (a UV cooling curve) during quiescence in response to accretional heating during the outburst, when the accretion rate is at least 2-3 orders of magnitude higher than in quiescence; (2) radial

velocity amplitudes and thus masses of the white dwarf primaries, independent
of disk emission lines; (3) white dwarf rotational velocities (needed to test standard boundary layer models including its boundary layer temperature/structure,
shear mixing of matter having angular momentum (with conversion of rotational
kinetic energy into heat), all of which are critically influenced by the rotation
rate of the white dwarf; (4) detailed line profile analyses and hence chemical
abundances of the freshly accreted photosphere as a function of time following
the deposition of mass, energy, and angular momentum by the outburst (accretion) event of a dwarf nova; (5) the physics of the differential spinup of the
white dwarf by accretion, of the redistribution of accreted matter over the white
dwarf surface and of the lateral and vertical extent of boundary layer heating
following the dwarf nova accretion event and; (5) physical processes in the white
dwarf envelope (diffusion, convective mixing and dilution, dredgeup from deeper
layers, magnetic channelling etc.) which directly affect the surface abundances
of accreted elements. This review highlights and summarizes recent progress on
several of the above scientific questions in the area of accretion processes and
the exposed white dwarf in non-magnetic CVs.

2 Hubble Space Telescope FOS and GHRS Spectroscopy of White Dwarfs in Dwarf Novae

HST observations of several bright dwarf novae with exposed white dwarfs during quiescence have been carried out to date for three, bright, eclipsing systems,
OY Car, Z Cha, HT Cas, and three systems, U Gem, VW Hyi and WZ Sge, in
which their extremely broad Lyman-α turnovers and UV flux shortward of 1600Å
in quiescence are unambiguously due to the white dwarf. All of these systems
have *IUE* temperature determinations (cf. Szkody 1993; Sion 1991 and references
therein). However, except for U Gem and WZ Sge, *IUE* analyses yielded little
spectroscopic information about chemical constituents other than hydrogen or
any other photospheric characteristic. *IUE* studies of U Gem and WZ Sge in quiescence (Panek and Holm 1984, Kiplinger *et al.* 1989; Holm 1988, Sion, Leckenby
and Szkody 1990) revealed absorption features due to metals and utilized pure
hydrogen model atmospheres due to Wesemael *et al.* (1980). Abundances of metals were estimated by comparing observed line strengths with the far ultraviolet
equivalent width versus metal abundance tabulations of Henry, Shipman and
Wesemael (1985). Within the past two years however, the availability of state
of the art, high gravity, solar composition synthetic spectra using TLUSTY and
SYNSPEC (Hubeny 1988, 1994) has opened the way for major advances in the
analysis of both HST and IUE archival spectra.

2.1 Bare White Dwarfs Viewed Through Warm Absorbing Curtains

In the eclipsing dwarf novae, high time resolution observations of eclipse ingress
and egress can separate out ultraviolet flux contributions from the white dwarf,
hot spot, disk and boundary layer. Early high speed photometric studies of

these systems (Wood *et al.* 1986, 1989) had led to predictions that the bare white dwarf should dominate the flux in UV quiescent spectra. The first HST application of the high time resolution method was to OY Car by Horne *et al.* (1994). Their HST spectrum of the white dwarf (Horne *et al.* 1994) revealed a complex, bizarre pattern of absorption lines. They modelled the spectrum by having the white dwarf photosphere viewed through an absorbing slab or curtain of solar abundance gas in LTE. The parameters of their curtain model are the gas temperature T, the electron density n_e, the turbulent broadening parameter V_{turb} and the total hydrogen column density N_h. The Fe-peak absorbers in the slab mask the true absorption spectrum of the white dwarf photosphere. Such absorbing Fe curtains were first discussed previously by Shore (1992) in symbiotic systems. Horne *et al.* (1994) discussed this model extensively and unexpectedly found evidence of mach 6 motions in the supported absorbing material above the disk plane.

A similar analysis was applied to *IUE* low resolution SWP spectra of OY Car and Z Cha by Wade *et al.* (1994) using solar composition model atmospheres with the effect of an overlying "Fe curtain" and inclusion of the H_2 quasi-molecule. Their result for OY Car was within a factor of 2 with Horne *et al.* 's (1994) HST analysis and their best model fits for Z Cha indicated a significantly weaker, cooler curtain obscuring the white dwarf but with a similar high turbulent broadening parameter. As mentioned below, there is no evidence of such absorbing curtains in front of the exposed white dwarfs in the high inclination system WZ Sagittae, or in U Gem or VW Hydri, two systems with lower inclination.

2.2 Measurements of White Dwarf Heating/Cooling in Response to Dwarf Nova Accretion Events

A UV flux decline, if due to a decrease in flux from the accretion disk during dwarf nova quiescence, would contradict the standard disk instability theory which predicts the disk should brighten through the quiescent interval because the accretion rate is gradually increasing. But these flux declines could be accounted for if they are due to cooling of the white dwarf in response to the accretion event. HST observations of three systems have provided conclusive evidence of white dwarf cooling in response to the accretion event.

In U Geminorum a number of investigations demonstrated that the white dwarf in U Gem dominates the system light during quiescence in the far UV (Panek and Holm 1984; Kiplinger *et al.* 1989; Long *et al.* 1994). The work by Kiplinger *et al.* . 1989 presented inconclusive evidence of white dwarf cooling using *IUE*. However, Long, Sion, Szkody and Huang (1994) observed U Gem in quiescence with HST/FOS at 13 days and 70 days after a normal dwarf nova outburst. They found a flux decline of 28% between the two observations and, that the white dwarf had cooled from 39,400K down to 32,100K, between the two observations.

For the eclipsing system OY Car, Cheng *et al.* (1994) compared the white dwarf FOS spectra in early quiescence to the the spectra at later times in quiescence. They determined that the temperature of the white dwarf just after the outburst was 19,500K but had cooled to 17,400K about 3 months after outburst. They found a cooling of 8000K in a two week interval.

HST high speed UV photometry of the eclipsing system Z Cha by Wood *et al.* (1993) from outburst far into quiescence revealed no evidence of an extended boundary layer even after outburst. The temperature of the white dwarf was 17,600K at the decline of outburst and cooled to its normal quiescent temperature of 15,600K 16 days later.

In WZ Sagittae, the dwarf nova with the longest interoutburst time the temperature of the white dwarf just after the December, 1980 outburst was 30,000K (Holm 1988). Since then the white dwarf surface temperature has been declining and had cooled to 12,500K in 1989 (LaDous 1990) based upon *IUE* measurements. More recently the white dwarf appears to be somewhat hotter (see below).

Simulations of the thermal response of the white dwarf to the dwarf nova accretion event can be compared with such white dwarf cooling determinations. These simulations are briefly discussed in the final section.

2.3 HST Photospheric Analyses of the Exposed White Dwarfs in U Geminorum, VW Hydri and WZ Sagittae

The three other CV systems observed to date with HST, VW Hyi, WZ Sagittae and U Gem, represent ideal laboratories for exploring the physics of disk-boundary layer accretion and its effect on the central white dwarf because their accretion rates are extremely low during quiescence and their underlying white dwarfs not only dominate the system's light in the far UV but exhibit a clear UV turnover associated with the Stark-broadened Lyman-α absorption wing. Principal results achieved up to the present are summarized below.

U Geminorum: In addition to the FOS study of the cooling of the white dwarf during quiescence by Long, Sion, Szkody and Huang (1994), a pilot GHRS study seeking the white dwarf rotation rate was also carried out. Sion, Long, Szkody and Huang (1994a) obtained a pair of consecutive far ultraviolet GHRS exposures of the Si IV region of the dwarf nova U Geminorum in early quiescence on September 25, 1993, 8 days after its return to optical quiescence when the underlying white dwarf dominates the UV light of the system. Their GHRS observation revealed a fully resolved line profile for the resonance doublet of Si IV. The two spectra occurred at mid-exposure orbital phase 0.67 and at phase 0.87. The Si IV profile in the second observation is clearly redshifted relative to the first spectrum, due to orbital motion. If the profiles are associated with the white dwarf photosphere, then the best synthetic fits are consistent with T_{eff} = 35,000K-38,000K, Log g = 8, a rotational velocity of 50 to 100 km/s, with a modestly enhanced silicon abundance (1.3-2.3 times solar). These results suggest that at least in U Gem and perhaps in other similar dwarf novae, the missing boundary layer cannot be explained by rapid rotation of the white dwarf. Unfortunately, the γ-velocity of the system remains uncertain. If the γ velocity is

43 km/s (Friend *et al.* 1990), then a gravitational redshift of $\sim 50 - 60$ km/s is implied for the white dwarf. If the γ-velocity is 84 km/s (Wade 1981), then a gravitational redshift of only 10-30 km/s is indicated, which may imply that either the white dwarf has a low (0.5-0.6 $M_\odot$) mass, or an extended atmosphere bloated by the outburst 8 days earlier. Further GHRS observations of U Gem during cycle 4 of HST should remove this uncertainty.

In order to rule out a rapidly spinning accretion belt, a comparison was made between the observed Si IV profile and models in which a portion of the white dwarf surface is assumed to be hot (60,000 K) and rotating with v sin i of 5000 km^{-1}. As the fraction of the surface rotating at high velocity increases, the fits get worse because Si IV is weak at 60,000 K (compared to 35,000 K) and because of the very large Doppler width at this rotational velocity.

WZ Sagittae: Sion, Cheng, Long, Szkody, Gilliland, Hubeny and Huang (1995) analyzed two consecutive two Hubble FOS spectra of the exposed white dwarf in the ultra-short period, high amplitude, dwarf nova WZ Sagittae. Their spectra revealed a rich absorption line spectrum of neutral carbon and ionized metals, the Stark-Broadened Lymanα absorption wing, the H_2 quasi-molecular Lymanα "satellite" absorption line, and a double-peaked C IV emission line which is variable with orbital phase. The two spectra are displayed in figure 1. The dominance of C I features confirms the identification of C I first reported with low resolution *IUE* spectra by Sion, Leckenby and Szkody (1991). The present spectral type of the white dwarf is DAQZ3. A synthetic spectral analysis of the white dwarf yields T_{eff}= 14,900 K $\pm$250K, Log g = 8.0. The best fit is shown in figure 2. In order to fit the strongest C I absorption lines and account for the weakness of the silicon absorption lines, the abundance of carbon in the photosphere must be about 0.5 solar while the abundances of 1×10^{-1} solar, silicon abundance is 5×10^{-3} solar, with all other metal species appearing to be 0.1 to 0.001 times solar. The H_2 quasi-molecular absorption is fitted very successfully.

The photospheric metals in the WZ Sge white dwarf have diffusion timescales of fractions of a year and thus they must have been accreted long after the 1978 December outburst. The source of the most abundant metal, carbon, is considered. If the time-averaged accretion rate during quiescence is low enough for diffusive equilibrium to prevail, then the equilibrium accretion rate of neutral carbon is 7×10^{-16} $M_\odot$/yr and the total accretion rate is 2×10^{-12} $M_\odot$/yr. However our derived concentration of carbon is consistent with the theoretically predicted convective dredgeup of core carbon from its equilibrium diffusion tail if the white dwarf has a relatively thin hydrogen layer or if factors associated with its cataclysmic binary membership (e.g. rotation) impede the rapid diffusion of C expected in a single degenerate. Since the core carbon convective dredgeup mechanism operates effectively in, and is only associated with, single helium-rich white dwarfs, it seems unlikely that this same mechanism can be responsible for the carbon in WZ Sge. Finally, although the T_{eff} of the white dwarf is within 1000K of the the blue edge of the ZZ Ceti instability strip, the possibility that the quasi-periodic oscillations are non-radial g-mode pulsations of the white dwarf

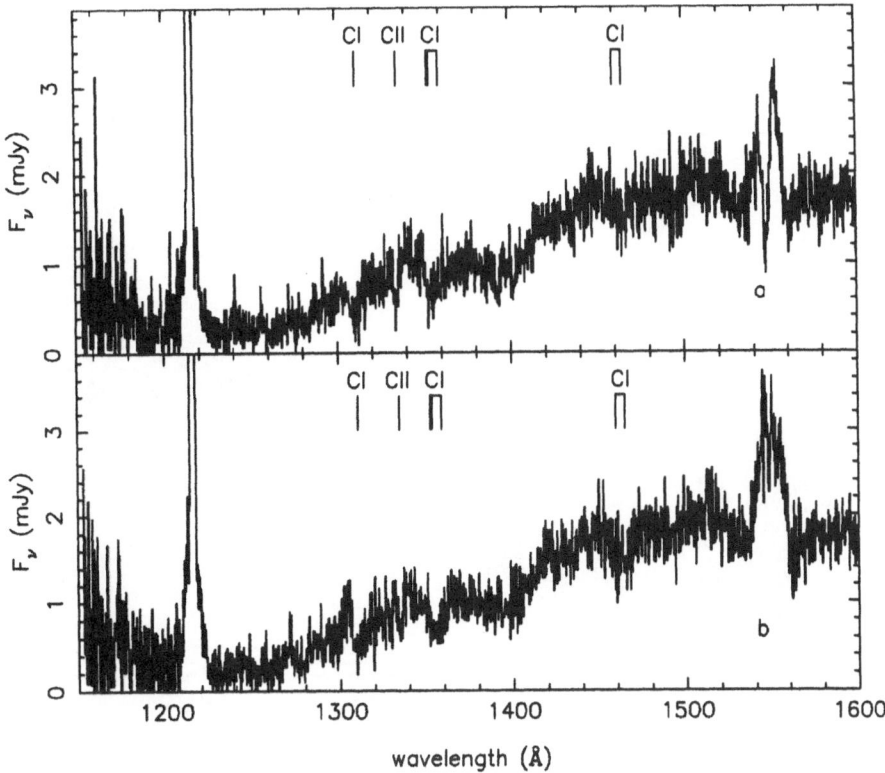

Fig. 1. The FOS spectrum of WZ Sagittae showing flux F_ν (mJy) versus wavelength (Å) for the first visit (top panel a) and the second visit (bottom panel b). The solid curve at the bottom of each panel is the error bars. Note the broad Lyα H2 quasi-molecular absorption at 1400 Å, the absorption line spectrum with the strongest C I features labelled, the emission at (1300 Å) in the panel b, the double-peaked emission feature at C IV (1550 Å) and its large variation in central absorption and the Stark-Broadened Lyα longward absorption wing. The emission at Lyα is geocoronal in origin. See the text for details.

is very remote. The periods are ultra-short, being a factor of 3-4 shorter than the shortest period ZZ Ceti pulsators.

VW Hydri: In an FOS study of the cooling of the white dwarf in the dwarf nova VW Hydri, Sion, Szkody, Cheng and Huang analyzed a far ultraviolet spectrum of the dwarf nova VW Hydri but a scheduling problem delayed the acquisition of the required second observation. The single observation occurred 10 days after the return to optical quiescence from a superoutburst of VW Hydri. The spectrum reveals a very strong Stark-broadened Lyman-α absorption

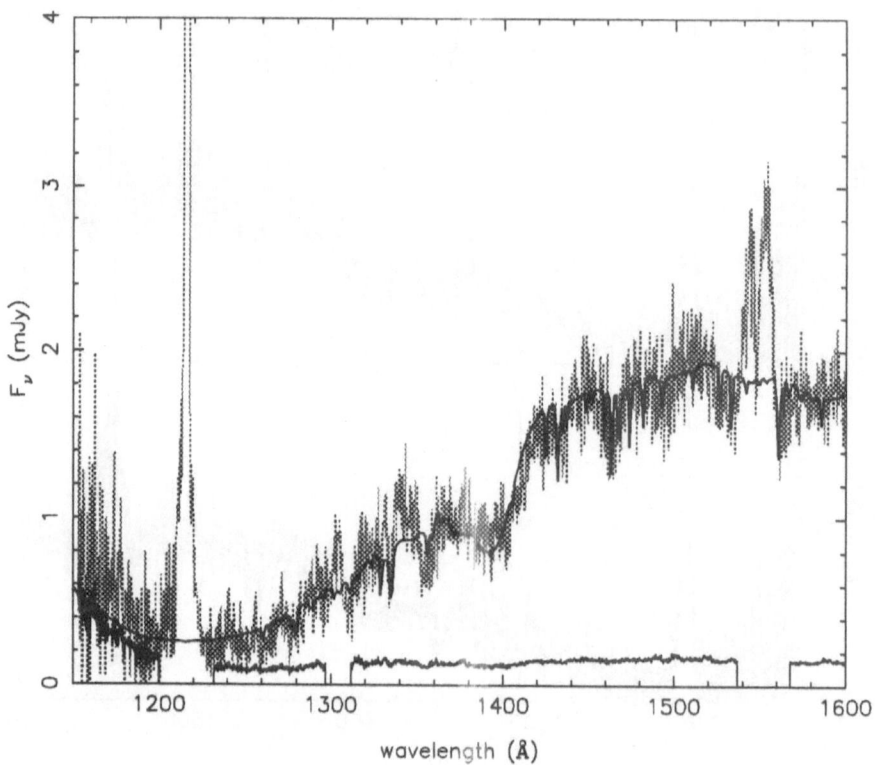

Fig. 2. The best-fitting ($\chi^2 = 1.51$) synthetic spectrum to the summed data for the two FOS spectra of WZ Sge. The error bars of the summed data are shown at the bottom. The three gaps in the error bar curve indicate the wavelength regions masked in the spectral fitting. Note the broad Lyα H$_2$ quasi-molecular absorption at 1400 Å, the rich absorption line spectrum with the strongest C I lines labelled, the emission at (1300 Å) in the panel b, the double-peaked emission feature at C IV (1550 Å) and its large variation in central absorption and the Stark-Broadened Lyα longward absorption wing. The emission at Lyα is geocoronal in origin. The physical parameters are T_{eff}= 14,900 K, Log g = 8, with Carbon being the the most abundant metal constituent in the hydrogen-rich atmosphere (from Sion *et al.* 1995, ApJ, 10 January issue, in press).

with narrow geocoronal emission, and a very rich metallic absorption line spectrum dominated by strong resonance absorption features of singly and doubly ionized silicon and carbon, the first solid identification of metallic absorption features arising in the accreted atmosphere of the white dwarf. They confirm the reported low resolution *IUE* detection of the underlying white dwarf photosphere by Mateo and Szkody (1984). A synthetic spectral analysis with hot, high gravity LTE model atmospheres yields a best fit model with the following parameters: $T_{eff}= 22,000K \pm 1000K$, Log g $= 8\pm0.3$, with chemical abundances of Oxygen $= 0.3$ x solar, Nitrogen $= 5$ x solar and all other heavy elements $= 0.15$ x solar (See figure 3.) Based upon their absorption line measurements in the subexposures at different orbital phases they find no conclusive evidence of equivalent width variations versus orbital phase. The elevation of the white dwarf surface temperature relative to previous largely *IUE*-based estimates of $T_{eff}(18\text{-}20,000K)$ could be due to the placement of our FOS observation so close to the end of a VW Hydri superoutburst. In the absence of any significant reduction of the white dwarf's core mass by past nova explosions, its lower limit cooling age is approximately 50 million years.

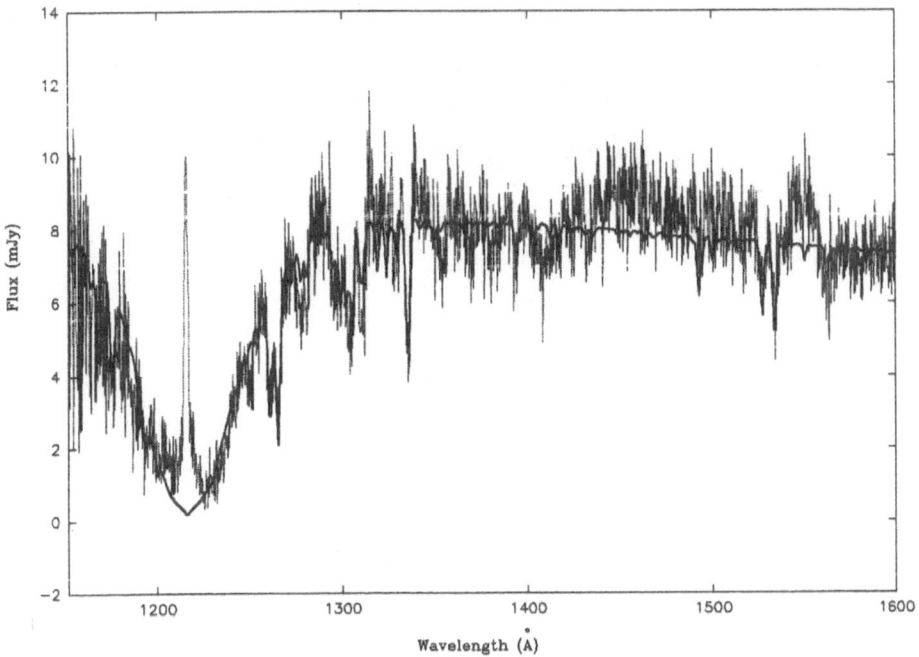

Fig. 3. The FOS spectrum of VW Hydri during quiescence. The summed data for the two sub-exposures is shown. Note the very broad Stark-broadened Lyman-α absorption with narrow airglow emission, the rich metallic absorption line spectrum dominated by strong lines of Si II, Si III, C II and C III.

A first attempt to determine the rotation rate of the white dwarf in VW Hydri was carried out with the Hubble GHRS by Sion, Long, Szkody and Huang (1994) They obtained a far ultraviolet spectrum in dwarf nova quiescence, covering the region of the Si IV (1393, 1402) resonance doublet. The broad, shallow Si IV doublet feature is fully resolved, has a total equivalent width of 2.8Å, and is the first metal absorption feature to be clearly detected in the exposed white dwarf. The synthetic spectral analysis, using a model grid constructed with the code TLUSTY (Hubeny 1988), resulted in a reasonable fit to a white dwarf photosphere with T_{eff}= 22, 000 ± 2000K, Log g = 7.5 ± 0.3, an approximately solar Si/H abundance, and a rotational velocity, V sin $i \simeq$ 600 km/s (see figure 4). This rotation rate, while not definitive because it is based upon just one line transition, is 20% of the Keplerian (breakup) velocity and hence does not account for the unexpectedly low boundary layer luminosity inferred from the soft X-ray/EUV bands(van der Woerd and Heise 1987; Belloni *et al.* 1991; Mauche *et al.* 1991, 1994) where most of the boundary layer luminosity should be radiated.

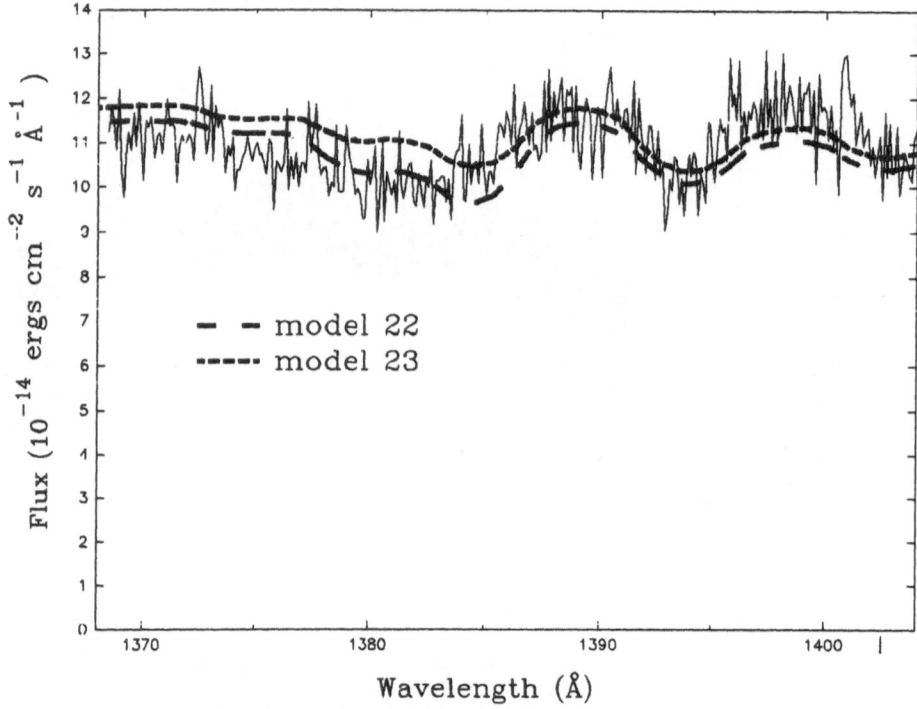

Fig. 4. A comparison between the observed GHRS spectrum of Si IV region and model fits to the region. The best fit white dwarf models with T_{eff}= 22, 000K ± 1000K, log g = 7.5±0.3, V sin $i \simeq$ 600 km/s are shown. Both models (22 and 23) have the following chemical abundances: Fe = 0.1 x solar, S = 0.5 x solar, all other heavy elements are 0.15 x solar except Al. In model 22, Al = 15 x solar while in model 23, Al = 5 x solar

Popham and Narayan (1995) have shown that the boundary layer luminosity must include the fraction of the rotational kinetic energy that goes into spinning up the white dwarf. Hence for a rotational kinetic energy $1/2\dot{M}\Omega_k(R_{wd})^2$ × R_{wd}^2 and accretion-imparted angular momentum $\dot{M} \times \Omega_{wd} \times R_{wd}^2$, Popham and Narayan (1995) have shown that the boundary layer luminosity should be:

$$L_{bl} = (GM_{wd}\dot{M}/R_{wd}) \times [1/2 - j\Omega_{wd}/\Omega_k(R_{wd}) + 1/2(\Omega_{wd}^2/(\Omega_k(R_{wd})^2))] \quad (1)$$

where j is the angular momentum accretion rate in units of the Keplarian angular momentum rate at the stellar surface ($j = \dot{J}/(\dot{M}\Omega_k(R_{wd})R_{wd}^2)$ and Ω_{wd} is the angular velocity of the white dwarf. For a white dwarf not rotating near breakup, $j \simeq 1$. The predicted boundary layer luminosity for a 0.6 $M_\odot$ white dwarf accreting at the rate 10^{-10} $M_\odot$yr and rotating at 600 km/s, corresponding to VW Hydri in quiescence, is $2 \times 10^{+32}$ ergs/s.

If the boundary layer area is equal to that of the white dwarf, then $T_{bl} = 24,000$K. These values are essentially identical to the photospheric luminosity and temperature determined in far ultraviolet photospheric analyses. If the boundary layer area is 10^{-3} of the white dwarf surface area, then $T_{bl} = 136,000$K. A preliminary conclusion based upon this single line measurement of rotation is that the significant under-luminosity of the VW Hydri boundary layer compared to the disk cannot be explained by rapid rotation of the white dwarf.

3 Theoretical Simulations of Accretional Heating of White Dwarfs in Dwarf Novae

There are three different principal sources of heat input into the white dwarf due to the dwarf nova outburst: (1) compressional heating due solely to the weight of the added matter; (2) accretion luminosity shining inward (associated with the kinetic energy of infall) and; (3) tangential accretion onto an equatorial belt with shear mixing. A comparison of the magnitudes of these three heat sources is useful.

The boundary layer downward irradiation of the white dwarf by a dwarf nova outburst has been considered semi-analytically by Pringle (1988) and with 1D hydrodynamics and radial accretion by Sparks, Sion, Starrfield and Austin (1993). Inclusion of the kinetic energy of infall of the accreting matter typically assumes that the radiated energy is equal to one-half of the gravitational potential energy of the accreted mass at the white dwarf surface; namely:

$$L_{acc} \approx 0.5GM_{wd}\Delta M/R_{wd} \quad (2)$$

This accretion luminosity, the accreted rotational kinetic energy rate, shines inward heating the white dwarf to a maximum temperature, T_{max}, in the outburst, down to a depth in the white dwarf atmosphere roughly equal to the depth in the original atmosphere where $T(M_r) = T_{max}$ (cf. Pringle 1988). Hence the higher the surface temperature to which the white dwarf is elevated at the peak of the outburst, the deeper is the radiative heat diffusion and thus the longer the thermal cooling timescale at the cessation of the accretion outburst episode.

A white dwarf will be heated by the weight of the accreted disk material in the dwarf nova outburst by simple adiabatic compression. Sion (1995) carried out a 1D quasi static simulation of heating due to periodic dwarf nova events.

The accreting material is assumed to rain down onto the stellar surface with essentially zero velocity (*i.e.*, it is assumed to "land softly") and hence completely excludes the effect of the kinetic energy of infall of the accreting matter. The implicit assumption is that all of the kinetic energy gained by the accreted matter in flowing from the donor star has been dissipated before this matter settles onto the surface of the star (at the same temperature and density as of the matter already at the white dwarf surface). The white dwarf undergoes pressure changes from model to model at any given M_r which is caused by readjustment of hydrostatic equilibrium. The white dwarf responds with (1) radius changes in the outer layers and (2) compensation for the weight of the newly added matter. However since r changes only very slightly for the largest part of envelope matter in mass fraction space M_r/M, most of the relative pressure change is due to compensation for the additional weight of the accreted gas (cf. Giannone & Weigert 1967). The change in total radius is also very small since the amount of mass added during a given dwarf nova outburst is also very small.

Hence compressional heating and heating due to downward accretion luminosity are therefore quite distinct; the kinetic energy of infall governs (dominates) the surface luminosity of a white dwarf in a dwarf nova via the necessity of the star to rid itself of a large heat source shining inward (*e.g.*, at $\dot{M} = 1 \times 10^{-9} M_\odot/\text{yr}$, $\dot{E}_k >> L_{wd}$) and is deposited in the the outermost surface layers. Compressional heating, on the other hand, affects the entire stellar structure (to varying degrees, as a function of mass fraction M_r/M). If the kinetic energy of infall were included, an accretion luminosity of 16 $L_\odot$ (corresponding to an accretion rate of $1 \times 10^{-8} M_\odot/\text{yr}$ onto a 0.6 $M_\odot$ white dwarf) would lead to a very high white dwarf surface temperature ($T_{eff} > 1 \times 10^5$K) compared to the undisturbed (initial) $T_{eff} = 27,000$K. Compressional heating, if operative over a long timescale, (and depending upon the accretion rate, white dwarf mass and outburst/quiescence timescale), could become comparable to or exceed the heating effect associated with accretion luminosity shining inward.

Shear mixing of the accreted matter into the white dwarf envelope, on the other hand, involves the conversion of rotational kinetic energy into heat thus providing a luminosity due to shear which emerges as surface luminosity (Sparks, Sion, Starrfield and Austin 1993). For a $0.93 M_\odot$ white dwarf accreting onto an equatorial belt of width $\pm 30°$ at a rate 2×10^{-8} for seven days; the maximum temperature of the white dwarf was $\sim 10^5$K. By comparison, the same model with radial (spherical) accretion and accretion luminosity shining inward at the surface, reached a peak temperature of 64,000 K. The higher temperature associated with shear mixing results largely from the fact that part of the energy continues as kinetic energy into the envelope before it is given up as an extra luminosity source due to shear. The accretion luminosity shining inward by kinetic energy of infall, on the other hand, is partly radiated away as high energy photons right at the surface.

Direct comparisons of simulations of the above three heating sources of CV degenerates with actual far UV cooling curves of exposed white dwarfs, have been discussed by Sion (1985), Pringle (1988), Sparks *et al.* (1993), Cheng *et al.*

(1994) and Sion (1995). In Sion (1985, 1995), compressional heating simulations are discussed in the context of CV degenerates in general while in the papers by Pringle (1988), Sparks *et al.* (1993) and Cheng *et al.* (1994), VW Hydri, WZ Sagittae and OY Car, respectively, are specifically considered.

4 Discussion

The large mirror of HST has led to the emergence of new breakthroughs on several fronts regarding equatorial accretion in non-magnetic cataclysmic systems : (a) the first rotational velocities for the underlying white dwarfs in cataclysmic variables; (b) the first white dwarf masses independent of disk emission lines; (c) the first chemical abundances for accreted atmospheres; (d) the first definitive evidence of white dwarf cooling in response to the dwarf nova accretion heating process in several systems. With the prospect of more high quality HST spectra on other CV systems and a growing HST archive, the stage is now set for further unprecedented detail on the structure of CV boundary layers, the diffusion and mixing of the accreted disk material, the white dwarf surface area and depth through which the accreted matter spreads or shear mixes and the detailed budget of the accretion energy. These developments will surely help to elucidate the detailed breakdown of the accretion energy budget, the true accretion geometry, how extensively the white dwarfs are heated laterally and with depth, why some systems have wind outflow during their outburst or high accretion phases, and the ultimate evolutionary effect of inward energy flow on the underlying white dwarf's evolution. The results summarized in this review mark only a bare beginning at best.

I am deeply grateful for the expertise, energy and enthusiasm of my collaborators on HST projects: Paula Szkody, Warren Sparks, Min Huang, Ivan Hubeny, Knox Long, and Fu-Hua Cheng. It is a pleasure to acknowledge stimulating discussions with many colleagues: Bob Popham, Richard Wade, Chris Mauche, Tom Marsh, Sumner Starrfield, Stephane Vennes, Janet Wood, Andy Silber, Craig Robinson, Keith Horne, Jim Liebert, Joe Patterson, and Bohdan Paczynski. This work is supported by NASA grant GO-3836.01-91A from the Space Telescope Science Institute, which is operated by the Association of Universities for Research in Astronomy, Inc., under NASA Contract NAS5-26555, and by NASA LTSA grant NAGW-3726.

References

Belloni,T., *et al.* 1991, ApJ, 246, L44

Cheng, F.H., Marsh, T.R., Horne, K., & Hubeny, I. 1994, in The Evolution of X-Ray Binaries, eds. S.S.Holt & C.S. Day, AIP Conf. Proc. 308, (New York: AIP Press), 197

Deng, S-B., Zhang, Z-Y., & Cheng, J-S.1994, A&A, 281, 759

Friend, M.T., Martin, J.S., Smith, R.C., & Jones, D.H.P. 1990, MNRAS, 246, 637

Giannone, P., & Weigert, A. 1967, Z.f.Astr, 67, 41

Henry, R.B.C., Shipman, H.L., & Wesemael, F. 1985, ApJS, 57, 145

Holm, A. 1988, in A Decade of Ultraviolet Astronomy with the IUE Satellite, Vol. I, (ESA: SP-281), p.229

Horne, K., Marsh, T.R., Cheng, F.H., Hubeny, I., & Lanz, T.1994, ApJ, 426, 294

Hubeny, I.1988, Comput.Phys.Comm, 52, 103

Hubeny, I.1994, Users Guide to TLUSTY & SYNSPEC, in press

Kiplinger, A., Sion, E.M., & Szkody, P.1989, ApJ, 366, 569

LaDous, C.1991, A&A, 252, 200

Long, K.S., Sion, E.M., Szkody, P., & Huang,M. 1994, ApJL, 424, L49

Lynden-Bell, D., and Pringle, J.E.1974, MNRAS, 168, 603.

Mateo, M., & Szkody, P. 1984,AJ, 89, 863

Mauche,C., Wade, R.,Polidan, R., van der Woerd, H., Paerels, F. 1991, ApJ, 372, 659

Mauche, C., Warren, J.K., Vallerga, J.V., Mukai, K., & Mattei, J. 1993, BAAS, 25, 863

Marsh, T., Horne, K., & Cheng, F.H. 1993, in Cataclysmic Variables and Related Physics, ed. O.Regev & G. Shaviv, Ann.Israel Phys.Soc., 10, 86

Panek, R., & Holm, A.1984, ApJ, 277, 700

Popham, R., & Narayan, R. 1995, ApJ, 10 March issue, in press

Pringle, J. 1988, MNRAS, 230, 587

Shore, S.N. 1992, in Nonisotropic and Variable Outflows from Stars, ed. L.Drissen, C. Leitherer, & A. Nota (ASP Conf.Ser.22), 342.

Sion, E.M., Leckenby, H., & Szkody, P. 1990, ApJL, 364, L41

Sion, E.M. 1985, ApJ, 297, 538

Sion, E.M.1991, AJ, 102, 295

Sion, E., Long, K.S., Szkody, P., & Huang, M.1994, ApJL, 430, L53

Sion, E.M., Cheng, F.H., Long, K.S., Szkody, P., Huang, M., Gilliland, R., & Hubeny, I. 1995, ApJ, 10 Feb. issue, in press

Sion, E.M., Szkody, P., Cheng, F.H., Huang, M. 1994, ApJ, submitted

Sion, E.M. 1995, ApJ, 438, in press

Sparks, Sion, Starrfield and Austin 1993, Ann.Israel Phys. 10, 96

Szkody, P. 1993, In Space-Based & Ground-Based Multi-Wavelength Astronomy, (ESA: SP-580), in press

Smak, J. 1993, Acta Astr, 43, 101

van der Woerd, H. et al. 1987, A&A, 182, 219

van der Woerd, H., & Heise, J.1987, MNRAS, 225, 93

Wade, R. 1981, ApJ, 246, 215

Wade, R.A., Cheng, F.H., Hubeny, I. 1994, BAAS, in press

Wesemael, F. et al. 1980, ApJS, 43, 159

Wood, J., Horne, K., Berriman, G., Wade, R., O'Donoghue, D., & Warner, B. 1986, MNRAS, 219, 629

Wood, J., Horne, K., Berriman, G., & Wade, R.A. 1989, ApJ, 341, 974

The Spatial and Kinematic Distributions of Field Hot Subdwarfs

Rex A. Saffer[1] *and J. Liebert*[2]

[1] Space Telescope Science Institute
3700 San Martin Drive
Baltimore, MD 21218 USA
[2] Steward Observatory
University of Arizona
Tucson, AZ 85721 USA

1 Introduction and Review

Hot subluminous O (sdO) and B (sdB) stars, or hot subdwarfs, are of interest in the study of the late stages of evolution of low- and intermediate-mass stars, since they are believed to evolve directly into white dwarfs. In both classes, the hydrogen and diffuse helium absorption lines are shallower and significantly more broadened than those of main sequence O and B stars, indicative of surface gravities between the main sequence and the white dwarf cooling sequence. The sdO stars have optical spectra characterized by the presence of strong He II λ4686, while in the sdB stars, the feature is lacking or is very weak, and the neutral helium lines show a broad range of strengths. At high dispersion in the ultraviolet, the spectra show many high excitation metal lines, such as N III, C IV, Si IV, and others, usually very sharp compared to the hydrogen and helium lines. The elemental abundances are modified by diffusion and, possibly, mass-loss processes in the hot, high-gravity atmospheres, with helium typically underabundant in the sdB stars and overabundant in the sdO stars, and with carbon and silicon underabundant in both classes. Nitrogen usually appears normal. The broadband colors are quite blue, for single stars, $(B-V) \sim -0.3$, $(U-B) \sim -1.0$, so that the detection efficiency for hot subdwarfs is very high in ultraviolet-excess color surveys. However, the occurrence of hot subdwarfs in binaries with bright main sequence or subgiant companions can redden the optical colors.

The first photometric identification of hot subdwarfs was made by Humason and Zwicky (1947). By the mid-1950's, a systematic spectroscopic survey of subdwarf candidates had been undertaken by J.L. Greenstein and G. Münch, and Münch (1958) performed the first quantitative spectroscopic analysis for the sdO star HZ 44. Greenstein (1960,1965) reported results of his spectroscopic survey, whose candidates were chosen from the large color-select surveys of Humason and Zwicky (1947; HZ), Feige (1958: F), and Iriarte and Chavira (1957) and Chavira (1958,1959: TON), on the basis of very blue color. However, beyond a few attempts (Sargent and Searle 1968; Newell 1969,1973; Norris 1970), the development of tools to derive accurate atmospheric parameters was slow, and progress was largely limited to characterizations of the stellar population through analysis of proper motions, and the construction of preliminary luminosity functions.

The first comprehensive analysis of hot subdwarfs and other faint blue stars was the seminal work of Greenstein and Sargent (1974; GS). Very briefly, through an improved quantitative analysis, the hot subdwarfs were identified as the field counterparts of the very faint extensions of metal-poor globular cluster (GC) blue horizontal branches, i.e., the extended horizontal branch (EHB). Most sdB stars had T_{eff} from 25 to 35,000 K and $\log g$ between 5 and 6, while most sdO stars appeared to be hotter than 40,000 K. Some sdO stars occupied a region of the H–R diagram near the hottest extension of the EHB, but they showed a considerably larger range in luminosity than the sdB stars. GS argued that the horizontal branch (HB) and EHB sequences were linked, both having helium-burning cores and hydrogen envelopes. The hot subdwarfs, however, had thin envelopes ($\lesssim 0.01\ M_\odot$) and could not sustain a hydrogen-burning shell, from which normal HB stars derive approximately half of their luminosities.

In the next decade, the spectra and photometry of a number of bright, well-observed hot subdwarfs were analyzed with increasingly sophisticated model atmosphere calculations (Baschek et al. 1972,1982a,1982b; Baschek and Norris 1975; Giddings and Dworetsky 1978; Kudritzki and Simon 1978; Kudritzki et al. 1980; Simon 1982; Bergeron et al. 1984; Heber et al. 1984; Schönberner and Drilling 1984; Wesemael et al. 1985; Lamontagne et al. 1985,1987; Heber 1986; Heber and Langhans 1986). Kudritzki (1976) published an extensive grid of model atmospheres for sdO stars, as did Wesemael et al. (1980) for pure hydrogen subdwarf and white dwarf atmospheres.

Concurrently, new theoretical evolutionary sequences of HB and EHB stars with improved treatments of convective overshooting and semiconvection became available (Caloi 1972,1979; Sweigart and Gross 1978; Castellani et al. 1983; Sweigart 1987). For helium core masses $\sim 0.5\ M_\odot$, and for very thin hydrogen envelope masses $\lesssim 0.01\ M_\odot$, the predicted effective temperatures and surface gravities agreed with observational estimates for the field sdB stars, as well as for a handful of GC EHB stars (cf. Heber et al. 1986).

Detailed calculations of radiative acceleration on helium and heavier elements showed that some elements could be supported against gravitational settling, while others could not (Lamontagne et al. 1985,1987; Michaud et al. 1985,1989; Bergeron et al. 1988). Observations contradicted the predicted abundances of silicon in sdB stars, which should have been supported but was found to be underabundant, and helium, which should have been completely depleted but was in general found to be underabundant only by a factor of ten. These results suggested that other particle transport processes, such as a weak stellar wind, operate in subdwarf atmospheres.

The Palomar Green (PG) Survey (Green, Schmidt, and Liebert 1986) provided a rich new hunting ground for hot subluminous stars. The sdB and sdO stars dominated the number counts, at 40% and 13% of the sample, respectively. Complementary programs in the southern hemisphere, the Edinburgh-Cape Survey (Stobie et al. 1988,1992) and the Hamburg Schmidt Survey (Engels et al. 1988), also have identified large numbers of hot subdwarfs at high galactic latitudes. The most recent compilation of Kilkenny, Heber, and Drilling (1988) lists 1225 spectroscopically identified hot subdwarfs.

Modern analyses of samples of hot subdwarfs drawn from color-select surveys have proceeded on a number of fronts. Newell (1973) and Greenstein and Sargent (1974) used broad-band, reddening-free photometric indices to estimate effective temperatures, which proved inadequate for the high temperature hot subdwarf atmospheres. Heber et al. (1984) and Heber (1986) supplemented absolutely calibrated IUE fluxes with ground-based broad-band and Strömgren photometry to obtain high precision estimates of $T_{\rm eff}$ for 18 southern hemisphere sdB stars. Surface gravities were determined from fine analysis of the Balmer and He I $\lambda 4471$ absorption line profiles. Most recently, four groups have analyzed large samples of hot subdwarf candidates drawn mainly from the PG Catalog:

- The Bonn and Kiel groups analyzed a total of 48 sdB stars (Moehler et al. 1990a,1990b; Theissen et al. 1993) and 15 sdO stars (Dreizler et al. 1990). In the sdB studies, effective temperatures were determined from reddening-free indices constructed from Johnson broad band, Greenstein multichannel, and Strömgren intermediate band photometry, predominately the latter, supplemented in a handful of stars by low-resolution IUE spectrophotometry. Surface gravities then were determined from equivalent widths and profile shapes of the Balmer lines. Helium abundances were determined from the equivalent width of He I $\lambda 4471$. The results were consistent with the interpretation of the field sdB stars as members of the EHB. Moehler et al. (1990b) derived $z_0 = 250$ pc for the population scale height, and a space density $n_0 = 1.0 \times 10^{-6}$ pc^{-3}, in general agreement with previous determinations (Heber 1986; Downes 1986; Green and Liebert 1988). Theissen et al. (1993) obtained similar estimates, $z_0 = 180$ pc and $n_0 = 1.9 \times 10^{-6}$ pc^{-3}.

- The Berkeley group analyzed a sample of 32 sdB candidates identified in their far-ultraviolet survey with the SCAP balloon-borne telescope (Bixler, Bowyer, and Laget 1991). Synthetic Strömgren b and y magnitudes were constructed from low- and intermediate-resolution followup spectroscopy, as well as quantitative measurements of the Balmer lines Hβ and Hγ. The observed quantities were compared with theoretical indices computed from the hot, pure hydrogen models of Wesemael et al. (1980) to determine $T_{\rm eff}$ and $\log g$, while the equivalent widths of the strong lines of He I and He II were compared with the models of Shipman and Strom (1970), Shipman (1971), and Auer and Mihalas (1972) to determine the helium abundances. The results were similar to those of the Kiel group, further confirming the identification of the field sdB stars with the EHB. The derived space density and scale height of the FUV-selected sample were 3.3×10^{-6} pc^{-3} and 240 pc, respectively.

- The Tucson group studied a sample of ~ 100 sdB candidates drawn from the

PG Survey (Saffer 1991; Saffer et al. 1994, SBKL). Using a homogeneous data set of high signal-to-noise (S/N) optical spectroscopy at intermediate spectral resolution (5Å FWHM) with spectral coverage $\lambda\lambda 3650-5200$, SBKL performed a fine line profile analysis of the Balmer lines $H\beta$ through H9 and all detected He I and He II lines. The resulting parameter estimates were very precise, due largely to the high S/N and homogeneity of the data set, as well as the sensitivity to atmospheric parameter variations of the large number of Balmer and helium lines analyzed. Saffer (1991) previously had applied completeness corrections derived from the sample to the entire sdB component of the PG Survey and estimated a population scale height of 285 pc.

- The Copenhagen group continues analysis of the complete sample of PG sdO stars to a limiting PG B magnitude of 15.6, as well as the complete sample of KPD sdO stars (Downes 1986). The technique is similar to that of Bergeron et al. (1992) and SBKL in that the line profiles of He I, He II, and H/He II blends in high S/N, intermediate resolution optical spectra are compared with the predictions of non-LTE model atmosphere theory. Preliminary results of the analysis of 21 helium-rich sdO stars (Thejll et al. 1994) reveal significant differences with the analysis of Dreizler et al. (1990) for 5 stars in common, with the former having estimated surface gravities systematically higher by 0.4 to 1.0 dex in $\log g$, and effective temperatures from 5,000 to 10,000 K higher. The discrepancy could be due to differences in the models and/or the fitting procedure, but is not yet known. The results of both studies remain the same: The PG sdO stars appear to be at large distances from the Galactic plane, with a scale height in excess of 600 pc, significantly larger than previous estimates of the sdB scale height.

2 Hot Subdwarf Spatial and Kinematic Distributions

2.1 Improved EHB Scale Height and Midplane Density Estimates

Observations and Reductions Spectroscopic observations of a large sample of stars drawn from the PG and KPD Surveys have recently been obtained in several observing runs at the Multiple Mirror Telescope (MMT) and the Steward Observatory 2.3-m reflector. The instrumentation, observing strategy, data reductions, and descriptions of representative spectra of both sdB and sdO stars are described in detail elsewhere (SBKL; Thejll et al. 1994).

Our observing efforts were concentrated in three fields, each surveying approximately 1200 square degrees and complete to a limiting PG B magnitude of 15.6. The first two fields were centered on Galactic coordinates $(l_{II}, b_{II}) = (85°, -44°)$ and $(211°, +44°)$, while the third field was centered on the north Galactic pole. Candidates were selected for observation if they had PG spectral classifications of sd, sdB, sdB-O, sdOA, or HBB, i.e., objects showing broad, shallow Balmer absorption. The complete sample comprises more than 250 EHB candidates, by far the largest surveyed to date.

Estimation of Stellar Parameters As described in detail by Saffer (1991), Bergeron et al. (1992), and SBKL, reliable atmospheric parameter estimates may be derived from detailed line profile analysis of the absorption lines of hydrogen and helium in optical spectra of intermediate resolution ($\sim$5Å FWHM). The grid of theoretical spectra, composed only of hydrogen and helium, encompasses $20{,}000\ K \leq T_{\mathrm{eff}} \leq 40{,}000\ K$, $4.0 \leq \log g \leq 6.0$, and $0.0 \leq N(\mathrm{H})/N(\mathrm{He}) \leq 0.1$, in steps of 5,000 K, 0.5, and 0.33, respectively. Briefly, observed and theoretical line profiles are normalized, line by line, to a linear continuum, and the atmospheric parameters are determined from a non-linear least squares minimization procedure. Individual internal error estimation follows straightforwardly from counting statistics, which dominate in the high S/N CCD spectra. Typical 1σ errors for T_{eff}, $\log g$, and $N(\mathrm{H})/N(\mathrm{He})$ are 500 K, 0.08 dex, and 0.003 by number, respectively.

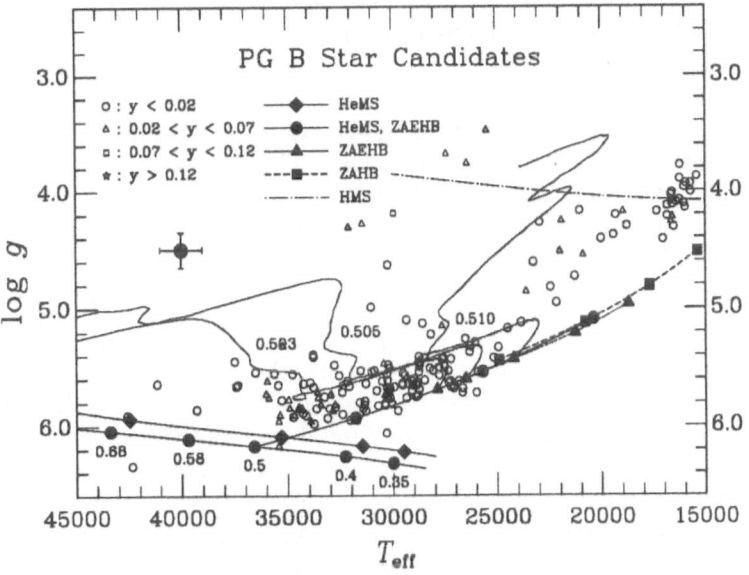

Fig. 1. Distribution of the complete sample in the T_{eff}–$\log g$ plane.

Distribution of Atmospheric Parameters The distribution of the sample in the T_{eff}–$\log g$ plane is shown in Figure 1. The majority ($\sim$75%) of the candidates form a tight locus bounded below by the theoretical zero-age extended horizontal branch (ZAEHB) sequences of Sweigart (1987) and Caloi (1972,1979) for 0.5 $M_{\odot}$ helium-burning cores having overlying hydrogen-rich envelopes with masses $\lesssim 0.01~M_\odot$. Rising from the ZAEHB, tracks for three values of the total stellar mass demonstrate the evolution of effective temperature and surface gravity during and subsequent to the core helium burning phase. After core helium exhaustion, evolution is governed by the mass of the hydrogen-rich envelope. For masses $\lesssim 0.01~M_\odot$, a hydrogen-burning shell cannot be sustained at all, and the stars move into the region occupied by lower luminosity sdO stars. For envelope masses between 0.01 and 0.02 $M_\odot$, stars may briefly ignite hydrogen burning in a thin shell, but they are unable to ascend the AGB and likewise move rapidly into regions occupied by sdO stars.

As shown in Figure 1, the scattering of stars just above the ZAEHB and extending to effective temperatures as high as 40,000 K is consistent with evolution following core helium exhaustion. The compact EHB distribution conclusively rules out the hypothesis that the bulk, indeed, even a modest fraction, of field EHB stars are formed by mergers of low mass helium degenerates in common envelope phases in close binary systems (cf. Iben and Tutkov 1986; Iben 1990). Such scenarios predict a broader range of helium-burning core masses, from 0.3 to 0.7 $M_\odot$, and this cannot be supported by the observed distribution.

About 20% of the candidates form two clumps at temperatures cooler than 24,000 K and with surface gravities lower than $\log g = 5.0$. They probably belong to the blue horizontal branch (BHB) or main sequence. The last distinct group of stars has temperatures higher than 30,000 K and gravities between the main sequence and the ZAEHB. These stars could be in a post-core helium burning phase, or even in a post-AGB phase, as evolutionary tracks for both phases pass through this region of the T_{eff}–$\log g$ plane. The stars at $T_{eff} \sim 25,000~K$ and $\log g < 3.5$ also could be in post-EHB or post-AGB phases. Further analysis with appropriate atmospheric compositions is in progress for the low-gravity stars and will be reported elsewhere. However, the current analysis is sufficient to exclude them from the confirmed EHB sample retained for further analysis.

EHB Distances and Population Scale Height EHB distance estimates may be made given estimates of the stellar mass and the reddening along the line of sight. The confirmed EHB stars form a well-bounded sequence along the theoretical ZAEHB tracks, and it seems well justified to assume a mass of 0.5 $M_\odot$ for all stars in the sample. The temperature and gravity determinations permit estimation of intrinsic Strömgren $(b-y)$ and Johnson $(B-V)$ colors from the model spectra. Reddenings (and total extinction at y and V) follow from comparison with observed colors in the literature, especially the large compilation of hot subdwarf Strömgren photometry of Wesemael et al. (1992). Stars without published colors were assigned reddening and extinction values by interpolating between neighboring objects. Finally, following the prescription of Saffer (1991), the distances d and heights $z = d \sin b$ above the Galactic plane were calculated.

For a stellar population with a barymetric density distribution, the differential counts obey

$$dN \propto (\sin b)^{-3} z^2 e^{-z/z_0} dz, \tag{1}$$

where z_0 is the population scale height and b is the Galactic latitude on which the sample field is centered. The maximum of the distribution lies at $z = 2z_0$. Figure 2 shows the differential z-histograms of the confirmed EHB stars in each of the 3 complete fields. For the two fields at mid-Galactic latitudes, the z-axes have been stretched to $z' = z/\sin b$, in order to compare directly with the field at the north Galactic pole. For each of the observed distributions, the maximum occurs at $z \gtrsim 1$ kpc, or $z_0 \gtrsim 500$ pc. In the past, estimates of the EHB scale height from formal fits to such differential distributions have ranged from 180 to 240 pc. However, while each of those samples was complete, the limiting magnitudes were bright, $y \sim 14.2$, and the total number of stars was small, about one dozen. It now is clear that falling counts in the last z-bins of those distributions were due to incompleteness at the relatively bright limiting magnitudes of the samples, leading to low estimates of the population scale height.

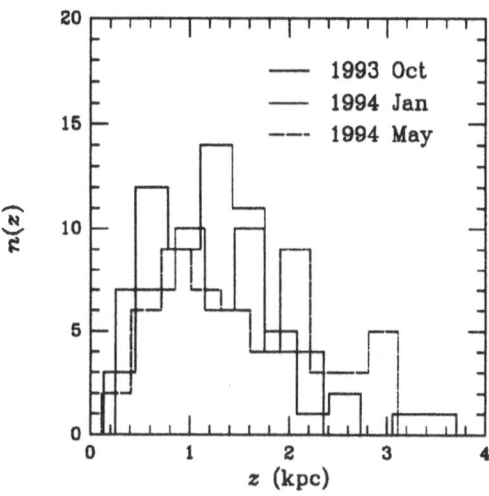

Fig. 2. Differential z-histograms for the complete sample.

An Improved EHB Midplane Density Estimate In the course of the observing program described above, we also observed and analyzed all but 2 of the complete sample of 31 sdB stars discovered in the Kitt Peak Downes Survey (KPD; Downes 1986). The KPD survey was similar to the PG Survey both in its use of wide-field Schmidt camera imaging in broadband filters, and in its UV-excess selection criteria, but it lies in the Galactic plane and is well-suited for the determination of the midplane space density. As in the high-latitude sample, some stars were misclassified: 0311+4801 is found to be a hot DA white dwarf; 0549+1642 shows only weak, broad lines of neutral and singly ionized helium and could be a hydrogen-deficient subdwarf or cataclysmic variable; 0721-0026 and 2022+2033 have low surface gravities, $\log g \sim 4$. Finally, 2220+5126 was partially resolved on the spectroscope slit, and it was possible to extract individual spectra. Both stars proved to be EHB stars, as did one system in the high latitude sample. In both cases, the observed unresolved magnitudes were deconvolved according to the estimated atmospheric parameters. The binary separations are much larger than the stellar radii, even when in the red giant branch (RGB) phase, so that each EHB star has evolved as a single star unperturbed by its companion and may fairly be included in the complete sample.

Individual reddenings again were determined from a comparison of intrinsic colors predicted from the synthetic spectra and the observed colors, followed by estimates of distances and absolute magnitudes M_V. We note here that Downes' (1986) M_V estimates were derived by assuming a mean intrinsic color for the sample, determining reddenings from the observed colors, then assigning distances from reddening vs. distance graphs. The dispersion was large, $3 \lesssim M_V \lesssim 7$, while our precise estimates of T_{eff} and $\log g$, together with an estimate of 0.5 $M_\odot$ for the stellar mass, lead to a narrower range of estimated absolute magnitudes, $3.2 \lesssim M_V \lesssim 5.1$. Using the $1/V_{max}$ method of Schmidt (1975) and summing over the sample, we find for the field EHB space density $n_0 = 7.5 \times 10^{-7}$ pc^{-3}, compared with estimates of 1.5–2 and 4×10^{-6} by Downes (1986) and Heber (1986), respectively. Assuming a core helium-burning lifetime $\tau = 1.5 \times 10^8$ yr, we estimate the EHB birthrate as $n_0/\tau = 5 \times 10^{-15}$ pc^{-3} yr^{-1}. Current best estimates of the local white dwarf birthrate range from 0.5–2 $\times 10^{-12}$ pc^{-3} yr^{-1}, that is, *the contribution to the local white dwarf birthrate from the EHB channel is just 0.25–1%*, compared to previous estimates of about 2%.

Our lower estimate for the space density is due to two effects: 1) Our high S/N spectroscopy and precise atmospheric analyses efficiently identify non-EHB stars, both for obvious spectral misclassifications and for low gravity. Excluding them from the complete sample eliminates their contribution to the space density. 2) Our precise estimates of M_V have a significantly smaller dispersion and brighter mean magnitude than those of Downes (1986). On average, our brighter M_V estimates correspond to larger volumes V_{max}, and the contributions $1/V_{max}$ to the midplane density are therefore smaller. The two effects work in the same direction and account for our smaller estimate.

PG sdO Proper Motions Saffer (1991) has analyzed PG sdO radial velocities in an attempt to estimate the population velocity dispersions, especially the w-component perpendicular to the Galactic plane, which is strongly correlated with population age. The sample comprised some 100 sdO stars with accurate measured radial velocities and was distributed nearly uniformly in solid angle on the celestial sphere. The analysis required various direction cosine-weighted terms to vanish upon summation over the sample. However, it was found during Monte Carlo simulations that those terms vanished only for sufficiently large sample sizes, well in excess of 1000 objects. The actual sample of only 100 objects had a derived w-dispersion $\sigma_w = 31.9 \pm 1.0$ km s^{-1}, where the quoted error was propagated from the very accurate measured echelle velocites. Taken at face value, this suggested that the population is significantly older than the bulk of the old disk stars. However, the simulations showed that the sampling error was about 10 times larger than the internal error, so that the estimated w-dispersion was not a precise indicator of population age.

To improve the situation, we have used the Space Telescope Guide Star Catalog (GSC) to measure proper motions for 56 of the sdO stars in the radial velocity sample (Saffer, Williamson, Roberts, and MacConnell; in preparation). For each object, 512×512 frames were extracted from optical disks scanned from the GSC and the POSS. The measured positions for all objects detected in both frames then were used to calculate plate solutions and relative proper motions. No attempt was made at reduction to an absolute frame. We found that of the 56 objects so analyzed, only 4 objects had detected proper motions in excess of the detection threshold of 0.030 arcsec yr^{-1}. These were 0155+383, 0823+546, 0934+553, and 0952+518, with proper motions of 0.037, 0.039, 0.047, and 0.112 arcsec yr^{-1}, respectively. If we assume membership in the old disk and a generous transverse velocity of 50 km s^{-1}, these translate to distances of 285, 270, 224, and 95 pc, respectively. Comparing with the confirmed EHB z-histograms of Figure 2, such a large lower bound on the distances of the sdO sample would be consistent with a population scale height similar to that of the EHB stars, of order 500 pc or larger. This also would be consistent with preliminary estimates of the sdO scale height in excess of 600 pc by Dreizler et al. (1990) and Thejll et al. (1994).

3 Discussion

Our estimate of the EHB scale height, $z_0 \gtrsim 500$ pc, is at least twice that of previous determinations. We attribute the difference to the relatively bright limiting magnitudes and small sizes of the previously analyzed samples. Our estimate for the EHB midplane density, $n_0 = 7.5 \times 10^{-15}$ pc^{-3}, is smaller than previous estimates by a factor of 2–4. This difference we attribute to spectral misclassification in the complete sample of Downes (1986), and to our smaller dispersion in the estimated absolute magnitudes. Our large value for z_0 implies that the population is significantly older than the bulk of the stars in the old part of the thin disk. A comparison may be drawn with the population of EHB stars recently discovered in the populous Galactic cluster NGC 6791 (Kałuży and Udalski 1992; Liebert, Saffer, and Green 1994, LSG). NGC 6791 is among the oldest and most metal-rich open clusters in the Galaxy, with an age of about 8 Gyr and [Fe/H] $\gtrsim$ +0.2. Horizontal branch stars in such a metal-rich population all would be expected to lie in a red clump near the RGB. However, LSG found that 3 EHB stars and one sdO star are probable members of the cluster. As there are some 20–22 red HB clump stars in the cluster, approximately 15% of the RGB progenitors of the HB stars managed to lose nearly all of the hydrogen-rich envelope and move to the EHB after igniting core helium burning. The cluster turnoff mass exceeds 1.1 $M_\odot$, so that an unusually large amount of mass loss must have occurred on the RGB.

Metal-poor globular cluster stars routinely lose 0.25–0.35 $M_\odot$ on the RGB and form blue horizontal branches, but the RGB stars in NGC 6791 must have lost ~ 0.6 $M_\odot$ to evolve to the EHB, where they are observed to lie on 0.5 $M_\odot$ core-helium burning tracks. Such extreme mass loss might occur only in extremely old and metal-rich populations, where mass loss rates could be enhanced, and where low turnoff masses would limit the amount of envelope to be lost. As suggested by LSG, the birthrate of EHB stars in the field could be explained if only the most metal-rich fraction of disk stars are able to form EHB stars. However, the hypothesis that EHB progenitors are among the most metal-rich *and* oldest stars in the disk runs counter to the expected relation between age and metallicity (Twarog 1980; Edvardsson et al. 1993). Still, it is worth noting that NGC 6791 lies ~ 1 kpc from the disk midplane. Sandage (1988) refers to it as a "thick disk" cluster.

References

Auer L.H., Mihalas D. 1972, ApJS, 24, 193

Baschek B., Sargent W.L.W., Searle L. 1972, ApJ, 173, 611

Baschek B., Norris J. 1975, ApJ, 199, 694

Baschek B., Kudritzki R.P., Scholz M., Simon K.P. 1982a, A&A, 108, 387

Baschek B., Höflich P., Scholz M. 1982b, A&A, 112, 76

Bergeron P., Fontaine G., Lacombe P., Wesemael F., Crawford D.L., Jakobsen A.M. 1984, AJ, 89, 374

Bergeron P., Wesemael F., Michaud G., Fontaine G. 1988, ApJ, 332, 964

Bergeron P., Saffer R.A., Liebert J. 1992, ApJ, 394, 228

Bixler J.B., Bowyer S., Laget M. 1991, A&A, 250, 370

Caloi V. 1972, A&A, 20, 357

Caloi V. 1979, A&A, 221, 27

Castellani V., Chieffi A., Tornambé A. 1983, ApJ, 272, 249

Chavira E. 1958, *Bol. Obs. Tonantzintla y Tacubaya*, No. 17

Chavira E. 1959, *Bol. Obs. Tonantzintla y Tacubaya*, No. 18

Dreizler S., Heber U., Werner K., Moehler S., de Boer K.S. 1990, A&A, 235, 234

Downes R.A. 1986, ApJS, 61, 569

Edvardsson B., Andersen J., Gustafsson B., Lambert D.L., Nisson P.E., Tompkin J. 1993, A&A, 275, 101

Engels D., Groote D., Hagen H.J., Reimers D. 1988, PASPC, 2, 143

Feige J. 1958, ApJ, 128, 267

Giddings J.R., Dworetsky M.M. 1978, MNRAS, 183, 265

Green R.F., Schmidt M., Liebert J. 1986, ApJS, 61, 305

Green R.F., Liebert J. 1988, in *The Second Conference on Faint Blue Stars*, IAU Coll. 95, (Schenectady:L. Davis Press), eds. A.D.G. Philip, D.S. Hayes, J. Liebert, p. 261

Greenstein J.L. 1960, in *Stellar Atmospheres*, Stars and Stellar Systems, Vol. VI, ed. J.L. Greenstein (Chicago Press), p. 676

Greenstein J.L. 1965, in *Galactic Structure*, Stars and Stellar Systems, Vol. V, eds. A. Blaauw and M. Schmidt (Chicago Press), p. 361

Greenstein J.L., Sargent A.I. 1974, ApJS, 28, 157 (GS)

Heber U. 1986, A&A, 155, 33

Heber U., Hunger K., Jonas G., Kudritzki R.P. 1984, A&A, 130, 119

Heber U., Kudritzki R.P., Caloi V., Castellani J., Danziger J., Gilmozzi R. 1986, A&A, 162, 171

Heber U., Langhans G. 1986, *New Insights in Astrophysics*, Proceedings of a Joint NASA/ESA/SERC Conference, ESA SP-263, p. 279
Humason M.L., Zwicky F. 1947, ApJ, 105, 85
Iben I. Jr. 1990, ApJ, 353, 215
Iben I. Jr., Tutukov A.V. 1986, ApJ, 311, 753
Iriarte B., Chavira E. 1957, *Bol. Obs. Tonantzintla y Tacubaya*, No. 16
Kałużny J., Udalski A. 1992, AcA, 42, 29
Kilkenny D., Heber U., Drilling J.S. 1988, SAAO, Circular 12
Kudritzki R.P. 1976, A&A, 52, 11
Kudritzki R.P., Simon K.P. 1978, A&A, 70, 653
Kudritzki R.P., Gruschinske J., Hunger K., Simon K.P. 1980, Proceedings of the 2nd-year IUE Conference, Tübingen
Lamontagne R., Wesemael F., Fontaine G., Sion E.M. 1985, ApJ, 299, 496
Lamontagne R., Wesemael F., Fontaine G. 1987, ApJ, 318, 844
Liebert J., Saffer R.A., Green E.M. 1994, AJ, 107, 1408 (LSG)
Michaud G., Bergeron P., Wesemael F., Fontaine G. 1985, ApJ, 299, 741
Michaud G., Bergeron P., Heber U., Wesemael F. 1989, ApJ, 338, 417
Moehler S., Richtler T., de Boer K.S., Dettmar R.J., Heber U. 1990a, A&AS, 86, 53
Moehler S., Heber U., de Boer K.S. 1990b, A&A, 239, 265
Münch G. 1958, ApJ, 127, 642
Newell E.B. 1969, PhD dissertation, Australian National University
Newell E.B. 1973, ApJS, 26, 37
Norris J. 1970, PhD dissertation, Australian National University
Saffer R. 1991, PhD dissertation, University of Arizona
Saffer R., Bergeron P., Koester D., Liebert J. 1994, ApJ, 432, 351 (SBKL)
Sandage A. 1988, in *Calibration of Stellar Ages*, (Schenectady:L. Davis Press), ed. A.G.D. Philip, p. 48
Sargent A.I., Searle L. 1968, ApJ, 152, 443
Schmidt M. 1975, ApJ, 202, 22
Schönberner D., Drilling J.S. 1984, ApJ, 278, 702
Shipman H.L. 1971, ApJ, 166, 587
Shipman H.L., Strom S.E. 1970, ApJ, 159, 183
Simon K.P. 1982, A&A, 107, 313
Stobie R.S., Morgan D.H., Bhatia R.K., Kilkenny D., O'Donoghue D. 1988, in *The Second Conference on Faint Blue Stars*, IAU Coll. 95, (Schenectady:L. Davis Press), eds. A.D.G. Philip, D.S. Hayes, J. Liebert, p. 493
Stobie R.S., Chen A., O'Donoghue D., Kilkenny D. 1992, in *Variable Stars and Galaxies*, ASP Conference Series, Vol. 30, (San Francisco:ASP), ed. B. Warner, p. 87
Sweigart A.V. 1987, ApJS, 65, 95
Sweigart A.V., Gross P.G. 1978, ApJS, 36, 405
Theissen A., Moehler S., Heber U., de Boer K.S. 1993, A&A, 273, 524
Thejll P., Bauer F., Saffer R., Liebert J., Kunze D., Shipman H.L. 1994, ApJ, 433, 819
Twarog B.A. 1980, ApJ 242, 242
Wesemael F., Auer L.H., Van Horn H.M., Savedoff M.P. 1980, ApJS, 43, 159
Wesemael F., Holberg J., Veilleux S., Lamontagne R., Fontaine G. 1985, ApJ, 298, 859
Wesemael F., Fontaine G., Bergeron P., Lamontagne R., Green R.F. 1992, AJ, 102, 203

Fundamental Properties of Magnetic White Dwarfs in CVs

Dayal T. Wickramasinghe

The Astrophysical Theory Centre
School of Mathematical Sciences
Australian National University
Canberra ACT 0200
Australia

1 Introduction

Model atmospheric studies of the spectra of magnetic white dwarfs is at present not at the level of sophistication that is required for determining gravities and masses (see Wickramasinghe 1994), mainly due to our lack of knowledge of opacities and line broadening in strong magnetic fields. It is therefore important that other more direct methods, such as those that can be applied to magnetic white dwarfs in binary systems, be exploited for estimating masses.

White dwarfs are found in the close interacting binaries known as the cataclysmic variables (CVs). About 200 CVs are known and of these about 70 are classified as being magnetic (MCVs). Magnetic CVs divide into the Intermediate Polars (IPs) and the AM Herculis type systems (AM Hers) (Wickramasinghe 1989). In the AM Hers, the magnetic field of the white dwarf is strong enough to lock it into synchronous rotation with the orbit. These systems do not have accretion discs. On the other hand, the magnetic white dwarfs in the IPs rotate more slowly than the orbital period and an accretion disc is generally present. They have lower fields than the white dwarfs in AM Hers.

If the orbital inclination is sufficiently high, partial or total eclipses can be seen of the various components of a CV, such as the accretion disc, the hot spot, the boundary layer and the white dwarf. If it is possible to disentangle contributions from these components (in particular the boundary layer and the white dwarf), it is in principle possible to use eclipsing systems to determine radii and hence the masses of the white dwarfs. This method has been applied with some success to the CVs known as the dwarf novae, but the errors in the mass determinations are generally large.

The situation with the magnetic CVs (AM Hers) is, however, quite different in that there is no disc or boundary layer. Eclipses can be seen of the bare white dwarf (and the accretion hot spots on its surface) and by careful analysis it is possible to determine accurate radii and hence masses for the white dwarfs.

The white dwarfs in the AM Hers provide an excellent opportunity for studying magnetic fields and field structure using methods which are either unavailable for isolated white dwarfs or which cannot be used effectively except in a few cases. Firstly, Zeeman spectroscopy can be used more effectively since the rotation periods are known (typically less than 4 hrs) and phase dependent information (spectra and polarisation) can be used to study field structure. Secondly the presence of accretion shocks enable fields to be determined in localised regions on the white dwarf surface using cyclotron spectroscopy (Wickramasinghe 1990).

In this paper I will briefly review results on the masses and magnetic fields of white dwarfs in the AM Herculis type systems.

1.1 Magnetic fields

The known isolated magnetic white dwarfs have fields in the range 0.1 - 1000 MG and appear to be almost uniformly distributed per decade in field strength in the range 1 - 1000 MG. They are rare among isolated white dwarfs (2 - 3 %) and are found routinely in optical surveys of white dwarfs. In contrast, magnetic white dwarfs are found in abundance among the cataclysmic variables. The majority are discovered from x-ray surveys through the detection of hard x-rays from accretion shocks located on the surface of the white dwarf near its magnetic poles. It is estimated that about 40 % of all CVs are magnetic with the majority belonging to the AM Herculis subclass. The high proportion of magnetic systems among the known CVs is clearly the result of the strong bias towards discovering such systems in X-ray surveys. The distribution of magnetic field strengths in CVs is difficult to establish over an extended field range, although there are some clear differences compared with isolated white dwarfs (Wu, Wickramasinghe and Li 1995) which require explanation.

For a given orbital period, an accretion disc forms at sufficiently low magnetic fields, and the radiation from the disc dominates the radiation from the white dwarf, so that direct magnetic field determinations (e.g. from Zeeman spectroscopy, as in isolated white dwarfs) are difficult. Some of these (the IPs) show X ray and/or optical modulations at the spin period of the white dwarf (typically 0.1 the orbital period) indicating magnetic accretion. It is estimated that the magnetic fields in IPs are typically less than 10 MG although no direct measurement are available. At somewhat higher fields, the white dwarf is locked into synchronous rotation with the orbit and accretion occurs directly onto the white dwarf surface without the formation of an accretion disc. In these higher field systems (the AM Hers) the magnetic field can be measured directly by two independent methods.

The AM Her systems exhibit high and low states corresponding to different levels of mass transfer. During a low state, the radiation from the photosphere of the white dwarf dominates at optical and UV wavelengths and it is possible to use Zeeman spectroscopy to measure magnetic fields. Since the orbital periods (and therefore the spin periods) are in the range 80 mins - 4 hrs, it is possible to carry out phase resolved spectroscopic and polarimetric studies to constrain the magnetic field structure. Such studies have generally shown that the magnetic field distributions do not correspond to a centered dipole indicating that higher order multipole components are present (see Wickramasinghe and Wu 1991). The modelling procedures generally assume, as a first approximation to more complex field structures, decentered dipoles. These generally show better agreement with phase dependent spectroscopic data.

A second method for determining magnetic fields is through the detection of cyclotron lines in optical and near IR spectra. These originate from the accretion shocks at the surface of the white dwarf (Wickramasinghe 1989), and appear to be a general characteristic of these systems. In general, accretion occurs onto two localised regions (spots) on the surface (located approximately at the foot points of a closed field line) and the detection of cyclotron lines enable the field to be determined at these spots. The two spots are expected to be close to the magnetic poles since the material couples onto field lines far away from the white dwarf. The cyclotron lines thus enable the fields to be determined near the magnetic poles, thus providing an independent method of probing the global field structure. The two methods have in general given consistent results (see Wickramasinghe and Martin 1985, Wickramasinghe et al. 1989)

The fields at or near the magnetic poles derived for the AM Hers from a combination of Zeeman and cyclotron spectroscopy are given in Table 1 (Wickramasinghe 1990). The large differences in field strength at the two poles is clear evidence for non-dipolar field distributions.

Another striking feature of Table 1 is that all AM Herculis type systems have fields in a very narrow range (8 - 80 MG), in stark contrast to what is seen in isolated magnetic white dwarfs. The absence of lower field AM Hers can simply be explained by the fact that discs develop in these systems and mask their magnetic nature (as already noted, the IPs are believed to harbour these lower field white dwarfs). The absence of white dwarfs with fields in excess of 80 MG has been a subject of much discussion, and cannot be explained by selection effects.

One possible explanation has recently been proposed by Wickramasinghe and Wu (1993). It is based on the assumption that most CVs with AM Her type (or larger) fields, first come out of the common envelope phase of evolution at orbital periods that are greater than about 10 hrs. It can then be shown that systems with fields in excess of a certain critical value (roughly the observed maximum for AM Hers) will synchronise quickly and the orbit will shrink thereafter, not by magnetic braking, but by orbital angular momentum loss by the less efficient mechanism of gravitational radiation. Such systems will not come into contact in a Hubble time scale and therefore never be seen as an active AM Her. A prediction of this model is that these strongly magnetised systems should be found among the pre CVs. They may, of course, be as rare as magnetic white dwarfs are among isolated white dwarfs, and be difficult to detect.

2 Mass determinations

Masses of the white dwarfs in the AM Herculis type systems have been determined by measuring radial velocity variations of absorption lines from the photosphere of the red star, of narrow emission lines attributed to the heated surface of the red star, and from eclipse light curves of the white dwarf by the red star. There are large uncertainties in interpreting the radial velocity curves of the narrow emission lines, since these may be distorted by contributions from the mass transfer stream.

The radial velocity methods using lines from the companion star can be expected to give accurate results only for the brightest systems where high resolution observations have been possible of absorption lines (in particular of the NaI lines (8183 - 8195 Å)). The errors in mass determinations using this method are dominated by uncertainties in the orbital inclination which is usually estimated independently from polarisation studies. The most accurate velocity data are for AM Her itself by Young and Schneider (1979) who gave the mass of the white dwarf as $M_{wd} = 0.6 M_\odot$ for an orbital inclination of 40°. Other similar studies have been reported by Mukai and Charles (1987).

Of the 54 known AM Hers, six are eclipsing, and of these only two have so far been studied at sufficiently high time resolution to enable radii and masses to be determined. The eclipse method was first applied to UZ For by Bailey and Cropper (1991) using photometry obtained with a time resolution of 1 sec with the Ango-Australian Telescope. The data through eclipse ingress and egress are shown in figure 1 together with model fits. The sources contributing to the light curves are the white dwarf and a hot spot at the base of the accretion funnel which emits optical cyclotron emission. The data show the intensity first declining gradually as the white dwarf begins to cross the limb of the companion star. A rapid decline in intensity then follows as the spot (typically of linear size $0.1 R_{wd}$) is eclipsed by the companion. Following the spot eclipse, the white dwarf becomes the dominant source of light and its intensity continues to decline until it is finally totally eclipsed by the companion star. The opposite sequence of events occurs at eclipse egress.

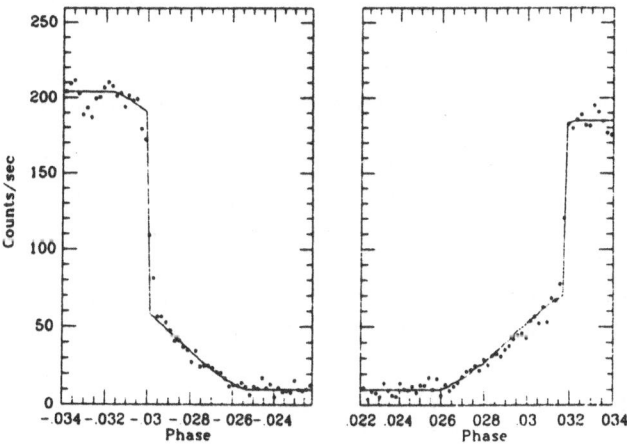

Fig. 1. Eclipse light curve of the white dwarf and hot spot of UZ For by the companion star

Table 1. Magnetic fields of AM Hers from Cylotron and Zeeman spectroscopy

System	Period Minutes	Field (pole 1) MG	Field (pole 2) MG
EF Eri	81.02	12	
DP Leo	89.8	32	59
VV Pup	100.4	32	56
V834 Cen	101.5	23	
1H1752+08	12.8	7	
MR Ser	113.6	24	
BL Hyi	113.7	22	
ST Lmi	113.89	12	
EK Uma	114.5	47	
AN Uma	114.8	36	
RE2107-0518	125	36	
UZ For	126.5	53	75
RE1938-4612	140	56	
AM Her	185.6	14	24
BY Cam	201.9	41	

The modelling procedure assumes a Roche lobe filling secondary. The mass ratio q ($= M_{rd}/M_{wd}$), where M_{rd} is the mass of the companion star, then determines the dimensions of the secondary relative to the orbital separation a. The duration of the white dwarf eclipse yields the orbital inclination i (for given q) and model fits to the eclipse ingress and egress light curves yield the ratio R_{wd}/a, where R_{wd} is the white dwarf radius. The emphrical relationship between the orbital period and the mass of the secondary (Patterson 1984) provides a final constraint, and these together then yield the absolute value of the radius of the white dwarf and therefore its mass from the mass — radius relationship.

Bailey and Cropper (1991) estimated $M_{wd} = 0.68 \pm 0.6 M_\odot$ for UZ For. A similar study of DP Leo yielded $M_{wd} = 0.71 \pm 0.5 M_\odot$ (Bailey et al. 1993). A source of error in the modelling procedure was the use of the simplifying assumption that the emission from the surface of the white dwarf is isotropic. It is estimated that the inclusion of limb darkening will decrease the mass estimates by 10 % and with this correction the most probable values for the masses of the white dwarfs in UZ For and DP Leo are $0.61\ M_\odot$ and $0.64\ M_\odot$ respectively. Thus the masses of the white dwarfs in the best studied systems do not appear to show significant deviations from the mean mass found for isolated white dwarfs.

3 Conclusions

Studies of the fundamental properties of magnetic white dwarfs in binaries are are at present at a rudimentary stage, but suggest that there are no significant differences between the masses of magnetic white dwarfs in CVs and the mean mass of isolated (non-magnetic) white dwarfs. The magnetic field distribution of white dwarfs in CVs is, however, significantly different from the field distribution in isolated stars, in that there are no systems observed with fields in excess of 80 MG.

References

Bailey, J. and Cropper, M.: Mon. Not. Roy Astr. Soc. **253** (1991) 27

Bailey, J., Wickramasinghe, D. T., Ferrario, L., Cropper, M. and Hough, J.: Mon. Not. Roy. Astr. Soc. **261** (1993) L31

Mukai, K. and Charles, P. A.: Mon. Not. Roy. Astr. Soc. **226** (1987) 209

Patterson, J.: Ap. J. Supp. **45** (1984) 517

Wickramasinghe, D. T.: Proceedings of the IAU General Assembly Joint Discussion **16** (Astrophysical applications of powerful new atomic databases) (1994) (in press)

Wickramasinghe, D. T.: in *Polarised Radiation of Circumstellar Origin*, Eds Coyne et al. (University of Arizona press). (1989) 3.

Wickramasinghe, D. T.: in *Spectral Line Shapes*, American Institute of Physics Conference proceedings **216** (1990) 574

Wickramasinghe, D. T. and Martin, B.: Mon. Not. Roy Astr. Soc. **212** (1985) 353

Wickramasinghe, D. T., Ferrario, L., and Bailey, J.: Ap. J. Letts **342** (1989) L35

Wickramasinghe, D. T. and Wu, K.: Mon. Not. Roy Astr. Soc. **253** (1991) 11P

Wickramasinghe, D. T. and Wu, K.: Mon. Not. Roy Astr. Soc. **266** (1993) L1

Wu, K., Wickramasinghe, D. T. and Li, J.: Proc. Astron. Soc. Aus. (1995) (in press)

Young, P. and Schneider, P. S.: Ap. J. **230** (1979) 502

Searching for Binary White Dwarfs: A Progress Report

Angela Bragaglia

Osservatorio Astronomico di Bologna, via Zamboni 33, I–40126 Bologna, Italy

1 Introduction

Extensive observational efforts have been made in order to discover the elusive progenitors of type Ia Supernovae. Among possible candidates, systems composed of 2 White Dwarfs (WDs), or "double degenerate" (DD) systems have been proposed (Iben & Tutukov 1984; Webbink 1984), as they naturally explain basic Sne Ia's characteristics, like the lack of hydrogen in their spectra, their uniformity and presence in elliptical galaxies.

Up to now no evidence[1] has been found for DD systems with the right properties to be SNe Ia's precursors (*i.e.* close enough to merge in less than one Hubble time and massive enough to reach the Chandrasekhar mass upon merging). This has given new life to the so-called "single degenerate" model, involving cataclysmic or symbiotic binaries (Munari & Renzini 1992; for a review see Renzini 1994).

Reliable data now exist for ∼ 130 WDs, but we still need to increase this number and better discriminate in some dubious cases of possible binaries for a decisive observational test (see Yungelson et al. 1994).

2 Past surveys: null results are a result too

Several searches have been dedicated to close DD's: i) Robinson & Shafter (1987) observed 44 WDs and did not find any DD with orbital period less than 3^h; ii) Foss, Wade & Green (1991) presented null results for periods less than 10^h for a sample of 25 DA WDs; iii) most of data for our survey were collected in 1988-1990; preliminary results were published in Bragaglia et al. (1990, BGRD) only for DA WDs and Bragaglia et al. (1994) for DB's too. As a result of further observations and reclassification of a few objects (Bragaglia, Renzini & Bergeron 1995, BRB) we have now multiple spectra available for 53 DAs and 31 DBs, none

---

[1] Finley has presented at this workshop two possible very close DD's; rapid rotation, though, may still be the best way to explain their spectra.

of which appears to be a close DD system. The published surveys [2] have several objects in common; all in all, data for 132 WDs (101 DAs, 31 DBs) are available. Note that in these samples *there are* confirmed binaries: the two WD+red dwarf pairs 0034–211 and 0419–487 (BGRD, BRB), and the two DD systems 0135–052 (L 870–2, Saffer, Liebert & Olzsewski 1988) and 0956–666 (BGRD, BRB). To these we must add those WDs which have too low a mass (M $\leq$ 0.4 $M_\odot$) to represent the final outcome of single-star evolution (Bergeron, Saffer & Liebert 1992; BRB). As shown in Fig. 1, they represent about 8-10 % of the observed samples. Apart from the AM CVn - like objects, the DD systems known up to now (more are being detected in the Hamburg Schmidt survey, or in the Common Proper Motion program by Oswalt et al., see this workshop) are just a few. A list of references is given in Tab. 1.

Table 1. Known/suspect DD systems. A: Visual pairs; B: spatially unresolved pairs

Reference	WDs
A: Greenstein 1986	6 visual pairs
A: Sion et al. 1991	15 additional visual pairs, among which L 151-81 A,B (DA+DB)
B: Saffer et al. 1988	0135-052: $M_1 \simeq M_2 \simeq 0.5\ M_\odot$, $P_{orb} = 1^d.5578$
B: BGRD	0957-666: $M_1 = 0.335\ M_\odot$, $P_{orb} = 1^d.15$
B: Bergeron et al. 1990	G4-34; GD 402; GD 387 (unresolved DA + DC)
B: Bergeron et al. 1992	G 62-46 (magnetic DA + DC)
B: Liebert et al. 1993	LB 11146 (DA + magnetic DA)

None of the mentioned known or suspect DD system may be assumed to represent a good example of SN Ia precursor since they are too wide and/or their combined mass is too low. There might still be some margin if sub-Chandrasekhar masses can produce Ia-like explosions via double detonation.

3 The BGRD survey revisited and increased

BGRD presented results for 54 "spectroscopically confirmed DA" WDs; after a fit of all spectra with model atmospheres (BRB) this number decreased since 7 hot subdwarf were misclassified as DAs in the McCook & Sion (1987) Catalog. Apart from that, we felt that the BGRD survey was in some ways still insufficient because: a) only DAs were observed, while there could be the possibility that the strong interactions resulting from a double common-envelope evolution leave preferably a non-DA type WD (Bragaglia et al. 1991); b) several of the WDs were observed only twice, while multiple observations increase the probability of detecting a radial velocity difference due to orbital motion; c) WDs observed only in 1985 lacked a completely satisfying wavelength calibration, even if precise enough to exclude the high values of ΔV_r's expected for the very close systems we were looking for.

---

[2] Tytler & Rubenstein (1989) did not pursue their search. They didn't give particulars, but reported no close DD in their sample of about 120 DAs/hot subdwarfs.

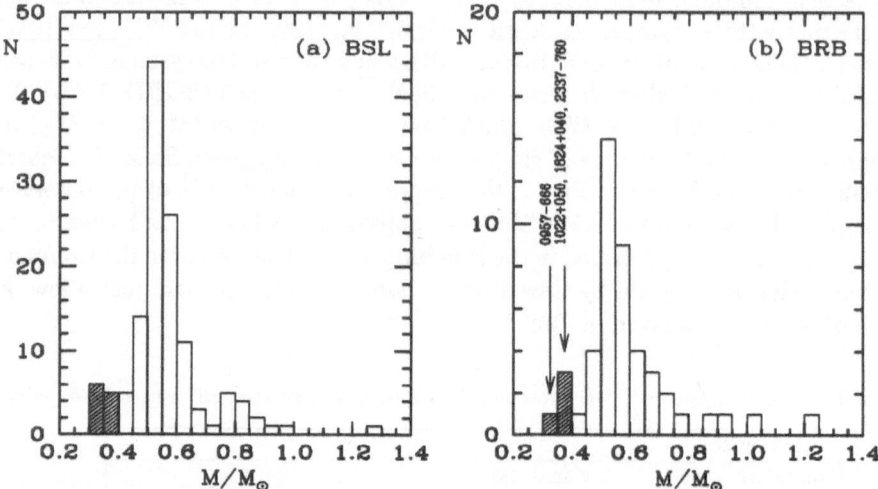

Fig. 1. Distribution of masses for the samples of (a) Bergeron, Saffer & Liebert (BSL, 129 WDs), and (b) Bragaglia, Renzini & Bergeron (BRB, 46 WDs) of DA WDs; WDs lighter than 0.4 $M_\odot$ are shaded

With this in mind, in 1990-1991 we added 31 DB type WDs to our sample; observations were made at the 3.6m ESO telescope, equipped with EFOSC1, at the 1.5m ESO telescope, using a Boller & Chivens spectrograph, and at the 90" of the Steward Observatory at Kitt Peak, also mounting a Boller & Chivens spectrograph. The strategy was the same adopted for the DAs: we took 2 to 5 spectra of each WD, and cross-correlated them to get differences in radial velocities. We did not get any positive DD detection, but 6 WDs showed hints of ΔV_r at the 2σ level (Bragaglia et al. 1993); one of them has been reobserved, and the other are scheduled for further measurements.

In 1993-1994 we reobserved 8 WDs, using the ESO 1.5m telescope. Two of them were "suspect" binaries (1845+019, 2130-047), two have a very low mass (1824+040, 2337-760), two lacked reliable wavelength calibration (0050-332, 1953-011) and finally two had only 2 spectra each (1615-154, 2149+021). We did non detect any ΔV_r in the case of 1615-154, 1953-011, 2130-047 (one of the 6 "suspect" DBs) and 2337-760, a very low mass WD for which we took 9 spectra. Since the wavelength coverage used in the last runs comprises several sky emission lines and bands, we were able to further check the precision and zero point of our wavelength calibration: errors in this can at most account for differences of about 30 kms^{-1} between spectra (in BGRD we estimated our total error as 40 kms^{-1}).

As indicated in Tab. 2, several WDs show some hint of orbital motion. One of them (0346-011, GD50) has very broad and shallow lines, making the values derived very imprecise, and in most other cases there is only one discrepant spectrum showing high relative velocity. In any case *we did not find any ΔV_r*

larger than about 100 kms^{-1}. Furthermore at least the 4 low mass star must be in binary systems (one is the confirmed DD 0957-666). So we can put a firm lower limit of 5% of DDs in our total sample, none a good "candididate progenitor" of type Ia SNe.

Table 2. DA and DB WDs observed for the radial velocity programme from 1985 to 1994. The number of usable spectra for each object is given in parenthesis, along with indication of suspect (RV?) radial velocity differences between spectra. A colon indicates uncertainty. To be added are also the two binaries composed by a DA WD + Red Dwarf, 0034−211 (2) and 0419−789 (15).

Observed DA White Dwarfs			
0047−524 (13)	0050−332 (7, RV?)	0135−052 (3)	0229−481 (2)
0255−205 (2) .	0310−688 (2)	0320−537 (2)	0343−007 (2)
0346−011 (10, RV?:)	0446−789 (3)	0549+158 (2)	0612+177 (3)
0651−020 (2)	0701−587 (3)	0732−427 (2, RV?)	0740−570 (2)
0839−327 (6)	0850−617 (2)	0954−710 (8)	0957−666 (26, DD)
1022+050 (6, RV?)	1042−690 (2)	1052+273 (2)	1053−550 (3)
1236−492 (2)	1257−723 (2)	1323−514 (2)	1407−575 (3)
1422+095 (2)	1451+006 (3)	1524−749 (2)	1555−089 (2)
1615−154 (2)	1620−391 (4)	1709−575 (2)	1824+040 (16, RV?)
1834−781 (2)	1845+019 (12, RV?)	1953−011 (4)	2007−219 (2)
2007−303 (9)	2014−575 (3)	2039−202 (3)	2039−682 (2)
2116−560 (3:)	2149+021 (9, RV?)	2204+071 (1)	2232−575 (2)
2309+105 (2)	2337−760 (11)	2359−434 (2:)	
Observed DB White Dwarfs			
0000−170 (3, RV?)	0002+729 (4, RV?)	0100−068 (5)	0112+104 (4)
0138−559 (2)	0308−565 (4)	0435+410 (3)	0437+138 (3)
0716+404 (5)	0840−262 (4)	0853+163 (4)	1026−056 (2)
1046−017 (2)	1107+265 (2)	1403−010 (2)	1444−006 (1)
1542+182 (3)	1614+270 (2)	1645+325 (4, RV?)	1709+230 (2)
1822+410 (4)	1940+374 (5)	2130−047 (8, RV?:)	2144−079 (4, RV?:)
2147+280 (2)	2222+683 (2)	2229+139 (4)	2234+064 (3, RV?:)
2253−062 (3)	2310+175 (5)	2316−173 (3:)	2328+510 (2)
2354+159 (2)			

4 So what?

We have not made yet any detailed computation of the DD frequency implied by our findings. But, if we have found *zero* DDs likely to later become Ia's (*i.e.* with $P_{orb} < 10^h$), then their frequency from simple Poisson statistics should be less than about 2 %. If the WD space density is about 4×10^{-4} pc^{-3} (Fleming et al. 1986), then $N_{DD}(P< 10^h) < 10^{-3}$ pc $^{-5}$. This has to be compared with the number of DDs needed to reproduce the observed galactic SN Ia rate (0.003

yr^{-1}, van den Bergh & Tammann 1991) $N_{DD,req}(P < 10^h) \sim 10^{-4}$ pc^{-3}. So we seem to be rather short of progenitors.

Should we think to move (back) to "single degenerate" systems? Maybe, but perhaps not all is lost for the DD model for SNIa's if we all have looked at the "wrong" orbital periods, assuming that most DDs form at longer P, then drift to shorter ones due to gravitational wave radiation. *If* they formed already al low P they would evolve quickly and we would find very few of them, if at all. This has implications on the previous evolution through the common envelope phases, since it involves a large orbital shrinkage and, using the current parameterization, implies a value for $\alpha_{CE} < 1$.

See Renzini (1994) for an extended discussion on the contradicting indication that after all α_{CE} may be > 1. Looking at the computations done by Iben (1990) and Yungelson et al. (1994), in the case $\alpha_{CE} = 1$ there should be an almost equal number of DD systems with P between 3 and 10 hours and between 10 hours and 1 day. Instead, all known or suspect DD systems found up to now have rather long periods; since shorter periods are easier to detect, this would favour larger values for α_{CE}. At the moment we are left in this unsatisfactory situation regarding DD systems as SN Ia precursors.

Our future plans comprise reobserving all binary candidates in our sample, computing in detail detection probabilities for the entire sample and deriving accurate space density of close DDs, to be compared with the SN Ia rate for a hopefully final observational test to the DD model for SNe Ia's.

References

Bergeron P., Greenstein J., & Liebert J. 1990, ApJ, 361, 190

Bergeron P., Ruiz M.T., & Legget S.K. 1993, ApJ, 407,733

Bragaglia A., Greggio L., Renzini A., & D'Odorico S. 1991, "Supernovae", ed. G. Woosley, p. 599

Bragaglia A., Greggio L., Renzini A., & D'Odorico S. 1990, ApJ, 365, L 13

Bragaglia A., Greggio L., & Renzini A. 1994, MemSAIt, 65, 411

Bragaglia A., Renzini A.,& Bergeron P. 1995, ApJ, in press

Greenstein J. 1986, AJ, 92, 867

Fleming T.A., Liebert J., & Green RF 1986, ApJ, 308, 176

Foss D., Wade R.A., & Green R.F. 1991, ApJ, 374, 281

Iben I. 1990, ApJ, 353, 215

Iben I. & Tutukov, A.V. 1984, ApJS, 54, 335

Liebert J., Bergeron P., Schmidt G.D., Green R.F., & Saffer R.A. 1993, ApJ, 418, 426

McCook G.P., & Sion E.M. 1987, ApJS, 65, 603

Munari U., & Renzini A. 1992, ApJL, 397, L87

Renzini, A. 1994, IAU Coll. 145, held in Xian, in press

Robinson E.L., & Shafter A.W. 1987, ApJ, 332, 296

Saffer R.A., Liebert J., & Olzewski E.W 1988, ApJ, 334, 947

Tytler D., & Rubenstein E. 1989, IAU Coll. 114, ed. G. Wegner, p. 524

Yungelson L.R., Livio M., Tutukov A.V., & Saffer R.A. 1994, ApJ, 420, 336

Webbink R.F. 1984, ApJ, 277, 355

Iron– and Nickel Abundances of sdO Stars

Stefan Haas[1], Stefan Dreizler[1], Ulrich Heber[1], Thomas Meier[1] and Klaus Werner[2]

[1] Dr. Remeis–Sternwarte, Universität Erlangen–Nürnberg, D-96049 Bamberg, Germany
[2] Institut für Astronomie und Astrophysik der Universität, D-24098 Kiel, Germany

1 Introduction

The sdO stars form a spectroscopic sequence which is very inhomogeneous with respect to their atmospheric parameters. They are spread out over a large area in the $(T_{\mathrm{eff}}\text{-}\log g)$–plane. Their effective temperatures range from 40 000 K up to 90 000 K and the surface gravities from $\log g = 4.0$ to $\log g = 6.5$. The helium abundance as well as the hydrogen abundance show large variations resulting in a classification into several subclasses by Green et al. (1986) and Moehler et al. (1990). Moreover, a large variety in depletion and enrichment of several metals is observed indicating different nuclear evolutions (Dreizler 1993). Regarding the evolutionary status of the sdOs one can distinguish two subtypes: On the one hand the "luminous" sdOs of low surface gravity lying close to the post-AGB tracks, and on the other hand the "compact" ones, i.e. of high surface gravity, situated close to the post-EHB tracks. Several evolutionary schemes have been suggested up to now, but it is still difficult to find one in which the sdOs fit satisfactorily (see Heber (1992) for a review).

To improve this situation we have selected during the last years 23 bright $(V \leq 13\,\mathrm{mag})$ sdOs for quantitative high resolution spectroscopy in the ultraviolet and optical spectral range. Their positions in the $(\log T_{\mathrm{eff}}\text{-}\log g)$–diagram are shown together with theoretical post-AGB and post-EHB evolutionary tracks in Fig. 1. Those stars for which the analysis of iron and nickel lines is under way are marked by filled symbols. A status report is given in section 3.

2 NLTE model atmospheres

Due to the high effective temperatures of the sdO stars non-LTE effects become very important. Thus the recent progress in analysing sdOs goes hand in hand with the development of more sophisticated NLTE-codes during the last years.

The model atmospheres for the analyses presented here were calculated using a NLTE Accelerated Lambda Iteration (ALI) code described in detail by Werner & Dreizler (1995). Basic assumptions are plane-parallel geometry, radiative and hydrostatic equilibrium. The treatment of line blanketing of the iron group elements (Sc–Ni) became possible, when Dreizler & Werner (1993, DW93) adapted Anderson's (1989) method to the ALI-algorithm to construct for each ionization stage a generic ion representing the whole iron group.

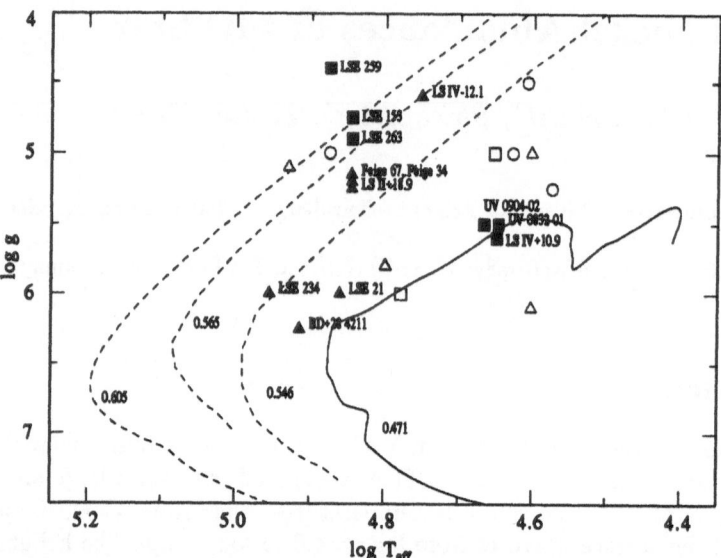

Fig. 1. Positions of the sdO programme stars in the (log T_{eff}-log g)–plane. He-strong, He-weak and hybrid (He- and H-strong) sdOs are marked with squares, triangles and circles, respectively. Theoretical evolutionary tracks are labeled with the respective stellar mass in solar units (Schönberner 1983 dashed lines and Dorman et al. 1993 solid line, a redward loop was omitted for clarity).

In a first step, all real levels (of all iron group elements in the same ionization stage) with similar excitation energy are combined to model bands according to the abundance ratio of the species relative to iron. Within such a band, a Boltzmann distribution for the population of the levels is assumed. To complete the generic ions, radiative and collisional transition data are calculated as described by DW93. The atomic data came from Kurucz' (1991) line list, but were restricted to line transitions between levels identified in laboratory experiments. Each generic ion consists of 6 to 7 bands. In a second step, the number of frequency points representing the complex cross sections is reduced to an appropriate amount ($\sim 10^3$) for model atmosphere calculations by using the opacity sampling technique as described by Anderson (1991) and DW93.

We have implemented two improvements concerning the complexity of the generic ions. First, the complete line list of Kurucz (1991) is included now, i.e. also those theoretically predicted levels as well as all accompanying line transitions are contributing to the model bands and the complex cross sections. We now account for the line blanketing of more than $1.5 \cdot 10^6$ lines of Sc through Ni in 5 ionization stages. Second, we apply radiative bound-free cross sections of Fe provided by the Opacity Project instead of using hydrogenic ones.

Furthermore – as an alternative way to reduce the huge number of frequency points of the complex cross sections – the construction of distribution functions is optional at our disposal. Since the correlation of overlapping distributions constitutes a serious problem inherent to this method, we scatter the steps of the step functions randomly over the respective frequency intervals following the idea outlined by Anderson (1989). NLTE calculations for checking this approach and a direct comparison to the opacity sampling technique are under way.

The influence of the approximations in our approach on the atmospheric structure and observable quantities (line profiles, UV and EUV fluxes) was studied by creating two additional generic ions. The first one was used to test the influence of the number of bands by increasing it from 7 to 20. The second one allows the validity of merging all iron group elements into one generic ion to be tested by creating a separate model atom for nickel, the second most abundant element within the iron group in the sun. The test calculations were carried out for $T_{eff} = 90\,000$ K and $\log g = 6.0$ with each of our four generic model atoms. Within the iron group the abundance ratios relative to Fe were always kept at the solar values. The calculations with the separate Ni model atom were performed with Ni/Fe = 1, in order to study even extreme abundance ratios which are reported for the sdO star Feige 67 (Becker & Butler 1994). Detailed standard model atoms for H and He as well as simplified versions of our C, N, O models (Werner & Heber 1991) were included. Abundances of He, C, N and O were kept solar and [Fe/H] = 1 and [Fe/H] = 10, respectively, was used. As an example of our results, we depict in Fig. 2 the ionization structures of the generic ions consisting of 20 bands (left hand side) and those of the separate Ni ions together with those of the generic ions representing the remaining iron group elements (right hand side). This comparison indicates that the ionization structure of the different model atoms has nearly the same shape for all ions. Thus the usage of generic ions representing the whole iron group is justified to account for the effects of the metal line blanketing on the atmospheric structures. However, due to the differences of the ionization fractions between the Ni and the generic ions (up to a factor ~ 10), it seems reasonable to apply single model atoms for line formation calculations and abundance determinations – at least for Fe and Ni, the both most abundant iron group elements – to achieve more reliable data. The increase in numbers of bands has negligible effects on the fluxes. A detailed description of the tests and the results will be published elsewhere (Haas et al. in prep.).

3 Abundance determination

ESO-CASPEC spectra as well as IUE-SWP high resolution spectra are available for all of our programme stars. The IUE spectra, which were used for the determination of iron and nickel abundances presented here, have been carefully re-reduced with STARLINK–IUEDR software.

The calculations of the theoretical spectra for each star were performed in two steps. First, we computed the atmospheric structure by applying the generic ions which contain the complete line list of Kurucz (1991) to account for the blanketing effect of the iron group elements. Second, line formation calculations were performed with the Ni ions and the generic ions consisting of the remaining iron group elements. The basic atmospheric parameters (T_{eff} and $\log g$) were taken from previous analyses carried out by various authors (for references see Table 1), but also models for higher and lower temperatures have been computed. For

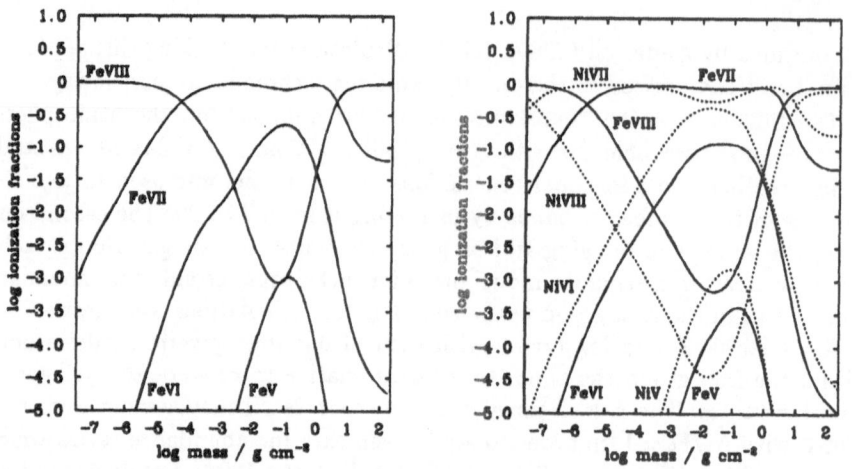

Fig. 2. Ionization structures of different iron group model atoms for $T_{eff} = 90\,000$ K and $\log g = 6.0$. **Left:** Generic ions with 20 bands. **Right:** Ni ions (dotted lines) versus generic ions representing the remaining iron group elements (solid lines).

each parameter set we varied the [Fe/H]-ratio, but kept Fe/Ni = 1 fixed. However, to achieve reliable estimates of the nickel abundances one has to change the Fe/Ni ratio, too. These calculations are in progress. Thus, we confine ourselves to present estimates of the iron abundances together with upper limits of the Fe/Ni ratios in Table 1. Since our iron group model atoms are based on a sta-

Table 1. Preliminary iron abundances in solar units and estimated upper limits of the Fe/Ni ratios. The solar Fe/Ni ratio is 20.

star	type	T_{eff}	$\log g$	[Fe/H]	Fe/Ni	references (T_{eff}, $\log g$)
BD +28°4211	He–weak	82 000	6.2	0.1	≤ 20	Napiwotzki (1993)
LS IV −12°1	He–weak	56 500	4.55	$1 \ldots 5 \cdot 10^{-3}$		Dreizler (1991)
LS II +18°9	He–weak	70 000	5.2	10	~ 3	this paper
Feige 67	He–weak	70 000	5.2	10	~ 3	Bauer & Husfeld (1994)
Feige 34	He–weak	70 000	5.2	10	~ 3	Becker & Butler (1994)
LSE 21	He–weak	72 500	6.0	$0.1 \ldots 0.5$	~ 20	this paper
LSE 234	He–weak	90 000	6.0	1	~ 20	this paper
LSE 153	He–strong	70 000	4.75	$0.1 \ldots 1.0$		Husfeld et al. (1989)
LSE 263	He–strong	70 000	4.9	$1 \ldots 3$		Husfeld et al. (1989)
LSE 259	He–strong	75 000	4.4			Husfeld et al. (1989)
UV 0904–02	He–strong	46 500	5.55	0.1	> 9	Dreizler (1993)
UV 0832–01	He–strong	44 500	5.55	0.5	~ 20	Dreizler (1993)
LS IV +10°9	He–strong	44 500	5.55	0.5	~ 20	Dreizler (1993)

tistical approach we didn't rely on single iron or nickel lines for the abundance estimates but compared the line strengths of all lines visible in the UV spectra with calculated ones. To demonstrate the quality of our computed UV spectra, we display in Fig. 3 a comparison with an observed spectrum of BD +28°4211. Note that we find a subsolar iron abundance ([Fe/H] = 0.1), whereas DW93 fitted the same spectrum with a solar iron abundance. The difference can be traced back to the improved T_{eff} determination by Napiwotzki (1993).

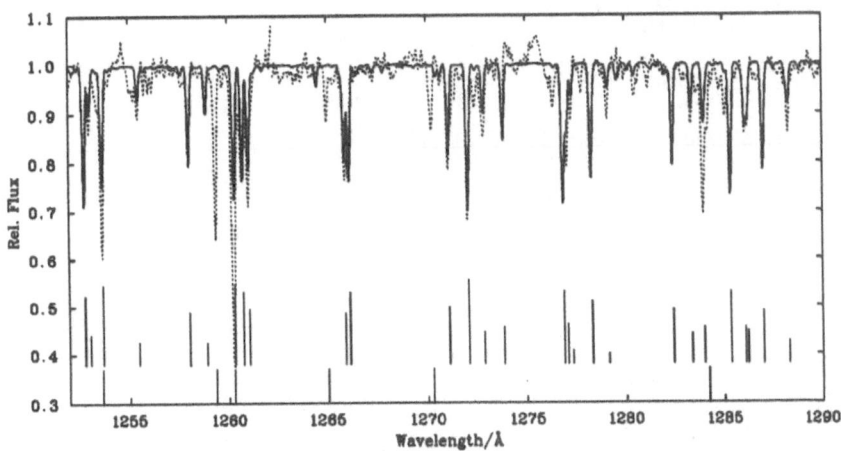

Fig. 3. Part of the UV spectrum of BD +28°4211 (dotted curve, coaddition of 30 high resolution SWP spectra), compared with the synthetic spectrum (solid curve, folded with a Gaussian, FWHM=0.15 Å) from a line blanketed NLTE model ($T_{eff} = 82\,000$ K, $\log g = 6.2$ and [Fe/H]=0.1). Positions of Fe VI lines are marked by solid lines with heights corresponding to the $\log gf$-value. Interstellar S II and Si II lines as well as Ni IV lines are marked by solid lines below.

Acknowledgement:
We acknowledge DFG support through grants He1356/17-1 and We1312/6-1.

References

Anderson L.S. 1989, ApJ 339, 558
Anderson L.S. 1991, NATO ASI Series C, Vol.341, Kluwer, p. 29
Bauer F., Husfeld D. 1994, A&A in press
Becker S.R., Butler K. 1994, A&A in press
Dorman B., Rood R.T., O'Connell R.W. 1993, ApJ 419, 596
Dreizler S. 1991, PhD thesis, University Kiel
Dreizler S. 1993, A&A 273, 212
Dreizler S., Werner K. 1993, A&A 278, 199 (DW93)
Green R.F., Schmidt M., Liebert J. 1986, ApJ Suppl. 61, 305
Heber U. 1992, Springer, Lecture Notes in Physics 401, p. 233
Husfeld D., Butler K., Heber U., Drilling J.S. 1989, A&A 222, 150
Kurucz R.L. 1991, NATO ASI Series C, Vol.341, Kluwer, p. 441
Moehler S., Richtler T., de Boer K.S., Dettmar R.J., Heber U. 1990, A&AS 86, 53
Napiwotzki R. 1993, PhD thesis, University Kiel
Schönberner D. 1983, ApJ 272, 708
Werner K., Heber U. 1991, NATO ASI Series C, Vol.341, Kluwer, p. 341
Werner K., Dreizler S. 1995, "Model Atmospheres" in Computational Astrophysics II,
    eds. R.P. Kudritzki, D. Mihalas, K. Nomoto, F.-K. Thielemann, submitted

The HST White Dwarf Project: V 471 Tauri and Procyon B

Harry L. Shipman

University of Delaware, Newark, DE 19711 USA

1 Introduction

Almost exactly one decade ago, we met for the Fifth European Workshop on White Dwarf Stars in Kiel. At that time, much discussion about the Hubble Space Telescope (HST) emphasized extragalactic research and "key projects". A number of us who attended the Kiel workshop in 1984 felt that our small science approach to white dwarf stars was in danger of being squeezed out of the HST's science program. Consequently we decided to form a consortium and apply for HST observing time as a group. The consortium eventually gathered an accretion disk of 20-plus members. We thus became practioners of another way of doing science, described here as "mass science".

Ten years later, the results are beginning to emerge. Since much data and interpretation will be published elsewhere (Provencal and Shipman 1995; Shipman et al. 1995a, 1995b, 1995c), only a brief summary of one of the highlights is presented here, to give readers the flavor of the kind of white dwarf science we can do with HST. In this paper I will provide some reflections on the style of doing science which emerged from the consortium approach, and its strengths and weaknesses, as a follow-up to the project description presented in Toulouse some years ago (Shipman 1991). This paper also contains a result, presented at the conference, which shows what our proposed observations of Procyon B can do.

2 The HST White Dwarf Consortium

Is mass science a good way of doing science? In the past several years a trend towards larger collaborations has developed. Science managers in the U.S. have tended to encourage astronomers to band together in large teams. In some cases, a multiwavelength approach is advocated; entire conferences have been devoted to the topic of multiwavelength astrophysics. From my perspective, I think that mass science has some significant weaknesses and I do not think it should become a standard way of doing business.

The mass-science approach did have some value in planning an initial set of observations of white dwarf stars with a new spacecraft and a new capability. Experience has shown that small- science telescope proposals generally are rather specifically targeted to one star or a small group of stars. I planned the initial two years of HST observations of white dwarfs as including a reasonable variety of target types. It was not easy to determine which observations would pay off rather nicely and which sort of observations would simply produce small increments in the science which had been explored previously with IUE. What's happened is that some of our targets turned out to be scientifically very rich – V471 Tau, described later in this paper, is an example. Other observations were not as productive. For example, the garden- variety DB star LDS 749 B remains a garden-variety DB star, even after we observed it with the power of HST. Because we observed a large variety of stars, the mass science approach seems to have produced some rather valuable guides as to what sort of follow-up observations will be most successful. Perhaps we also helped everyone in the community convince telescope allocation committees that HST could do some good white dwarf science.

Our mass science approach also succeeded in garnering a 25.5 hours of space-craft time, approximately 1 for our field. Whether a small science approach would have obtained the same amount of resources is a question which is difficult to answer. Would a dozen individual proposers have obtained the same amount of resources to devote to white dwarf stars? We can't experiment. We can't go back to 1985 and find out whether a different route would have led to an allocation of 2.5 spacecraft hours, 250 spacecraft hours, or to the 25.5 spacecraft hours which we actually had. I believe that it is quite possible that the astronomical landscape in 1985 was such that a small science approach would have failed. However, circumstances are different now. Our team proposal failed to win approval in cycle 3. Now that we know what HST can do, a small-science approach seems to make more sense. I also think that the climate in all of astrophysics has changed and I would recommend a small-science approach to a new instrument.

However, there are some distinct disadvantages in the execution of the mass-science approach. Precisely because we had chosen to observe a wide variety of stars, there were many targets on the list which I had never worked on and which I was not intimately familiar with. But as Principal Investigator (PI), I had to make a number of decisions which might better have been made by others. For example, the failure of Side 1 of the GHRS detectors created a big problem. Many of our original observing plans used the low-resolution grating of GHRS and the side 1 detector so that one exposure could obtain data on nearly 300 A of the spectrum. The two alternatives were to either use the FOS, which provides us with IUE-level resolution, or to try to pick the right 30-A spectral windows and use the higher resolution capabilities of the GHRS. Which wavelengths should we pick? In some cases team members had provided reasonably specific suggestions. In others I simply had to do the best that I could.

A second disadvantage appeared in the data-reduction stage. HST is a sufficiently complex spacecraft that it seemed to make sense to have one person

– usually my postdoc – do most of the work in data reduction. We had agreed
that our first papers would be a preliminary reconaissance of most of the cycle
1 data. This job took longer than would have been desirable. For example, my
postdoc Scott Roby came to me with spectra which apparently had absorption
lines in them with flat bottoms and square sides. Neither of us believed that
such features could be real – and as it turned out, they weren't. We and other
HST investigators discovered the badly calibrated diodes in the FOS which, because each FOS diode samples four wavelengths, produces absorption or emission
features with boxcar profiles. Because of the volume of data which we had do
deal with, these papers took a fair amount of time to emerge (Shipman et al.
1995a, 1995b). The HST data- reduction process has now become considerably
simplified, but that didn't help us in cycle 1.

While we can't rewrite history, I believe that our experience with the mass-
science approach to doing astronomy should be taken into consideration if any
similar approach, with a future instrument, is considered. It might sound good
at the outset, but there are a number of problems along the way, at least for this
type of project.

I now turn to one of the highlights of the HST observations of white dwarf
stars in cycles 1 and 2.

3 V 471 Tauri

For well over a decade, some of us on the HST team have sought evidence for
the accretion of material onto the white dwarf component of the variable star V
471 Tauri. V 471 is a member of a very small class of close but detached binaries
(Vennes and Thorstensen 1994) in which a late-type main sequence star orbits
very close to a white dwarf companion. Other members of this class include Feige
24, EUVE0720-317), and several others (see Barstow et al. 1993 for some more
examples of systems). This class of stars is interesting because it can serve as
an accretion laboratory; the signatures in these weakly accreting systems may
provide more direct indications of accretion dynamics than in systems where
accretion rates are greater. In addition, these systems may prove to be the key
to our understanding of the origin of cataclysmic variables.

The HST observations of V 471 Tau have been briefly described in Shipman
et al. (1995a) and will be presented in more detail in Shipman, Schieble, Roby,
and Sion (1995c). Since these publications exist, I will simply summarize the
results here. We observed the C IV 1548+1550 A doublet, the Si IV 1394+1402
A doublet, and the He II Balmer alpha line at 1640 A. Our original intent was to
obtain data on the entire 300-A spectral range from 1200-1800 A in two GHRS
observations; the failure of side 1 of the detector meant we had to choose some
spectral regions to observe, and we were lucky. We made sixteen 58.8 second
integrations, thus obtaining eight observations for each rotational phase of the
white dwarf star. Our observations were obtained at orbital phase 0.75 so that the
contamination of our spectra by the wind from the K dwarf would be minimized.

The time-averaged spectra clearly show the existence of a strong emission feature at C IV, a weak absorption feature at Si IV, and broad, diffuse emission at He II. When the spectra are separated into time-resolved integrations taken at different rotational phases of the white dwarf, we can begin to determine just where it is that the accretion is taking place. A naive model of the system, supported by the dark-pole model of the X-ray variations, would suggest that most accreting material is associated with the poles of the white dwarf.

The time-resolved spectra support this interpretation, to some extent, though there are some considerable differences from a naive expectation of photospheric absorption lines at the poles and pure hydrogen atmospheres at the equator. The C IV emission features are strongest when the magnetic equator faces us and are seen at rotational phases 0.19, 0.30, and 0.74. (The dark, magnetic poles face our line of sight at X-ray phases 0.0 and 0.5.) While at first glance this might seem surprising, the velocity of this emission (-80 km/s) is close to the rotational velocity of the white dwarf, suggesting that the emission feature comes from gas located just above the accreting pole. The Si IV data show what we expected, an absorption doublet seen predominantly when the dominant accretion pole is face-on, near phase zero. The He II data shows emission at only two phases, 0.22 and 0.33, with negative velocities.

Further HST observations of this system would be highly desirable. A really interesting question is whether the features we see are persistent with time, or whether they vary. Ground based studies suggest that the intrabinary H-alpha lines seen in this system show variations on time scales varying from months to years.

4 Procyon B

Procyon B has been one of the most fascinating and one of the most elusive white dwarf stars. It was discovered in the nineteenth century as part of the "astronomy of the invisible", detailed measurements of nearby binary stars which led to the discoveries of many stellar companions including Sirius B. A precise mass determination by Strand (1951) demonstrated that this star was sufficiently massive and too faint to be a main sequence star. Subsequently, Irwin et al. (1992) have measured a much more precise mass. Procyon B's mass is one of the best-determined white dwarf masses.

But besides its mass, little else is known about Procyon B. It is about 10,000 times fainter than Procyon A, and most of the time the separation is a minimal 2". Eggen and Greenstein's (1965) classic tabulation of the first comprehensive list of white dwarf stars lists Procyon B as a white dwarf and provides a rather uncertain visual magnitude of 10.7. Various observers have tried to obtain better data from the ground, and until recently success was elusive. Consequently we made an effort to observe this star during cycle 1 of our HST project. The spherical aberration dispersed the signal from Procyon B among a sufficiently large number of pixels in the WF/PC so that we were not able to identify

Procyon B in the WF/PC frames. A second observation is scheduled with the refurbished telescope.

But in the meantime, ground-based observers have developed better instruments which can handle strong contrasts. Walker et al. (1994) used the High-Resolution Camera and the CFHT to obtain V=11.3, B-V = +0.26, and B-I = +0.62. Unfortunately they do not quote any error bars. Based on the number of significant figures quoted in their result, the non-standard nature of their photometric bands, and past experience in analyzing photometry of this sort, I assume errors of 0.1 mag for this analysis.

I first determine the temperature and radius of Procyon B from these data. A temperature scale set by stars in this temperature range with known T(eff) has been determined by Bergeron (1988) and by Bergeron et al. (1995). The B-V color leads to T(eff) = 7,800 K with an estimated uncertainty of 700 K. At this level of precision, it does not matter that we do not know whether Procyon B is a DA or DZ star. The indicated radius in solar units is log R = -2.00 +/- 0.12, consistent at the 1-sigma level with the value of -1.88 expected from the mass-radius relation.

This result is significant for two different areas of astrophysics. First, as our host Volker Weidemann pointed out in a lecture in the last Kiel workshop in 1984, our basic understanding of white dwarf stars rests on a mass-radius relation which has very few data points to calibrate it with. No one seriously expects the theory of stellar degeneracy to be wrong, but most theories of such stature are based on a richer observational comparison than we have obtained so far. What's at stake is far more than Chandresekhar's Nobel Prize (which he deserves for many astrophysical achievements in addition to the theory of stellar degeneracy!). So far, the Nobel Prize is still safe.

We can also use these results to calibrate white dwarf cooling ages. Claver (1995) discusses this topic in much more detail in the context of his observations of star clusters. Procyon A+B constitute a "cluster" with two members, and a known age of 1.5 - 2 Gyr based on the observation that Procyon A is leaving the main sequence. Wood's (1992) most recent models give a white dwarf cooling age of 1.6 - 2.5 Gyr. Thus the Procyon B results confirm, in broad terms, the cooling ages of white dwarf stars.

Still another result, which should be interpreted considerably more cautiously, is that we can use these data to determine the mass of Procyon B's progenitor. The comparison between the cooling age of Procyon B and the total lifetime of the system suggests that the nuclear burning lifetime of Procyon B's progenitor is less than 0.5 Gyr, which suggests that the progenitor mass exceeds 2.5 solar masses.

These results are certainly very preliminary and should be considered as illustrations of what can be learned from more precise photometry and spectroscopy of Procyon B.

I thank Lauretta Nagel of the Space Telescope Science Institute, the Technical Associate who is associated with this program, for all of her help in translating our scientific needs into the rather complex set of instructions which need to be uploaded to HST. We also thank Hugh Van Horn for his encouragement throughout this ten-year initiative. Support for this work was provided by NASA through grant numbers GO-2593.01-87A and GO- 3816.01-87A from the Space Telescope Science Institute. The National Science Foundation's support of H.S. during the long planning stage of this project in the 1980s was invaluable in bringing this project to fruition.

References

Barstow, M., et al., 1993, in White Dwarfs: Advances in Observation and Theory, M.A. Barstow ed., NATO ASI C Vol 403, Kluwer, p. 433.

Bergeron, P. 1988, Ph.D. Thesis, Universite de Montreal.

Bergeron, P., Saumon, D., and Wesemael, F. 1995, ApJ (in press).

Claver, C.F., and Winget, D.E. 1995, this volume.

Eggen, O.J., and Greenstein, J.L. 1965, ApJ, 141, 83.

Irwin, A.W., Fletcher, J.M., Yang, S.L.S., Walker, G.A.H., and Goodenough, C. 1992, PASP, 104, 1992

Provencal, J., and Shipman, H. 1995, this volume.

Shipman, H.L. 1991. In: White Dwarfs, Vauclair, G. and Sion, E. (eds.), Kluwer, Dordrecht, p. 369-378.

Shipman, H., et al. 1995a, AJ, (in press; scheduled for March 1995)

Shipman, H., et al. 1995b, AJ, (in press; scheduled for March 1995)

Shipman, H., Schieble, M., Roby, S., and Sion, E. 1995c, manuscript circulating among co-authors.

Strand, K. Aa. 1951, ApJ, 113, 1.

Vennes, S., and Thorstensen, J.R. 1994, Astrophys.J. Letters, in press.

Walker, G.A.H., Walker, A.R., Racine, R., Fletcher, J.M., and McClure, R.D. 1994, PASP, 108, 356.

Wood, M. 1992, ApJ 386, 539.

The Interacting Binary White Dwarfs

J. L. Provencal

Department of Physics and Astronomy, University of Delaware, Newark, DE 19716

1 Introduction

Interacting binary white dwarfs (IBWDs) are a special class of ultra-short period binaries, which provide an exciting avenue to explore nucleosynthesis, binary star evolution, and stellar structure. IBWD models contain two helium white dwarfs of extreme mass ratio (Faulkner, Flannery & Warner 1972, hereafter FFW). The less massive but more distended component, an evolved object, is transferring material to the primary via an accretion disk, allowing us to directly view the processed byproduct of stellar nucleosynthesis and constrain models of stellar nuclear reactions (Marsh & Horne 1991). IBWDs mark an endpoint of binary star evolution, and can unravel the history of binary systems, much as solitary white dwarfs represent the end of single star evolution. Also, the IBWDs lie in a temperature range, $\approx 25000 \, K$ (Patterson et al. 1992), where helium is partially ionized and provides an efficient mechanism for driving pulsations (Winget & Fontaine 1982). If we can find even one pulsating white dwarf within a binary system, we can use asteroseimological techniques (Winget et al. 1991) to unlock its structural secrets.

At present, six objects form the IBWD family: the notorious prototype AM CVn (HZ-29), EC1533-1403, recently discovered during the Edinburgh-Cape Survey, CR Boo (PG1346+082), V803 Cen, CP Eri, and GP Com (G61-29). Several characteristics are shared by all family members. The first, and most important, is a complete lack of hydrogen. The spectra of all known IBWDs are dominated by broad lines, either in emission or absorption, of neutral helium. All six are short period photometric variables, ranging from 3000 to 200 seconds. CR Boo, V803 Cen and CP Eri also undergo large 3-4 magnitude outbursts on a timescale of several days. Each IBWD's characteristics are summarized in Table 1.

Despite years of investigation, many questions concerning IBWDs remain. Just one, GP Com, has had its orbital period unambiguously identified. Only circumstantial evidence supports the binary nature of the others. We do not understand the mechanism responsible for the photometric variations, although we assume they represent the orbital periods, where within the system the vari-

Table 1. The Interacting Binary White Dwarf Systems

Name	Temp (K)	Magnitude (B)	Period (s)	Amp (mmi)	Spectrum
AM CVn	25000	13.9	1011.4	11-< 3	absorption
			525.6	11	
			350.4	3.2	
EC15330-1403		13.6	1119	≈18	absorption
PG1346	20000(h)	13.6-17.3	1471.3	55(l) < 5(h)	absorption (h)
			1492.8	32(l) < 7(h)	weak emission (l)
V803 Cen	20000(h)	13.4-17.4	1600	20	absorption (h)
	15000(l)		175		emission (l)
CP Eri		≈ 16.7 − 21	1800	68(l) 16(h)	absorption (h)
G61-29	≤10000 (disk)	15.8	2790	< 3	emission

ations arise, or what the timescale for period change is. These questions must be considered before we can understand the IBWD's place in stellar evolution and white dwarf formation.

2 The IBWDs

IBWDs can be further subdivided into two groups: those which demonstrate dwarf-novae like eruptions, and those that do not. Space is limited, so I cannot discuss each IBWD, but I will highlight one from each group.

AM CVn is the best studied, and perhaps most controversial, IBWD (see Provencal et al. 1994 and Patterson et al. 1993 for overviews). Since its discovery, debate has brewed over the actual periodicities present in the light curve and their long term stability. Our overriding goal is the precise identification of exactly what frequencies are detected in the Fourier transforms (FTs), in hopes of using the photometric evidence to choose whether 951.3 or 1902.5 μHz represents a physical process occurring in AM CVn, and if a stable frequency exists in this system.

With the exception of the 988.7 μHz peak, AM CVn's FT is dominated by a remarkably stable distribution of power, at 1902.5, 2853.8, 3805.2, 4756.5 and 5707.8 μHz (fig. 1). Each frequency, except 988.7 μHz, is a multiple of 951.3 μHz, but no power is detected, to a limit of 1 mma, at 951.3 μHz itself. In 1990, AM CVn was the prime target of the Whole Earth Telescope (Nather et al. 1990), eliminating alias artifacts due to periodic gaps and improving the FT's noise level. We uncovered new low amplitude power at 654 and 1249 μHz, and found fine structure, with a frequency splitting of 20.77 μHz, associated with the 1902.5 and 2853.8 μHz peaks.

The 20.77 μHz fine structure's discovery allows us to address the long term stability of AM CVn's photometric variations. We created an O-C diagram for the 1902.5 μHz peak, taking the fine structure into account (Provencal et al. 1994). After much trial and tribulation, discussed in appropriate detail in

Provencal et al. (1994) we found $\frac{\dot{P}}{P} = +1.71 \times 10^{-11}$ s s^{-1}, corresponding to a timescale of 10^6 years. This $\dot{P}$ is much larger than predicted by gravitational radiation (FFW).

AM CVn remains a complicated object. The new power at 654 and 1249 μHz, the fine structure associated with only a few of the dominant peaks, and the lack of power at 951.3 μHz warns that interpreting the dominant peaks as a series of harmonics of an unseen fundamental is too simplistic. Something more complex is going on in this system. The 988.7 μHz power's amplitude variability suggests it arises via a different mechanism than the 1902.5 μHz frequency. We believe 988.7 μHz is the best candidate for the orbital period, while the 1902.5 μHz and associated peaks arise on the mass accretor, either via rotation or pulsation.

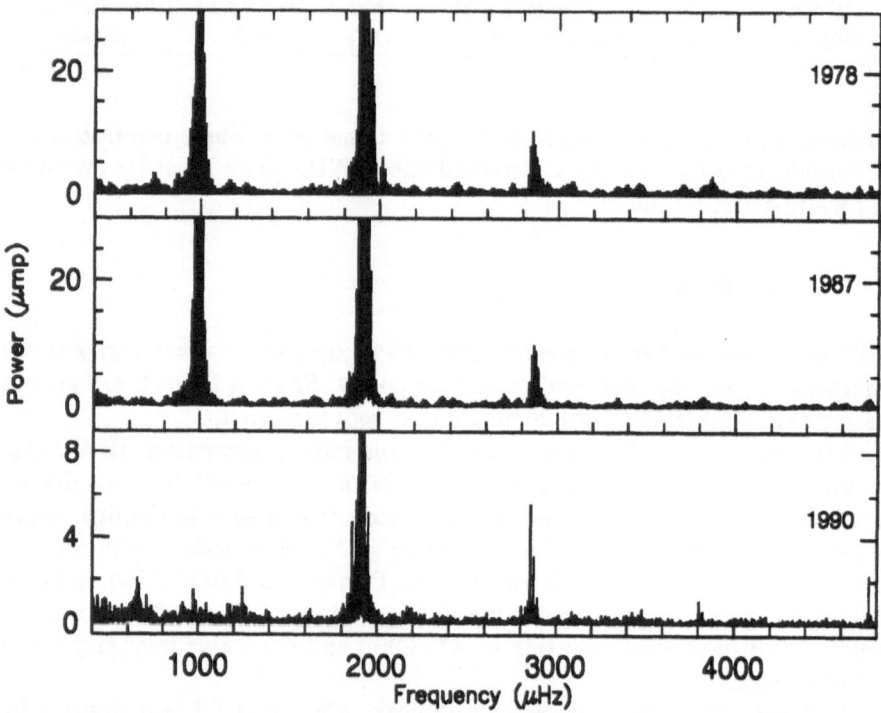

Fig. 1. Three seasonal FTs of AM CVn. The dominant power is at 988.7, 1902.6, 2853.9, 3813.9, 4756.4, and 988.7 μHz. Note changes in y-scale to accommodate the dynamic range of each FT.

In 1988, the eruptive IBWD CR Boo was the very first WET target. We were interested in the short period variations at 1490 μHz reported by Wood et al. (1987), but the large magnitude outbursts presented reduction and analysis obstacles, which forced us to divide the data into two groups by magnitude (Provencal et al. 1994b). The low state (dim) FT (fig. 2) is dominated by power at 679.670 ±0.004μHz and 669.887±0.008μHz. We also find power at

1359.58±0.01μHz, the first harmonic of 679.670 μHz. There is no evidence of a harmonic of 669.89 μHz.

The high state (bright) power spectrum contains an unresolved band centered at 673 μHz and a series of higher order harmonics. We find two dominant peaks, at 672.99±0.04 μHz and 686.12 μHz. Because these two peaks are aliases of each other, we cannot determine which is real, or if additional frequencies are present. However, neither of these frequencies match the observed low state power, and we place upper limits of about 5 and 7 mma for the low state 679 and 669 μHz peaks, respectively.

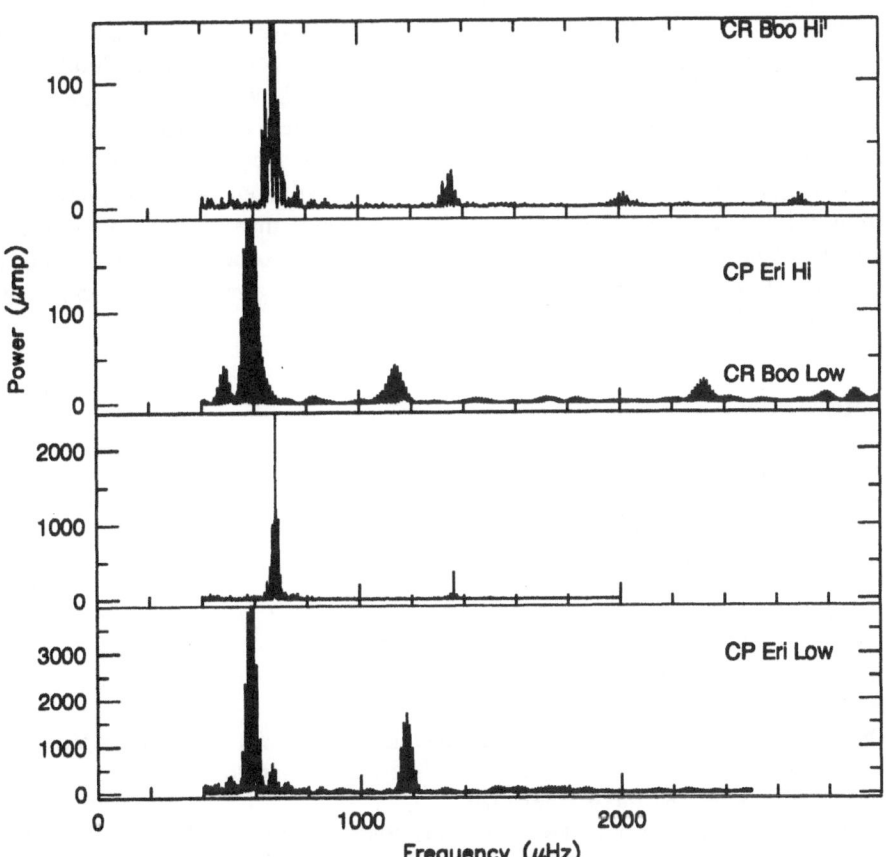

Fig. 2. PG1346+082 and CP Eri in high state and low state. The high state FTs are dominated by a fundamental and a series of harmonics, while the low state power spectra contain only a fundamental and one harmonic. Note changes in y-scale to accommodate the dynamic range of each FT.

Figure 2 also compares the behavior of two IBWDs in high and low state. Both display a complex fundamental frequency and a series of harmonics in high state, but a slightly different fundamental frequency and first harmonic in low state. The high frequency power's amplitude variability closely mimics a source at constant absolute brightness seen against a fluctuating background (Provencal et al. 1994b). It is fair to conclude that the higher order harmonics are associated with the outburst, perhaps arising in the disk. Clearly, similar mechanisms are at work in these objects.

3 Conclusions

IBWDs are complicated, with much work still remaining to be done. We need accurate models of helium disks in an IBWD environment. We do not know if such disks are stable over long periods, or if tidal forces always create elliptical, precessing disks. There seems to be a behavioral trend with photometric period, but our sample is small. We must find more IBWD systems, although there is no simple criteria which set these objects apart from solitary DBs. Finally, we must explore the past and future history of IBWDs. The secondary is believed to be ≈ 0.04 solar masses. If this star followed normal main sequence evolution, we should be well into the carbon/oxygen core. Finally, recent evidence (Clemens 1994) suggests that different white dwarf flavors (hydrogen vs helium) have different origins. Because the secondary is degenerate, mass transfer in IBWDs should not stop until the secondary is completely absorbed, leaving behind a solitary helium white dwarf. IBWDs could be the progenitors of a significant fraction of field helium white dwarfs.

References

Clemens, J. C.: Ph.D. thesis (1993) University of Texas
Faulkner, J., Flannery, B. P., & Warner, B.: ApJ **175** (1972) L79
Marsh, T. R., & Horne, K.: ApJ **366** (1991) 535
Nather, R. E., Winget, D. E., Clemens, J. C., Hansen, C. J., Hine, B. P.: ApJ **361** (1990) 309
Patterson, J., Sterner, E., Halpern, J. P., Raymond, J. C.: ApJ **384** (1992) 234
Patterson, J., Halpern, J., & Shambrook, A.: ApJ **419** 803
Provencal, J. L., et al.: ApJ *submitted* (1994)
Provencal, J. L., et al.: *in preparation* (1994b)
Winget, D. E., & Fontaine, G.: in Pulsations in Classical and Cataclysmic Variable Stars, ed. J. P. Cox & C. J. Hansen (Boulder:Univ. Colorado Press), 46
Winget, D.E., et al.: ApJ **378** (1991) 326
Wood, M. A., Winget, D. E., Nather, R. E., Hessman, Frederic V., Liebert, J., Kurtz, D. W., Wesemael, F., & Wegner, G.: ApJ **313** (1987) 757

GD1401 and GD984: X-ray Binaries with Degenerate Components

Irmela Bues and Turgut Aslan

Dr.Remeis-Sternwarte Bamberg, Sternwartstr. 7, D-96049 Bamberg

1 Introduction

During a spectrophotometric survey for blue stars in the solar neighbourhood (Rupprecht 1983; Rupprecht, Bues 1983) two very blue objects, GD984 ($m_V = 13.^m93, \mu = 0.^"14/y$) and GD1401 ($m_V = 14.^m51, \mu = 0.^"20/y$), with a positive (R-I)-index were detected. In the two colour diagrams (U-B)/ (B-V) and (u-b)/(b-y) both objects are located very close to the black body line, GD984 in the extremely hot WD-region, GD1401 in the region of cataclysmic variables. In the combined two colour diagrams (U-V)/(R-I) and (u-b)/(R-I) the positions are shifted far below the WD region, thus suggesting the binary structure of both objects. That is why we investigated these stars with photometry and spectra since 1982 to check on variability in colours and spectral features. Preliminary results for GD1401 were obtained by Müller (1989).

Data in the EUV and X-ray region were taken by ROSAT in September 1990 during its All-Sky-Survey (Barstow et al.,1993) and are listed in Table 1. Our investigation with optical and IUE spectra started independently. The results are compared to the X-ray data afterwards.

Table 1. Fluxes (counts/sec) for GD984 and GD1401 measured by ROSAT

object	PSPC	S1	S2
GD984	1.745 ± 0.195	0.230 ± 0.013	0.612 ± 0.022
GD1401		0.012	0.023 ± 0.007

2 Analysis of GD1401

During 3 observing periods of 5 nights in 1981, 1985 and 1987 colour measurements in the UBVRI, Strömgren and JHK-system have been obtained with the ESO 1m telescope. In addition, 120 blue plates of the Bamberg Southern Sky Archive have been investigated for variability. On a long term scale, a variability in the blue of $0.^m3$ is likely, whereas the ESO observations show $0.^m1$ for the V magnitude, $0.^m3$ in the y magnitude; (R-I) remains constant at $0.^m78$ as well as (H-K) at $0.^m32$. The latter values definitely belong to a very cool star, later than K7.

Spectra in the blue (3800-5100Å), taken in 1986, 1987, 1990 and 1993 with the ESO 1.52m telescope, show DA structure with Balmer lines in absorption. 5 IUE low dispersion spectra also give no hint of any interaction between the stars, a pure continuum with strong Lyα- absorption is observed. Red spectra (1990 and 1993) around the Hα position, yield weak metallic line and band features together with Hα in absorption only.

A model atmosphere analysis of the IUE and blue spectra, including line profile calculations, gave the best fit for a DA of $T_{eff} = (21000 \pm 1000)$K, log $g = 7.0 \pm 0.25, \epsilon_{He} < 10^{-3}$. With the colour measurements in V, the contribution of the secondary being $0.^m1$ in V, and a computed bolometric correction, $M_V = 10.^m68$ and the distance amounts to 65 pc. With these values the observed X-ray flux cannot be reproduced for reasonable values of the HI column density of the interstellar medium. It should be due to the secondary, for which our comparison of colours and fluxes with computations by Allard et al.(1992) results in an M4V star, although no emission components have been observed in the visible.

3 Analysis of GD984

24 spectra were taken in the years 1986 to 1994 with the ESO 1.52m telescope (+ B&C + CCD) in the blue and red region, as well as 7 UV-spectra (low dispersion) with the IUE satellite. Magnitudes and colours for GD984 were observed in the Johnson UBVRI- and Strömgren uvby-system with the ESO 1m telescope. (R-I) varies by 0.06 magnitudes.

The optical spectrum of GD984 in Fig.1 shows broad Balmer lines with emission components, strongest in Hα. Weak HeI 4471Å and HeII 4686Å are also present, they vary by a factor of 3 in equivalent width.

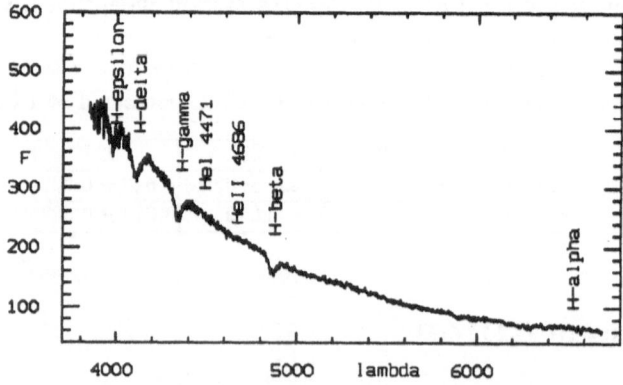

Fig. 1. An optical spectrum of GD984 (1990/12/13); Calibrated flux

The UV-spectra show a number of highly ionized species of ions (see Fig.2): SiIII, SiIV, FeV,NV, etc. in absorption and also HeII 1640Å in absorption with a central emission component. The presence of both SiIII and SiIV give a guess for a cooler companion, e.g. a subdwarf: The ionization equilibrium SiIII/SiIV yields $T_{eff} = 30000K - 40000K$, if $\log g$ is ~6, in accordance with the UV slope.

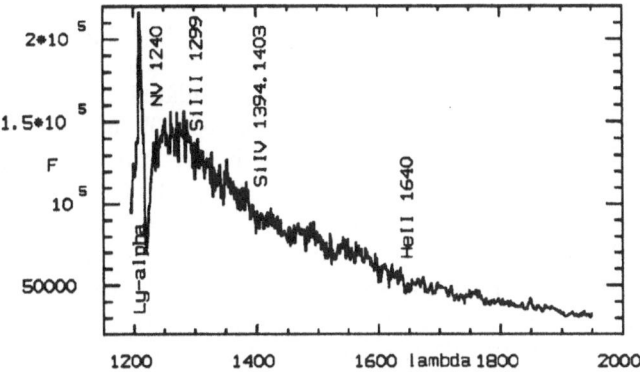

Fig. 2. UV spectrum of GD984 (1991/8/18) from IUE; Flux in cgs

An independent analysis of blue spectra with pure hydrogen model atmospheres by Kidder (1991) gave $T_{eff} = (45800 \pm 3700)$K, $\log g = 7.90 \pm 0.43$. Modelling of the X-ray region alone with stratified as well as homogenous H+He model atmospheres by Barstow et al.(1993) failed to reproduce the gradient. In both cases GD984 was assumed to be a single white dwarf.

With the assumption of the binary character of the system, Aslan (1994) calculated H+He+C model atmospheres in LTE for the temperature range $T_{eff} = 30000K...64000K$, $\log g = 6.0, 6.5, 7.0, 7.5, 8.0, 8.5$ and variable mixtures of hydrogen, helium and carbon. Even small amounts of carbon affect the flux in the X-ray region which is shown in Fig.3. Curve I is a model atmosphere with $\epsilon_{He} = 1 * 10^{-6}$ and no carbon (Napiwotzki, priv. comm.), Curve II a model atmosphere with $\epsilon_{He} = 0$ and $\epsilon_C = 4 * 10^{-6}$.

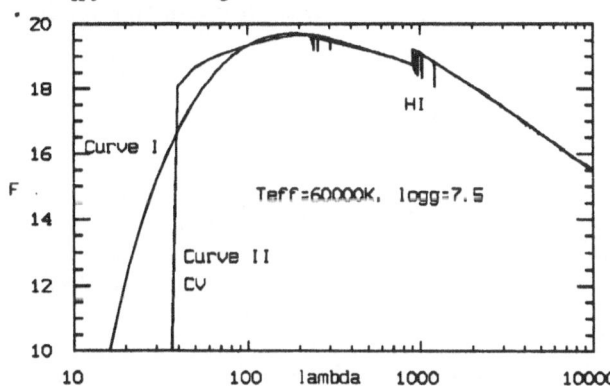

Fig. 3. Flux (log scale) from model atmospheres with and without carbon (see text)

The CV absorption edge at 36.47Å is blocking most of the flux at shorter wavelengths, the gradient of the continuum is provided by the absorption of the ground state. The ROSAT cameras PSPC, S1 and S2 are working in this wavelength-region, so that the presence of carbon is crucial to understand the ROSAT data.

For the carbon abundance of both stars we derive an upper limit of $\epsilon_C \leq 4 * 10^{-6}$ from the IUE spectra (CIV resonance lines 1548Å and 1551Å). With abundances $\epsilon_{He} = 10^{-3}$, $\epsilon_C = 4 * 10^{-6}$ and calculated H+He+C model atmospheres for $T_{eff} = 54000K, 56000K, 58000K, 60000K, 62000K, 64000K$, log $g = 7.5, 8.0, 8.5$ we get theoretical count rates in the ROSAT cameras. It was assumed that both components contribute equally to the flux at $\lambda = 5500$Å and the interstellar HI column is $N_H = 10^{18} cm^{-2}$. Comparison of observed and calculated ROSAT counts yields $T_{eff} = (60000 \pm 2000)K$, log $g = 8.0 \pm 0.25$ for the hot white dwarf. Part of the spectrum in the IUE region is provided by a hot subdwarf with parameters $T_{eff} = (32000 \pm 2000)K$, log $g = 6.5 \pm 0.5$, $\epsilon_{He} = 0.01, \epsilon_C = 10^{-6}$. The (R-I) colour of $0.^m25$ and the Hα region is mainly due to a very cool M5 star in agreement with model fluxes of Allard et al.(1992). The distance of the whole interacting system is $d = (136 \pm 25)$pc.

4 Conclusions

From our model atmosphere analyses of GD1401 and GD984 we learn that two white dwarfs in the direction of the Galactic South Pole ($l = 173^0, b = -76^0$ and $l = 167^0, b = -75.^01$) with masses of 0.33 $M_\odot$ and 0.63 $M_\odot$ belong to multiple systems. The wide binary system of GD1401 is explained in the whole spectral region, from X-ray to 1μm by a normal DA and dM star, whereas for the hot DA white dwarf in GD984 traces of carbon are present, the cool secondary is a main sequence object, yet a third component, providing the variable helium features , is needed. The assumption of a hot accretion disk would account for the helium and the emission components, but not for the metal absorption lines in the IUE region. That is why we prefer a hot interacting subdwarf as the third component of the system.

References

Allard, F., Scholz, M., Wehrse, R., 1992, Rev.Mex.Astron.Astrofis. 23,203
Aslan, T.,1994, Diploma thesis, Friedrich-Alexander-Universität Erlangen-Nürnberg
Barstow, M.A., Fleming, T.A., Diamond, C.J., Finley, D.S., Sansom, A.E., Rosen, S.R., Koester, D., Marsh, M.C., Holberg, J.B., Kidder, K., 1993, MNRAS 264,16
Bues, I., Rupprecht, G., 1983, ESO Messenger 34,24
Kidder, K., 1991, Ph.D. thesis, University of Arizona
Müller, B.I., 1989, Ph.D. thesis, Friedrich-Alexander-Universität Erlangen-Nürnberg
Rupprecht, G., 1983, Ph.D. thesis, Friedrich-Alexander-Universität Erlangen-Nürnberg

The White Dwarf in AM Her

B.T. Gänsicke[1], K. Beuermann[1,2], D. de Martino[3]

[1] Universitäts-Sternwarte Göttingen, Geismarlandstr. 11, 37083 Göttingen, Germany
[2] MPE Garching
[3] IUE Observatory VILSPA ESA, PO Box 50727, 28080 Madrid, Spain

AM Her is the prototype of the *polars*, cataclysmic variables characterized by their strongly magnetized white-dwarf primary. During quiescense, the white dwarf should contribute significantly to the observed UV-flux and be detectable by its strong Lyα absorption. The contradicting temperature estimates for the white dwarf of 50000 K (Szkody et al., 1982, ApJ 257, 686) and 20000 K (Heise & Verbunt, 1988, A&A 189, 112) derived from *IUE* low-state spectra were hitherto interpreted by the cooling of the accretion-heated polecap or by different levels of activity between the individual low states. We obtained three SWP exposures with *IUE* on Sept 21, 1990, simultaneous with the *ROSAT* survey coverage of AM Herculis, and completed our dataset with all reliable quiescense spectra from the *ULDA* archive. The UV flux (Fig. 1b) varies over the phase consistently with the hard X-ray flux (Fig. 1a), the UV and X-ray modulations are $\sim 3.6 \times 10^{-11}$ and $\sim 10^{-11}$ergs cm^{-2}s^{-1}, respectively. The derived *flux-weighted mean temperatures* (Fig. 1c) follow tightly an orbital variation of sinusoidal form in phase with the UV flux. With the time the system had spent in low state previous to the observation ranging from 30 to 150 days, there is no evidence for cooling in AM Her. Possible cause for the UV modulation is heating by ongoing accretion at a low rate. The source radius (Fig. 1d) is compatible either with a white dwarf of 0.9 M$_\odot$ and non-uniform temperature or with a very cool unseen low-mass white dwarf with a large heated spot. This latter possibility can be excluded for distances $d > 100$pc.

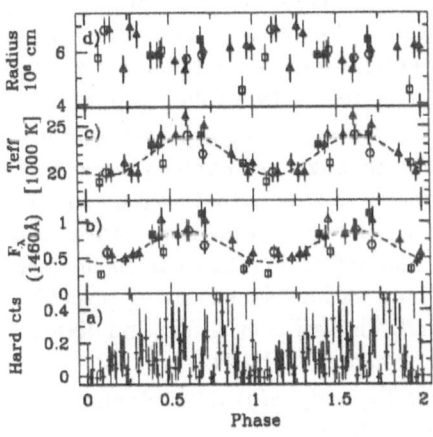

Fig. 1. (a) Hard X-ray lightcurve from the *ROSAT* all sky survey. (b) *IUE* lightcurve of AM Her for several low states: June 1980 (■), Sept. 1980 (△), Nov. 1983 (□), Sept. 1990 (○) and June 1992 (▲). The flux is given in 10^{-13}ergs cm^{-2}s^{-1}Å^{-1}, averaged over the interval 1420–1500 Å. (c) effective temperatures derived from fitting $\log g = 8$ pure-hydrogen white-dwarf models to the observed Lyα profile. (d) the radius of the emitting (circular) surface at distance 75 pc.

The Final Results of: A NLTE Analysis of the Helium-rich sdO Stars in the Palomar Green Survey of High Galactic Latitude Blue Objects

Peter Thejll[1], Rex Saffer[2], Franziska Bauer[3], Jim Liebert[4], Dietmar Kunze[3], and Harry Shipman[5]

[1] Niels Bohr Institute, Copenhagen, Denmark
[2] STScI, Baltimore, USA
[3] University of München, Germany
[4] Steward Observatory, Tucson Arizona, USA
[5] University of Delaware, USA

The NLTE analysis of the majority of the Helium-rich (Y>50% He) sdO stars in the Palomar Green survey (Schmidt, Green and Liebert, 1986), is finally completed (Thejll *et al.* 1994).

The main result is that the sdO stars, like the sdB stars in the PG survey (Saffer *et al.* 1994), are distributed above the plane of the galaxy with a scale height that indicates membership in the thick disk or halo population. Because of the relatively small number of stars involved (we analyze 21 stars out of a possible total in the PG survey of 25 - 30 helium rich sdO stars) the possibilities of reducing statistical errors by large number effects are ruled out - as opposed to the analysis of sdBs in the PG survey. The scale-height finding is in agreement with the recent analysis of the kinematics of hot subdwarfs (Thejll, 1993). We also find that the stars are distributed in the T_{eff} vs. $\log(g)$ plane in the area delimited by the Helium Main sequence (HEMS), the 0.5 $M_\odot$ extended horizontal branch (EHB) and the high luminosity post-asymptotic giant branch tracks.

The large scale height indicates that there is reason to look at evolution channels that produce stars at a mature stage - such as near-complete envelope loss of red giant branch stars during single star evolution of low mass stars (Thejll, Jørgensen and Flynn, 1992), or coalescence of close binary stars whose orbits have decayed over a long time thanks to gravitational wave radiation. A large age and old population kinematics is not consistent with a young population.

The distribution in the T_{eff} vs. $\log(g)$ plane shows, as has been discussed before (Greenstein and Sargent, 1974), that no single evolutionary stage easily can be assigned to the sdOs, but it is reasonable to require that even if evolution tracks criss-cross in the area covered, we must require that those stages that last a long time will be the best candidates for explaining the sdOs. Such stages include the Helium Main Sequence itself and various slow-evolving post-EHB stages that have recently been brought into discussion (Greggio and Renzini, 1990, Jørgensen *et al.* 1993a and 1993b, Dorman *et al.* 1994).

We do have to face the problem that our T_{eff} and $\log(g)$ values do not match those values found by Dreizler *et al.* (1990) in their analysis of a sample of PG stars, 5 of which coincide with our sample. The methods used by Dreizler *et al.* are very similar to ours but, for as-yet unclear reasons, they find generally smaller gravities than we do. For a fixed assumed mass this means that the derived distance is greater than we find - the scale height derivable from Dreizler *et al.* 's analysis, if it were to be extended to the whole PG survey sample of sdOs, such as we have done, would therefore lead to larger scale heights and therefore a greater implied age. This does not change our conclusions. The problem of understanding why the two works give different results is being analyzed, but there are no clear answers yet.

During the work on the helium-rich fraction of stars in the PG survey, it became evident that the majority of the stars categorized as sdO are hydrogen-rich. Typically these are stars with strong H+HeII lines and also showing HeII 4686 Å and sometimes HeII 4542 Å in addition. These stars have all been observed and will be analyzed when a new grid of H-rich H/He models has been completed, in collaboration with Dirk Husfeld of München.

Since H/He spectra are not very sensitive indicators of gravity when the HeI lines are absent, it will be necessary to have at hand as much additional information about the objects, as possible. We are in separate investigations acquiring parallaxes (from the HIPPARCOS satellite as well as from ground-based CCD parallax work), proper motions (with the Carlsberg meridian circle telescope on La Palma), and we have started an effort to use the occasionally occurring companion stars as distance indicators (Thejll, Jimenez, Saffer and Jørgensen 1993, and Thejll, Ulla and MacDonald, AA in press 1994). The luminosity class of the companion stars must be determined, as must the true physical nature of the binarity. We are gathering spectroscopic data on such companions, and detecting evidence for companions in ground-based infrared photometric work, details of which were presented as posters at this meeting.

References

Dorman, B., O'Connell, R.W., and Rood, R.T., 1994, ApJ Preprint.

Dreizler S., Heber U., Werner K., Moehler S., & de Boer K. S., 1990, AA 235, 234.

Greenstein J. L., & Sargent A. I., 1974, ApJSS 28, 157.

Green R. F., Schmidt M., & Liebert J., 1986, ApJSS 61, 305.

Greggio, L., and A. Renzini, 1990 ApJ 364, 35.

Jørgensen, U.G., and Thejll, 1993a, AA 272, 255.

Jørgensen, U.G. and Thejll, P.A., 1993b, ApJ 411, L67.

Saffer, R.A., Bergeron, P., Koester, D., Liebert, J., 1994, ApJ 432, 351.

Thejll, P.A., Jørgensen, U. G., and Flynn, C., 1992, in Proceedings of the 8th European White Dwarf Workshop, held in Leicester, M. Barstow (ed.).

Thejll, P.A., Jimenez, R., Saffer, R., and Jørgensen U., 1993, in Proceedings of the "Hot stars in the Halo" Saul J. Adelman, Arthur R. Upgren, and Carol J. Adelman (eds.), Cambridge University Press

Thejll, P., 1993, in proceedings of "Hot Stars in the Galactic Halo", Saul J. Adelman, Arthur R. Upgren, and Carol J. Adelman (eds.), Cambridge University Press

Thejll, P.A., Bauer F., Saffer R., Liebert J., Kunze D., Shipman H, 1994, ApJ, 433, 819.

³He- and Metal Anomalies in Subluminous B Stars

U. Heber[1] and L. Kügler[2]

[1] Dr. Remeis Sternwarte, Universität Erlangen-Nürnberg, Bamberg, Germany
[2] Institut für Astronomie und Astrophysik der Universität, D-24098 Kiel, Germany

High resolution optical spectra of 16 subluminous B stars (HBB, sdB and sdOB) and five main sequence ³He stars (Hartoog & Cowley, 1979, ApJ 228,229) obtained with the ESO CASPEC at the 3.6m telescope are analysed for the isotopic abundance ratio of helium. For 8 sudwarfs metal abundances are derived.

The He I 6678Å line has the largest isotopic line shift and is best suited amongst all optical lines to detect the ³He isotope by measuring its wavelength displacement. A prerequisite for such a line shift measurement, however, is to determine the stellar radial velocity from metal lines. Due to the scarcity of spectral lines about 1000Å need to be recorded simultaneously, including He I 5875Å. Its negligible isotopic line shift (0.04Å) allows an important consistency check to be made. Our CASPEC measurements fully confirmed the line shifts of the definite ³He stars 3 Cen A, α Scl and HR 2306 (Hartoog & Cowley, 1979) to within 0.02Å. However, no isotopic line shift is found for the probable ³He stars 33 Gem and 40 Gem. Four of our programme stars display significant line shifts of He I, 6678Å (0.43Å for PHL 25; 0.45Å for PHL 382; 0.45Å for SB 290 and 0.11Å for SB 459). The measurement for SB 459 is somewhat uncertain and therefore this star should be termed a possible ³He star, whereas PHL 25, PHL 382 and SB 290 are definite ³He stars. In fact, their isotopic line shifts are to our knowledge the largest ever measured for any star and indicate that in their atmospheres ⁴He is largely replaced by ³He.

Blue spectra (4000Å to 5000Å) were also recorded with CASPEC for most programme stars and allowed several weak metallic lines to be measured. Several spectral peculiarities are obvious. Some species are found to be too weak (or even absent) when compared to a normal B star. However, in a few cases we find species to have enhanced line strengths, e.g. Cl II and Ar II in PHL 382, PHL 25 and PHL 1434. For 8 stars of our samples abundances are derived differentially to the normal B star γ Peg. Three HBB stars show an enrichment of chlorine by factors ranging from 3 to 170. In addition Ar was found to be slightly enriched in PHL 382 and PHL 25 and Sr in PHL 1434. The helium deficiency of the programme stars is accompanied by large deficiencies of C and Si whereas N is almost normal. The Cl, Ar and P enrichment is probably caused by the radiative acceleration being large for these elements ("radiative levitation"). The observed abundance patterns of HBB and sdB stars are still a challenge to theoreticians in the field of diffusion processes. Moreover, the occurrence of the ³He-anomaly may allow to constrain the evolutionary time scale of sdB stars. Calculations of Vauclair et al. (1974, A&A 31, 381) indicate that the replacement of ⁴He by ³He is a slow process and requires some 10^8 years for main sequence stars. Calculations for HBB and sdB stars are needed.

HST Observations of the White Dwarf in V471 Tauri

Howard E. Bond,[1] E. M. Sion,[2] Karen G. Schaefer,[1] Rex A. Saffer,[1] and John R. Stauffer[3]

[1] Space Telescope Science Institute, Baltimore, MD USA
[2] Villanova University, Villanova, PA USA
[3] Center for Astrophysics, Cambridge, MA USA

V471 Tauri is a 12.5-hr eclipsing DA + K2 V binary, and a probable member of the Hyades cluster. It is the prototypical example of a post-common-envelope, pre-cataclysmic binary, and it offers a rare opportunity to determine a white-dwarf (WD) mass directly from radial-velocity measurements of a very "clean" system, and to learn more about the physics of common envelopes.

Unfortunately, it is not possible to obtain ground-based velocities of the WD, since it is outshone at optical wavelengths by the dK star. It has proven possible, however, to measure the velocities by using *HST* and the GHRS to obtain spectra in a 36 Å region centered on Lyman-alpha. This is a brief progress report, based on four spectra obtained in 1993 October. Details will be published elsewhere, and additional observations are scheduled for late 1994.

A Hubeny profile with $T_{\text{eff}} = 35,000$ K, $\log g = 8$ provides an excellent fit to the GHRS Lyman-alpha profiles. The corresponding cooling age from Wood's 1992 tracks is 6–7 Myr. Since the Hyades age is 700–800 Myr, the progenitor mass of the DA component must have been essentially that of the current Hyades red giants, $\sim 2.5 M_{\odot}$. *These conclusions rest on the assumption that there is no heating of the WD due to accretion of a portion of the dK's wind.*

The radial velocities obtained so far give a WD velocity amplitude of about $180 \, \text{km s}^{-1}$, and, since the dK velocities are already known, 1σ limits on the masses of the two stars can be set as follows: $M_{\text{DA}} = 0.76{-}1.06 M_{\odot}$ and $M_{\text{K2}} = 0.78{-}1.48 M_{\odot}$. Astrophysical considerations suggest that the lower ends of these ranges are most likely.

With this information, we can reconstruct the progenitor binary, following precepts reviewed by Iben & Livio (1993, PASP, 105, 1373). V471 Tau appears to be descended from a system with an initial period of ~ 5 years, and about $1.7 M_{\odot}$ was ejected during the common-envelope interaction. Since the parameters of the initial and final binaries are fully known, we can compare the initial orbital energy with the final energy plus the energy used to eject the common envelope, and thus estimate α_{CE}, the efficiency with which orbital energy went into ejecting material from the system. The drastic contraction to a period of 12.5 hr requires that α_{CE} must be very low, < 0.1.

This surprisingly low value of α_{CE} is *not* required if we drop the assumption that the cooling age of the WD can be obtained from its present T_{eff}; indeed our result can be taken as an argument that the WD is in fact much older, that the progenitor had a mass $> 2.5 M_{\odot}$, and that α_{CE} is of order 0.4–1 as indicated by other theoretical and observational considerations, as reviewed by Iben & Livio.

Hot Stars in Globular Clusters

Sabine Moehler[1], Ulrich Heber[2] and Klaas S. de Boer[3]

[1] Landessternwarte, Königstuhl, D-69117 Heidelberg
[2] Dr. Remeis-Sternwarte Bamberg, Sternwartstr. 7, D-96049 Bamberg
[3] Sternwarte d. Univ. Bonn, Auf dem Hügel 71, D-53121 Bonn

In the colour-magnitude diagrams of many globular clusters one can find gaps along the blue horizontal branch (BHB). One of the many explanations proposed for this long known phenomenon severely affects work done on hot subdwarfs in the field: The stars below the gaps may not be BHB but subdwarf B (sdB) stars produced by white dwarf mergers (Iben 1990, ApJ 353, 215). Such an origin for sdB stars could also apply to field sdB/sdOB stars but predicts a mass distribution very different than the classical extended HB (EHB) model (Heber 1986, A&A 155, 33) does. In order to find out which of the two models holds true we started observations of the BHB stars in the clusters M 15 and NGC 6752.

Low and intermediate resolution optical spectra of stars above and below the gaps are analysed for the atmospheric parameters T_{eff} and $\log g$ and spectroscopic masses are derived using ATLAS9 models of Kurucz (1992, IAU Symp. 149, 225). For a detailed description see Moehler et al. (1994a, in Hot stars in the galactic halo, ed. S.J. Adelman, Cambridge University Press, p. 217b; 1994b, A&A in press). The blue tail of the HB in M 15 extends to temperatures as high as 20000 K, and thus consists of B stars similar to the Horizontal Branch B-type (HBB) stars in the field. The stars below the gap in NGC 6752, however, show temperatures as high as 26000 K, which correspond to the temperatures shown by sdB stars in the field of the Milky Way.

A careful comparison with evolutionary models for the BHB reveals that the surface gravities for T_{eff} below 20000 K are systematically lower than predicted by canonical HB models even when luminosity evolution is accounted for. Also the resulting masses lie in general significantly below the values predicted by canonical HB theory. Non-canonical HB theory (i.e. the merger scenario of Iben) is also found to be inconsistent with our results. The sdB stars in NGC 6752, on the other hand, lie rather close to the zero age EHB. A comparison with data for other clusters shows that the contradictory results for the BHB stars are not limited to M 15 and NGC 6752 but represent a common phenomenon.

An increase in gravity by about 0.2 dex would erase the discrepancies both for the $\log g$–T_{eff} diagram as well as for the masses. However, an investigation into the analysis techniques as well as possible deficits in the model atmospheres for any systematic effects on the gravity determination was unsuccessful. Another explanation lies with systematic errors in photometric data and/or distances.

From Interacting Binary Systems to DB White Dwarfs

Jan-Erik Solheim

University of Tromsø, Auroral Observatory, N-9037 Tromsø, Norway

From V. Weidemann's studies of white dwarfs (Weidemann 1990) it is concluded that about 15 per cent may be a result of binary evolution – or binary mergers. The AM CVn stars constitute a small group of stars which appear to be in the last stage of mass transfer before evolving into single white dwarf stars. They show only helium and heavier elements in their spectra. The observed mass density of the AM CVn stars is comparable with the space density of DB stars (Warner 1995). The colours of the AM CVn stars are similar to the DB stars, but spectra are different in having wide asymmetric features in the visible spectrum indicating fast rotating disks (Patterson et al. 1993). We also observe narrow blue shifted UV lines indicating a hot wind (Solheim and Kjeldseth-Moe 1987; Solheim and Sion 1994). A low mass secondary and a precessing disk model explain the photometric periods observed in AM CVn itself (Patterson et al. 1993; Solheim et al. 1995).

Since the AM CVn stars show only helium and some traces of C, N, and O, we expect that the final product of a merger would be a DO or a DB white dwarf star with rapidly rotating outer parts. Such a system is discovered in GD 358 where the outer layers rotate 50 per cent faster than the deeper layers investigated by asteroseismology (Winget et al. 1994). Future Whole Earth Telescope campaigns will investigate if GD 358 is a special case, or if differential rotating outer layers are common for DB stars.

More work is needed to follow the AM CVn stars from binaries to DB's.

References

Patterson et al.: Astrophys. J. **419** (1993) 83
Solheim, J.-E., Kjeldseth–Moe, O.: Astr. Space Phys. **131** (1987) 785
Solheim, J.-E., Sion, E.: Astron. Astrophys. **287** (1994) 503
Solheim, J.-E. et al.: Astron. Astrophys. (1995) in preparation
Warner, B.: Astrophys. and Space Sci. (1995) in press
Weideman, V.: Ann. Rev. Astron. Astrophys. **28** (1990) 103
Winget, D. et al.: Astrophys. J. **287** (1994) 503

White Dwarfs in AM CVn Systems

J.-E. Solheim and C. Massacand

University of Tromsø, Auroral Observatory, N–9037 Tromsø, Norway

We assume that the AM CVn stars are mass transferring interacting binary white dwarf systems with orbital periods 15–50 minutes. The mass transfer creates an accretion disk. The primary stars are hot white dwarf stars with temperature greater than 50 000 K, detected in the far UV and EUV part of the spectrum when a disk spectrum is subtracted from the observed spectrum. The hot central star is the source of a strong wind which constitutes highly ionized particles (Solheim and Sion 1994, A&A, 287, 503). The secondary object must be a low mass ($< 0.1\ M_\odot$), low luminosity star, and its spectral features are completely swamped by the flux from the disk. In the case of AM CVn itself we deduce a mass of 0.05 $M_\odot$ and a radius 25 000 km if it is completely degenerate (Fig. 1). The other AM CVn stars have lower masses (Warner 1995, Astrophys. and Space Sci., in press) and larger radii and may be even closer to the brown dwarf sequence.

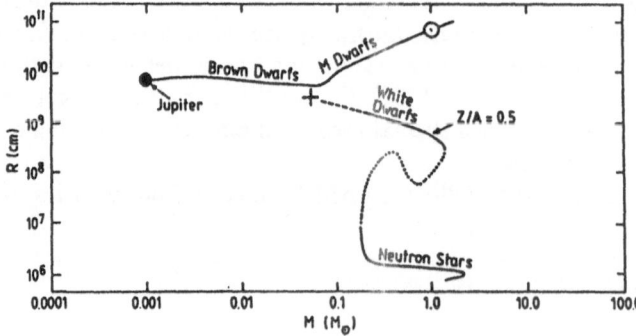

Fig. 1. Radius–mass relation for compact objects with surfaces. Depicted are the brown dwarf and M-dwarf branches, the white dwarf ($Z/A = 0.5$), and the neutron star family branches. The positions of the Sun and Jupiter are marked. The cross indicates the position of the secondary object in AM CVn. (Figure adapted from Burrows & Liebert, Rev. Mod. Phys. **65** (1993) 301.)

Hot Subdwarf Stars in Binary Systems

A. Theissen[1], S. Moehler[2], T. Bauer[1,4], U. Heber[3],
K.S. de Boer[1], and J.H.K. Schmidt[1]

[1] Sternwarte der Universität Bonn, Auf dem Hügel 71, D–53121 Bonn, Germany
[2] Landessternwarte Königstuhl, D–69117 Heidelberg, Germany
[3] Dr. Remeis Sternwarte Bamberg, Sternwartstr. 7, D–96049 Bamberg, Germany
[4] Observatory of the Sternwarte Bonn, Hoher List, D–54550 Daun, Germany

UV spectra observed with the IUE satellite of the binary stars PG 1718+519, PG 2110+127, and PHL 1079 (each consisting of a hot subdwarf B star and a cool main sequence star) are presented. Assuming that the UV region is not affected by the cool companions, models for the hot components of the binary stars are fitted to the UV, thereby enabling the separation of the fluxes of the cool and hot star. We so find for the effective temperatures of the hot stars 30 kK each. The optical spectra are then corrected for the contribution of the cool stars allowing precise determinations of the gravities from the cleaned Balmer profiles (being $\log g = 5.0$, 5.0, and 5.25, respectively).

Modelling the optical excess fluxes in Strömgren and Johnson photometry, effective temperatures of the cool companion stars are derived. The values lie within a range of 5750 to 4750 K in $T_{\rm eff}$, matching stars of spectral types G2 to K2. Having determined $T_{\rm eff}$ of both components, the disentangled spectra result in luminosity ratios of hot and cool component. For any reasonably assumed luminosity of the hot star (those corresponding to a mass of 0.3 $M_\odot$ – as a minimum for He–burning – or 0.5 $M_\odot$ – as suggested by horizontal branch morphology) we find that the cool companions are of subgiant nature and not main–sequence stars, as often assumed in the literature. Since binaries may form with components of similar masses, and hot subdwarf stars are in a late stage of evolution, it is reasonable that the companion star is somewhat evolved, too.

On 16 CCD–images obtained through Strömgren vby-filters under sub–arcsec seeing conditions we separated the cool companion star BD +28°4211 B from the standard subdwarf O star BD +28°4211 A (angular separation = 2.9 arcsec; see also Ulla et al., these proceedings). Component B is 4.62 mag fainter than A in the visual. An optical spectrum of component B is classified as G8, luminosity class V, thus having an absolute visual magnitude of $M_V = 5.5$. Allowing for some variation ($M_V = 4.5...5.5$), we calculate a distance of 1000 ± 230 pc. If component B were a physical companion, this distance would also apply to component A, and we could derive the absolute luminosity of A. Together with $T_{\rm eff} = 85$ kK and $\log g = 6.0$ (Napiwotzki, 1993, Ph.D. thesis, Kiel) for component A, we then calculate M = 13 $M_\odot$ for the hot star. This is not compatible with the subdwarf nature. Comparison with post AGB–tracks result in M = 0.53 $M_\odot$. Adopting the lower mass, a distance of about 170 pc is derived. This is consistent with the upper limit of 300 pc derived from the weak interstellar Ca II line (Greenstein, 1952, PASP 64, 256). Hence we conclude that the two stars form an optical pair only.

CCD Imaging, Optical Spectroscopy and UBVRIJHK Photometry of Cool Companions to Hot Subdwarfs

Ana Ulla[1,2], Peter Thejll[1], Cristina S. Hansen-Ruiz[1,3],
José Luis Rasilla[3], Armin Theissen[4], Jim MacDonald[5]

[1] Niels Bohr Institute, Blegdamsvej 17, DK-2100 Copenhagen, Denmark
[2] Laboratorio de Astrofísica Espacial y Física Fundamental (LAEFF-INTA), Apdo. 50727, E-28080 Madrid, Spain
[3] Instituto de Astrofísica de Canarias, E-38200 La Laguna, Spain
[4] Sternwarte der Universität Bonn, Auf dem Hügel 71, D-53121 Bonn, Germany
[5] Physics Department, University of Delaware, Newark DE 19716, USA

We are looking for stellar companions to very hot helium-rich subluminous stars by using infrared photometry, optical CCD imaging and spectroscopy and give some preliminary results in this poster. Companions can be used to indirectly measure the distance to the hot subdwarfs, and give clues about possible binary evolution scenarios for hot subdwarfs. We need to know the distances in order to test whether direct methods, based on non-LTE spectral analysis, are working correctly. Reliable distances to hot subdwarfs are also needed in order to determine their role in stellar evolution, in particular in connection with the evolution of helium-rich white dwarfs. We have therefore gathered and analysed IR (JHK) data, filtered CCD imaging and spectroscopy for over 57 hot subdwarfs in a search for IR excesses and companions. 43 of the stars are observed for the first time in the IR and 28 objects are found to have larger JH and K band fluxes than expected from a single hot star. Out of this study we can preliminary conclude: *i)* that the rate of IR excess for the sdOs and sdBs studied so far is about 50/50 for both classes, by simply counting the number of our targets with an excess. For the sdBs this is consistent with the 50% rate of binarity previously found by other authors - provided the IR excess we observe are due to companion stars, and for the sdOs it is a new result; *ii)* three new separable companions to sdBs seem to have been found: KPD20242+5303, PG2148+095, and BD+11 4571, by inspection of both their IR and CCD data; and *iii)* an analysis of the spectrum of the supposed physical companion to the sdO BD+28 4211 is given and it poses problems for the interpretation of other data.

Part VI
Pulsating White Dwarfs

The Multi-Periodic Pulsating PG1159 White Dwarf PG0122+200

G. Vauclair[1], B. Pfeiffer[1], A.D. Grauer[2], J.A. Belmonte[3], A. Jimenez[3], M. Chevreton[4], N. Dolez[1], I. Vidal[3]

[1] Observatoire Midi-Pyrénées, Laboratoire d'Astrophysique CNRS/URA285, 14 avenue E. Belin, F-31400, Toulouse, France
[2] Department of Physics and Astronomy, University of Arkansas at Little Rock, 2801 S. University Avenue, Little Rock, AR 72204, USA
[3] Instituto de Astrofisica de Canarias, Via Lactea S.N., E-38200 La Laguna, Tenerife, Spain
[4] Observatoire de Paris-Meudon, DAEC, F-92195, Meudon, France

1 Introduction

The PG1159 stars form a spectroscopic class a hot pre-white dwarf stars in an intermediate evolutionary stage between the planetary nebula nuclei and the white dwarf cooling sequence. Some are central stars of planetary nebula still embedded in their nebula, while other have lost their nebula and show up as isolated stars (naked PG1159). Approximately half of them are variable stars with periods between 400s and 2000s. The pulsations of PG1159-035, the first star of this class, were discovered by McGraw et al. (1979). Its name was given to this new class of variable stars also known as GW Virginis stars. Subsequently, pulsations were discovered in other stars of this type, either in naked ones: PG1707+427 and PG2131+066 (Bond and Grauer 1984), PG0122+200 (Bond and Grauer 1987), RXJ2117+3412 (Watson 1992; Vauclair et al. 1993) or in planetary nebulae nuclei: K1-16 (Grauer and Bond 1984; 1987), Lo-4 (Bond and Meakes 1990), NGC 1501, NGC 2371-2, Sand-3 and NGC 6905 (Bond and Ciardullo 1993).

The long period oscillations observed in the pulsating PG1159 stars are interpreted as non radial g-modes, with buoyancy acting as the restoring force. This interpretation is now strongly supported by the detailed analysis of the prototype star PG1159-035 (Winget et al. 1991) and of PG2131+066 (Kawaler et al. 1994), from extensive photometric campaigns with the Whole Earth Telescope (Nather et al. 1990).

In the case of the variable planetary nebulae nuclei (PNNV), CCD photometry technics is necessarily used to properly correct for the contamination by the nebula. The effective temperature of the central stars suffers from large uncertainties. It has been estimated for K1-16 (130 000K< T_e < 140 000K) and for Lo-4 ($T_e \approx$ 120 000K). There is still no estimate for the effective temperature of the four other PNNV. As a sub-class of variable PG1159 stars, the PNNV have longer periods than the naked ones. The dominant modes in Lo-4 and K1-16, the two best studied cases, have periods of 1800s-2000s and of 1500s-1700s respectively. The dominant mode periods vary from 1318s-1154s in NGC 1501, 983s in NGC 2371-2, 932s in Sand-3 to 710s-875s in NGC 6905.

For the pulsating naked PG1159 stars, fast multichannel photometry can be used. The effective temperature can also be determined with a better confidence. The hottest member of this sub-class is RXJ2117+3412: Motch et al. (1993) derived an effective temperature of 150 000K ± 15 000K and a surface gravity in the range log g=5.6 - 6.3 from NLTE model atmosphere analysis. This object is an intermediate case between the PNNV and the naked PG1159 variable stars as it shows an extended nebula. Its dominant pulsation period is 821s (Vauclair et al. 1993). PG1159-035 itself is the best studied object of its class. Werner et al. (1991) determined an effective temperature of 140 000K ±14 000K and a surface gravity of log g=7.0 ±0.5 from the NLTE model atmosphere and line profile fit to the optical spectrum. The effective temperature has been recently redetermined from a FOS-HST spectrum: the analysis confirms an effective temperature of 140 000K but with a better precision (3%: Werner and Heber 1993). The pulsations of PG1159-035 have been analysed with unprecedented details owing to a WET campaign (Winget et al. 1991). More than 100 frequencies were resolved. Winget et al.(1991) showed how the frequencies are distributed in triplets and quintuplets and interpreted the entire power spectrum in terms of ℓ =1 and ℓ =2 modes. From the identification of the modes, they were able to derive the mass, the rate of rotation and to give evidence for a stratified chemical composition. The time derivative of the dominant mode at 516s was measured.

The atmospheric parameters for PG1707+427 have been determined by Werner et al. (1991) from a fitting of the line profiles with synthetic spectrum calculated from NLTE model atmosphere. They derived an effective temperature of 100 000K and a surface gravity of log g=7.0. The pulsations, discovered by Bond and Grauer (1984), have been studied with more details from high signal/noise fast photometry (Fontaine et al. 1991) and extended observations (Grauer et al. 1992). While pulsation modes are found in the entire period interval between 335s and 940s, the maximum power is found in a doublet with periods of 447s and 449s. The remaining two pulsating naked PG1159 stars, PG2131+066 and PG0122+200 are much cooler with effective temperature of 80 000K and 75 000K respectively (Dreizler et al.1995). PG2131+066 pulsations are dominated by three modes with periods between 380s and 408s. A WET campaign was organized to study the pulsations of that star in details (Kawaler et al. 1994). PG0122+200 was discovered as a pulsating PG1159 star by Bond and Grauer (1987)(see also Grauer and Bond 1987). It is the only pulsating naked PG1159 star which has not yet been observed thoroughly during a WET campaign, the only limited follow-up observations, to our knowledge, being the ones described by Hill et al.(1987). With an effective temperature of 75 000K (Dreizler 1995), it is presently the coolest pulsating PG1159 star and it defines the red edge of the instability strip. For this reason it is an important object.

There is a clear correlation between the evolutionary phase and the period of the dominant modes: the periods decrease with effective temperature in the naked pulsating PG1159 stars, and the periods in PNNV are longer than in the naked PG1159. Such a correlation is in agreement with the idea that the PG1159 stars are pre-white dwarf stars still contracting towards the white dwarf cooling sequence. This is in contrast with the global periods/effective temperature correlation in ZZ Ceti stars (DAV) in which the periods of the largest amplitude modes increases with decreasing temperature. In the DAV white dwarfs, it is known that the most unstable modes should be those with periods comparable to the thermal time scale at the bottom of the convection zone, which increases as the convection zone deepens with decreasing temperature.

There are still some fundamental physical mechanisms missing in our present understanding of the pulsations in PG1159 stars. One plausible driving mechanism relies on the partial ionization of carbon and/or oxygen. Theoretical modeling of the g-mode instability (Starrfield et al. 1983, 1984; Stanghellini, Cox and Starrfield 1991) shows the sensitivity of the instability strip location in the H-R diagram to the He abundance. While the carbon and oxygen abundances (50% and 17% by mass respectively) derived by Werner et al. (1991) do support the idea that partial ionization of these two elements causes the instability, the large He abundance (33% by mass) produces a too cool instability strip. In addition, all theoretical stability analyses made the assumption of an homogeneous chemical composition. The detailed asteroseismological studies of PG1159-035 (Winget et al. 1991) and PG2131+066 (Kawaler et al. 1994) convincingly demonstrate that the outer layers, where the observed g-modes propagate, are not chemically homogeneous: the period distribution clearly shows the signature of a composition stratification. Such a stratification does naturally explain the different sign of period variation with time observed for different modes: while most modes do increase their period with time, trapped modes show a negative dP/dt, according to the detailed analysis by Kawaler and Bradley (1994). This is in good agreement with the negative dP/dt measured for the 516s mode in PG1159-035, which they find very close from a trapped mode in their best fit model. Finally, the proposed theoretical models do not provide any explanation to the fact that among twin PG1159 stars of similar atmospheric parameters (effective temperature, surface gravity and chemical composition) one does pulsate while the other one does not.

An alternative model, which attempted to explain the heavy elements enrichment in the outer layers, and the instability and filtering mechanisms as the consequences of radiative levitation (Vauclair 1990), seems ruled out by the extremely carbon/oxygen rich composition derived from the NLTE analysis of Werner et al. (1991). The diffusion mechanism cannot produce such a C/O rich composition: in the best cases, the predicted enrichment does not exceed a few times the solar abundances.

PG1159 stars are believed to be in a transition evolutionary phase between post-AGB and white dwarf stars. Their structure and composition contain informations on the late stages of evolution which are dominated by mass loss, a physical process still poorly understood. This is the reason why the pulsating PG1159 stars are so important. Asteroseismological study of the very few known stars of this class will reveal their internal structure with unprecedented details.

Among the pulsating naked PG1159 stars, PG0122+200 is the last one which has not been observed with WET. It is also the coolest presently known. As it defines the red edge of the PG1159 instability strip, its study may reveal important clues on the instability mechanism. The present paper describes the results of two-site campaigns of fast photometry (§2). The data analysis is presented in §3 and an interpretation is proposed in §4. We summarize our conclusions in §5.

2 Observations

Two campaigns of observations were organized in 1990. The first one, in October 1990, was a coordinated two-site campaign between the Steward Mt. Lemmon 1.5m telescope and the Roque de los Muchachos Nordic Optical 2.5m telescope, followed by observations from the Steward Mt. Bigelow 1.5m telescope. This first campaign allowed to accumulate 74.7 hours of fast photometry, with a time coverage of 26.4% over the 11 days spanned by the observations. The second campaign, in December 1990, was also a coordinated bi-site òne, between the Kitt Peak 1.3m telescope and the Haute Provence Observatory 1.9m telescope. It allowed to accumulate additional 30.8 hours of fast photometry, with a time coverage of 22.8 % over the 5 days of the campaign.

3 Data Analysis

The data were reduced according to the standard procedure: the sky background recorded in the third channel is used, after calibration with the two other channels, to substract the sky background contribution in the two channels recording PG0122+200 and the comparison star. The normalized light curves are then divided by polynomials to suppress the low frequencies introduced by the sky transparency variations during the night.

The light curves were then analyzed by two independent methods. An iterative sine wave fitting (ISWF), adapted from Ponman (1981), and followed by prewhitening, was performed at IAC. The details of this method is described in Belmonte et al. (1991). The other method, developed at OMP, is an improvement of the "CLEAN" deconvolution technics (Schwarz 1978; Roberts et al. 1987). The classical "CLEAN" technics assumes that the amplitude of a given mode may be directly read from the power spectrum. This assumption neglects the fact that the amplitude may be polluted by the aliases due to the neighbour modes. The error in amplitude and phase introduced by this assumption propagates in the following iterations. As a consequence the amplitudes derived by this method cannot be trusted in this case. A deconvolution method, intitled "Several Peaks Deconvolution Method" (SPD) has been devised, which deconvolves simultaneously all the peaks in a given range of frequencies of the Fourier Transform. A description of this method is given in Pfeiffer (1993, 1995).

We analysed the October campaign with both "ISWF" and "SPD" methods. The frequencies derived from both methods are listed in Table 1. No significant signal is present in the power spectrum below 2000 μHz and above 3000 μHz. The power spectrum is characterized by four frequency "bands" which peak around 2130 μHz (470s), 2225 μHz (450s), 2497 μHz (400s) and 2973 μHz (336s). The mode of oscillation at 2497.4 μHz (400.4s) seems to have remained the dominant one since the discovery in 1986 (Bond and Grauer 1987).

Table 1: Frequencies and power in PG0122+200 (October 1990)

ISWF			SPD			
f(μHz)	Π(s)	P(μmp)	f(μHz)	Π(s)	P(μmp)	
2130.3	469.4	2.5	2130.3	469.4	3.1	f_1
2144.9	466.2	0.5	2145.0	466.2	0.9	f_2
2150.3	465.0	0.6	–	–	–	
2221.6	450.1	1.6	2221.6	450.1	1.5	f_3^-
2225.1	449.4	2.2	2225.0	449.4	2.8	f_3
2227.8	448.8	1.8	2227.7	448.9	1.7	f_3^+
2486.5	402.2	0.6	–	–	–	
2490.6	401.5	1.1	–	–	–	
2493.6	401.0	6.8	2493.4	401.0	2.9	f_4^-
2494.6	400.9	2.1	2494.4	400.9	4.1	f_4
2497.4	400.4	22.1	2497.4	400.4	21.5	f_4^+
2970.4	336.6	0.9	2970.2	336.6	0.7	f_5^-
2973.7	336.3	4.0	2973.7	336.3	3.5	f_5

The follow-up observations performed in December on a period of time a little shorter than 5 days (2.4 μHz resolution in the power spectrum) do show qualitatively similar light curves and power spectra. Adding the October and December campaigns together would result in 105.5 hours of fast photometry on a time interval of roughly two months. With such a low coverage ($\approx$ 6.6%) the deconvolution technics become unstable. As a consequence, the addition of both sets of data, which provides a 0.2 μHz resolution on the frequencies, does not give the amplitudes with enough confidence. The addition was only performed to check the reality of close frequencies small amplitude peaks.

4 Discussion

4.1 The power spectrum

The power spectrum of the PG0122+200 is characterized by 10 frequencies (Table 1), distributed in four groups: a doublet at $f_1 = 2130.3$ μHz (469.4s) and $f_2 = 2145.0$ μHz (466.2s), a triplet around $f_3 = 2225.0$ μHz (450s), another triplet around $f_4 = 2494.4$ μHz (400s) with the dominant power at $f_4^+ = 2497.4$ μHz (400.4s), and a doublet around $f_5 = 2973.7$ μHz.

4.2 Fine structure and rotational splitting

Looking at the multiplets in Table 1, one finds that most of the components are separated in frequency by a somewhat constant value between 2.7 μHz and 3.5 μHz.

The triplet at f_3 is almost symmetrical in amplitude but not in frequency. If interpreted as a $\ell = 1$ mode split by rotation, the resultant triplet shows a retrograde mode displaced by -3.4 μHz while the prograde mode is displaced by +2.7 μHz. The 0.7 μHz difference in the m $\neq 0$ modes splitting is marginally significant. It cannot be explained by the influence of a magnetic field (Jones et al. 1989) which always increases frequencies: the frequency difference between the prograde mode f_3^+ and the mode m=0, f_3, should be larger than the difference between the retrograde mode f_3^- and f_3, which is the contrary of what is observed in this triplet.

The second well identified triplet f_4 is more intriguing. It is strongly asymmetrical in both frequency and amplitude. If it is also a $\ell = 1$ mode, f_4 would correspond to the m=0 component; the 2493.4 μHz (f_4^-) is then the retrograde mode with a frequency difference of only -1.0 μHz, of the same order as the resolution. However, in this case, we have performed a Fourier analysis of the joint October + December 1990 data, which gives a frequency resolution of 0.2 μHz. In this power spectrum, both f_4^- and f_4 are well separated. This confirms that both frequencies are probably real. The 2497.4 μHz (f_4^+) would correspond to the prograde mode separated from the m=0 mode by 3 μHz.

The two highest frequency peaks at 2970.2 μHz (f_5^-) and 2973.7 μHz (f_5) are separated by 3.5 μHz, a value very close to the separation found in the triplets. It suggests that these two frequencies could also be members of a triplet whose third member has not been detected with the signal/noise achieved in the present data. If one excludes the triplet f_4, with its very asymmetrical frequency fine structure, one finds that both the triplet f_3 and the doublet (triplet) f_5 point to an average separation of 3.2 μHz that we interpret as rotational splitting.

If 3.2 μHz is the splitting due to rotation, then the 14.7 μHz difference between the two lowest frequency modes f_1 and f_2 cannot be explained by rotational splitting of an $\ell = 1$ mode. Would these two peaks be part of a $\ell = 2$ quintuplet with the three other peaks missing, their expected frequency separation should be a multiple of 5.3 μHz, as a consequence of the dependence of the rotational splitting on ℓ in the asymptotic regime, which is not the case. These two frequencies more probably correspond to two modes of different ℓ values (one $\ell = 1$ and one $\ell = 2$). They cannot reasonably be two modes of $\ell = 1$ and different k, because of their period difference of only 3s. With an average value of 3.2 μHz as a tentative rotational splitting, one derives a rotation period of $\approx$ 1.8 days.

4.3 Period spacing

With only 10 periods of pulsation clearly detected in the power spectrum, an analysis of the period spacing by the methods frequently used: Fourier transform of the period distribution and Kolmogorov-Smirnov test, would be inconclusive. One did not attempt to perform these classical tests. One can look for period spacing hints in two different ways. The first one, suggested by Hill et al. (1987), considers that PG0122+200 should show a period spacing in the same ratio compared to PG1159- 035 than their respective dominant period ratio= 400.5s/516s. Such an argument relies on the assumption that both stars have the same structure. Repeating the Hill et al. (1987) argument leads to a predicted period spacing of 16.70s. One other way is to look at the periods of the observed modes and note that: $(\pi_4 - \pi_5)/(\pi_3 - \pi_4) = 64.6s/48.5s = 4/3$ ($\pm$ 0.2%) This suggests a period spacing of 16.16s.

Considering the period distribution given in Table 1, one may check how it fits with these period spacings. To do this comparison, we took the central mode of the triplets, assuming that they were the m=0 modes. It is not known a priori which of f_1 or f_2 is an $\ell = 1$ mode. Assuming that the observed modes are successive k modes of same ℓ (except one among f_1 or f_2), with some modes missing, we estimated the difference δk between the modes by dividing their period difference by the predicted period spacing. The chosen reference mode was the shortest period mode at 336.3s which we consider as the m=0 mode of a triplet, with the prograde mode missing, on the basis of its larger amplitude. We then calculated the predicted periods π_{Pred} with the two values of the period spacing, to be compared with the observed periods π_{Obs}. The result of the comparison is shown in Table 2. The relative differences between the two periods are indicated in percents. Period spacings of 16.70s or 16.16s both agree with the observed period distribution. With the 16.70s period spacing, the 469.4s mode (f_1) should be considered as a $\ell = 1$ mode while with 16.16s, the agreement is better if the 466.2s (f_2) is a $\ell = 1$ mode. However, if 16.70s or 16.16s is the true period spacing in PG0122+200, one cannot derive unambiguously the mass of the star because the luminosity is not known. As outlined in the introduction, the period/effective temperature correlation in PG1159 pulsating stars suggests a strong dependence of the pulsation periods on the luminosity, a consequence of the final contraction of the PG1159 towards the white dwarf cooling sequence. Taken at face value, such a small value for the period spacing would point to a high mass ($\approx$1 M$_\odot$) for PG0122+200.

Table 2: Comparison between observed and predicted period distribution in PG0122+200

$\Delta\Pi = 16.70$ s				$\Delta\Pi = 16.16$ s			
Δk	Π_{Pred} (sec)	Π_{Obs} (sec)	%	Δk	Π_{Pred} (sec)	Π_{Obs} (sec)	%
4	403.1	400.9	0.5	4	400.9	400.9	0
7	453.2	449.4	0.8	7	449.4	449.4	0
8	469.9	$\begin{cases} 469.4 \\ 466.2 \end{cases}$	$\begin{matrix} 0.1 \\ 0.8 \end{matrix}$	8	465.6	$\begin{cases} 469.4 \\ 466.2 \end{cases}$	$\begin{matrix} 0.8 \\ 0.1 \end{matrix}$

4.4 Mode trapping

An examination of the period distribution suggests that the observed modes in PG0122+200 are trapped modes separated by 3-4 normal modes (undetected). By comparing the period ratio of the presumably trapped modes with the trapping coefficients estimated for PG1159 stars by Kawaler and Bradley (1994), one infers that these modes could reasonably correspond to the trapping resonances i= 3, 4 and 5. The observed period ratios are : $\pi_5/\pi_4 = 336.3s/400.9s = .839$, $\pi_4/\pi_3 = 400.9s/449.4s = .892$, $\pi_4/\pi_2 = 400.9s/466.2s = .860$. These period ratios should be compared to the trapping coefficient ratios estimated by Kawaler and Bradley (1994): $\lambda_3/\lambda_4 = .850$, $\lambda_4/\lambda_5 = .861$, $\lambda_5/\lambda_6 = .887$. A consistent description emerges if one associates the observed modes of periods π_5 with i=3, π_4 with i=4 and π_2 with i=5. The trapping cycle corresponds to $\delta k=4$. The number of normal modes separating trapped modes also suggests that the He-rich/C+O transition zone is rather deep in PG0122+200. Computation of the trapping cycle in a representative model for PG1159-035 leads Kawaler and Bradley to a 3-4 modes trapping cycle for their thickest helium layer ($\approx$ 2% in mass). However, this cannot be directly compared to PG0122+200 which has a significantly lower effective temperature.

5 Summary and Conclusions

Fast photometry observations of the pulsating PG1159 star PG0122+200 has been obtained during two bi-site campaigns. The resulting power spectrum shows that PG0122+200 is a multiperiodic pulsator with at least 10 modes; the maximum power is in a mode with a period of 400s; the dominant mode seems to have remained the same since the discovery of PG0122+200 as a pulsating star.

The period distribution shows four domains of power. The fine structure in triplets and doublet is interpreted in terms of $\ell =1$ g-modes split by rotation. The inferred rotation period is of about 1.8 days. A period spacing of 16.70s or

16.16s seems compatible with the observed periods. Taken at face value, such a small period spacing would indicate a high mass ($\approx$ 1 $M_\odot$) for PG0122+200, if its structure were similar to the PG1159-035 one. However, in the absence of any estimate of the star luminosity, it is not possible to derive unambiguously the mass of PG0122+200, as the period spacing depends on both the total stellar mass and the luminosity. A parallax measurement of this faint ($m_B=16.13$) PG1159 star is urgently needed. Furthermore, the number of observed frequencies in PG0122+200 should be significantly increased to ultimately get a true measurement of its period spacing.

In addition, the structure of the period distribution strongly suggests that the observed modes are trapped modes, corresponding to trapping resonance i between 3 and 5. The separation of 4 normal modes between successive trapped modes indicates a rather deep transition zone between the outer He-rich layer and the C/O core. PG0122+200 is the only pulsating PG1159 star not yet observed with the Whole Earth Telescope (Nather et al. 1990). It is also the coolest pulsating PG1159 star presently known, defining the red edge of the instability strip. The present preliminary asteroseismological investigation of that star, based on bi-site photometric campaigns, is hopefully a starting point to a more detailed study.

References

Belmonte,J.A., Chevreton,M., Mangeney, A., Saint-Pe, O., Puget, P., Praderie, F., Alvarez, M. and Roca-Cortes, T. 1991, AA 246, 71

Bond, H.E. and Ciardullo, R. 1993, in "White Dwarfs:Advances in Observation and Theory", M. Barstow (Ed.), Kluwer Academic Publishers, 491

Bond, H.E. and Grauer, A.D. 1984, ApJ 279, 751

Bond, H.E. and Grauer, A.D. 1987, ApJ 321, L123

Bond, H.E. and Meakes, M.G. 1990, AJ 100, 788

Dreizler, S., Werner, K., Heber, U. 1995, these proceedings

Fontaine, G., Bergeron, P., Vauclair, G., Brassard, P., Wesemael, F., Kawaler, S.D., Grauer, A.D. and Winget, D.E. 1991, ApJ 378, L49

Grauer, A.D. and Bond, H.E. 1984, ApJ 277, 211

Grauer, A.D. and Bond, H.E. 1987, in "The Second Conference on Faint Blue Stars", IAU colloquium 95, A.G. Davis Philip, D.S. Hayes and J. Liebert (Eds.), L. Davis Press , Schenectady, New-York, USA, 231

Grauer, A.D., Green, R.F. and Liebert, J. 1992, ApJ 399, 686

Hill, J.A., Winget, D.E. and Nather, R.E., in "The Second Conference on Faint Blue Stars", IAU colloquium 95, A.G. Davis Philip, D.S. Hayes and J. Liebert (Eds.), L. Davis Press, Schenectady, New-York, USA, 627

Jones, P.W., Pesnell, W.D., Hansen, C.J. and Kawaler, S.D. 1989, ApJ 336, 403

Kawaler, S.D. and Bradley, P.A. 1994, ApJ 427, 415

Kawaler, S.D., O'Brien, M.S., Clemens, J.C., Nather, R.E., Winget, D.E., Watson, T.K., Yanagida, K., Dixson, J.S., Bradley, P.A., Wood, M.A., Sullivan, D.J., Kleinman, S.J., Meistas, E., Leibowitz, E.M., Moskalik, P., Zola, S., Pajdosz, G., Kzresinski, J., Solheim, J.-E., Bruvold, A., O'Donoghue, D., Katz, M., Vauclair, G., Dolez, N., Chevreton, M., Barstow, M.A., Kanaan, A., Kepler, S.O., Giovannini, O., Provencal, J.L. and Hansen, C.J. 1994, submitted

McGraw, J.T., Starrfield, S.G., Liebert, J. and Green, R.F. 1979, in "White Dwarfs and Variable Degenerate Stars", IAU colloquium 53, H.M. Van Horn and V. Weidemann (Eds.), Rochester, New-York, University of Rochester press, 377

Motch, C., Werner, K. and Pakull, M.W. 1993, AA 268, 561

Nather, R.E., Winget, D.E., Clemens, J.C., Hansen, C.J. and Hine, B.P. 1990, ApJ 361, 309

Pfeiffer, B. 1993, in Proceedings of the Second WET Workshop, Baltic Astronomy, 2, 538

Pfeiffer, B. 1995, unpublished PhD thesis, University Paul Sabatier, Toulouse

Ponman, T. 1981, MNRAS 196, 543

Roberts, D.H., Lehar, J. and Dreher, J.W. 1987, AA 93, 968

Schwarz, O.J. 1978, AA 65, 345

Stanghellini, L., Cox, A.N. and Starrfield, S.G. 1991, ApJ 383, 766

Starrfield, S.G., Cox, A.N., Hodson, S.W. and Pesnell, W.D. 1983, ApJ 268, L27

Starrfield, S.G., Cox, A.N., Kidman, R.B. and Pesnell, W.D. 1984, ApJ 281, 800

Vauclair, G. 1990, in "Progress of Seismology of the Sun and Stars", Y. Osaki and H. Shibahashi (Eds.), Lecture Notes in Physics, Springer-Verlag, 437

Vauclair, G., Belmonte, J.A., Pfeiffer, B., Chevreton, M., Dolez, N., Motch, C., Werner, K. and Pakull, M.W. 1993, AA 267, L35

Watson, T. 1992, IAU Circular N° 5603

Werner, K. and Heber, U. 1993, in "White Dwarfs: Advances in Observation and Theory", M. Barstow (Ed.), Kluwer Academic Publishers, 303

Werner, K., Heber, U. and Hunger, K. 1991, AA 244, 437

Winget, D.E., Nather, R.E., Clemens, J.C., Provencal, J., Kleinman, S.J., Bradley, P.A., Wood, M.A., Claver, C.F., Grauer, A.D., Hine, B.P., Hansen, C.J., Fontaine, G., Wickramasinghe, D.T., Achilleos, N., Marar, T.M.K., Seetha, S., Ashoka, B.N., O'Donoghue, D., Warner, B., Kurtz, D.W., Buckley, D.A., Martinez, P., Vauclair, G., Chevreton, M., Dolez, N., Barstow, M.A., Solheim, J.-E., Ulla, A.M., Kanaan, A., Kepler, S.O., Henry, G.A. and Kawaler, S.D. 1991, ApJ 378, 326

Asteroseismology of DA White Dwarf Stars

Paul A. Bradley

Los Alamos National Laboratory
Los Alamos NM 87545

Abstract: Following the recent mode identification by Clemens suggesting the hydrogen layer mass of the DAV (ZZ Ceti) stars is nearly the same, I make the extreme assumption that models with a *single* composition profile with a hydrogen layer mass of $1.5 \times 10^{-4} M_*$ may be able to fit the observed pulsation periods. Because I use only a single composition profile, this seismological analysis of eight DAV stars is preliminary, pending model fits for a range of hydrogen layer masses. These models duplicate the observed periods of several hot DAV pulsators quite well, when I allow for a small spread in total mass and effective temperature. My results confirm the dominant pulsation mode power in the DAV stars resides in $\ell = 1$ modes, although a few $\ell \geq 2$ modes are present. Assuming this hydrogen layer mass is typical for all DA white dwarfs, current theories for the evolution of stars from the asymptotic giant branch through the planetary nebula stage are at least qualitatively correct and will affect the ages and average mass of the DA white dwarfs.

1 Introduction

The hydrogen atmosphere (DA) white dwarfs are the most numerous (about 80 %) spectral class of white dwarf, hence they dominate the bulk properties of white dwarfs. The intense surface gravity ($\log g \approx 8$) causes the DA white dwarfs to have a surface layer of pure hydrogen overlying a thicker helium mantle, which rides on top of a C/O core. Because stellar evolution theory *cannot* predict the hydrogen and helium layer masses or the C/O composition profile, we must rely on observational means — such as asteroseismology — to constrain them. Knowing the hydrogen layer mass will provide clues for why the number fraction of DA white dwarfs changes with effective temperature (see Fontaine & Wesemael 1987, 1991; Koester & Finley 1992) and provide explanations for the dichotomy between DA and helium atmosphere (DB) white dwarfs. Finally, determining the C/O mass fraction in the cores of white dwarfs will allow us to constrain the $^{12}C(\alpha, \gamma)^{16}O$ reaction rate (Caughlan & Fowler 1988), which is still only known to within a factor of two in spite of considerable experimental and theoretical effort.

Fortunately, asteroseismology — the comparison of theoretical and observed properties of pulsating stars — offers a way to determine the internal structure of DA white dwarfs once we identify their observed pulsation modes. The 24 known pulsating DA white dwarfs are known as DAV (or ZZ Ceti) stars, and are located between $13,000$ K and $11,000$ K. Up to now, the DAV stars resisted seismological analysis because each star either has too few modes present to constrain the mode identification or their power spectra change radically on timescales of months to years. The only current constraints we have for the hydrogen layer mass is a lower limit of $\sim 10^{-13} M_*$ set by the absence of helium in the hottest DA white dwarfs (Barstow, Holberg, & Koester 1994) and an upper limit greater than $\sim 10^{-4} M_*$, where so much hydrogen would be able to fuse that it would produce more luminosity than observed (Iben & MacDonald 1985,1986).

Fontaine et al. (1992), Bergeron et al. (1993), and Bradley (1993) made the first suggestions that it might be possible to constrain the structure of DAV stars. In all cases, the authors assume a mode identification and then derive limits to the hydrogen layer mass. In most cases, they assume the shortest period mode is the first radial overtone ($k = 1$) nonradial g-mode with a spherical harmonic index of $\ell = 1, 2$, or 3. The thickest hydrogen layer mass occurs when they assume $\ell = 1$, typically $\gtrsim 10^{-5} M_*$, although Bradley (1993) found some examples where models with thinner hydrogen layers could not be ruled out. If they assume $\ell = 2$ or 3, the thickest hydrogen layer masses drops to $10^{-7} M_*$ or less.

Clemens (1994, 1995) reanalyzed the existing pulsation period data on the DAV stars and determined that *as a class*, the DAV stars show a remarkably uniform pulsation pattern. Instead of being scattered randomly, the pulsation periods fall into distinct "bands", suggesting that the pulsation mode pattern is due to a common set of modes. The observed period scatter is easily accounted for by differences in stellar mass and effective temperature. Clemens also suggested the hydrogen layer mass must be nearly the same for all the pulsating DA white dwarfs, and the best fit to the observed pulsation spectrum — using the then available models of Brassard et al. (1992a) and Bradley (1993) — occurs for $\ell = 1$ modes when the hydrogen layer mass is near $10^{-4} M_*$. The coarseness of these model grids precluded more accurate seismology, and in response, I am computing a grid of DA models specifically tailored to match the observed DAV star periods. Here, I provide preliminary fits of the observed pulsation spectra of several DAV stars to theoretically predicted periods derived from models with a common composition profile, varying only the mass and effective temperature.

The layout of the rest of this paper is as follows. First, in § 2, I briefly describe the evolutionary white dwarf models, the common DAV star composition profile, and pulsation analysis programs. § 3 contains a brief overview of the pulsation properties of DA white dwarf models. In § 4, I describe the structure of several DAV stars derived through asteroseismology. I close with conclusions and a summary in § 5.

2 Models and Pulsation Calculations

I use equilibrium models computed with the Rochester-Texas white dwarf evolution code ($\equiv$ WDEC) described by Lamb & Van Horn (1975), Wood (1992),
Bradley, Winget, & Wood (1993, hereafter BWW93), and Bradley (1995). These
models are essentially identical to those presented in BWW93 and Bradley
(1995). Here, I mention some features WDEC has for creating suitable DA
models and refer the reader to Wood (1992), BWW93, and Bradley (1995) for
additional details.

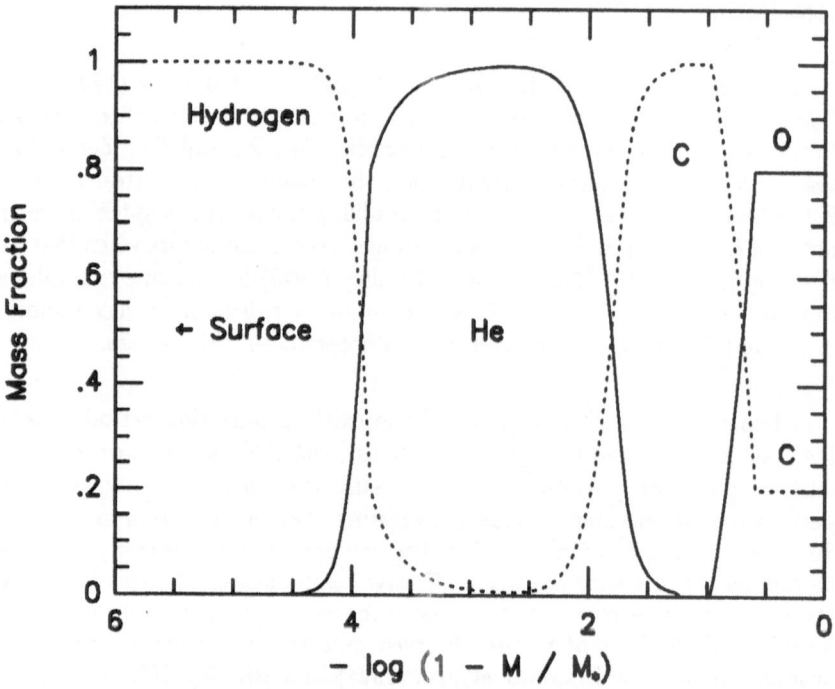

Fig. 1. The composition profile of my models

I use a series of DA model sequences with a common H/He/C/O profile (see
Fig. 1) for my seismological fits. The hydrogen layer mass is $1.5 \times 10^{-4} M_\star$, the
helium layer mass is $1.5 \times 10^{-2} M_\star$, and the core is 80 % oxygen. This oxygen-
rich core extends to $0.75 M_\star$, where I impose a linear change to a pure carbon
composition by $0.90 M_\star$ (see Fig. 1); this approximates the profiles computed
from stellar evolution models by Mazzitelli & D'Antona (1986) and D'Antona
& Mazzitelli (1989). My H/He and He/C transition zone profiles are computed
within the framework of diffusive equilibrium presented by Arcouragi & Fontaine
(1980). However, the He/C transition zone is somewhat steeper than diffusive

equilibrium predicts to avoid the unphysical problem of having helium present in the center of the model and because the strong degeneracy present in the region of the He "tail" inhibits diffusion, sharpening the profile.

The Brunt-Väisälä frequency is the critical quantity that determines the pulsation properties of g-modes, so I use the "Modified Ledoux" treatment of the Brunt-Väisälä frequency in the transition zones as described by Brassard et al. (1991) in these models. Including composition gradient terms in the Brunt-Väisälä frequency is responsible for mode trapping (see Brassard et al. 1992b for details), and it strongly affects the pulsation periods.

I solve the nonradial oscillation equations in the form described by Saio & Cox (1980) using an adiabatic pulsation analysis program that employs a fourth order Runge-Kutta-Fehlberg (RKF) integration scheme with error limiters (Kawaler et al. 1985). RKF has the ability to interpolate extra zones into an equilibrium model until it resolves the nodes of an eigenfunction, allowing RKF to obtain accurate solutions for high overtone modes. The adiabatic approximation neglects entropy and luminosity perturbations; thus, we cannot predict the pulsational stability of a mode. However, the adiabatic pulsation periods are quite accurate, since only the outermost layers are nonadiabatic.

3 Pulsation Properties of DA Models

Before describing the seismological results for several DAV stars, I will sketch the dependence of the theoretical pulsation periods on various structural parameters of the models. I refer the reader to Brassard et al. (1992a, b) and Bradley (1995) for more detailed explanations of the theoretical trends. From here on, I use the following nomenclature for mode identifications; (1,3) represents an $\ell = 1$, $k = 3$ mode.

Increasing the mass of a DA model acts to raise the Brunt-Väisälä frequency in the outer layers, causing the periods of all modes to drop. I see this effect observationally; the period of the (1,1) mode of G 226–29 is 109 s, while it is about 120 s for L 19–2 and GD 165. This tells me G 226–29 should be more massive than the other two stars. In addition, higher mass models have smaller period spacings between consecutive overtone modes, allowing me to differentiate between changes in stellar mass and changes in the hydrogen layer mass. Increasing either one causes the period of the (1,1) mode to drop, but increasing the stellar mass also causes *all* of the other $\ell = 1$ mode periods to drop as well. My model fitting of L 19–2 is an example where this difference comes into play.

The thickness of the hydrogen layer has a large effect on the pulsation spectrum of the DA models. The key to understanding the change in pulsation periods as the hydrogen layer mass changes lies in the Brunt-Väisälä frequency. The Brunt-Väisälä frequency of the H/He transition region is much larger than in the surrounding H or He layers because of the contribution due to the changing mean molecular weight; this dominates over the smaller Brunt-Väisälä frequency of hydrogen compared to helium. Thus, as I increase the mass of the hydrogen layer — i.e., push the region of large Brunt-Väisälä frequency deeper — I effectively increase the Brunt-Väisälä frequency in the model. The pulsation periods of the model varies as the inverse of the Brunt-Väisälä frequency, so they decrease with increasing hydrogen layer mass. Unlike changing the stellar mass, the average period spacing between consecutive overtone modes does not change.

The steep composition gradient of the H/He transition region also produces a rapid change in density that is able to partially reflect the energy within a pulsation mode. Here, I am most interested in the case where the mode can be confined between the surface and the H/He transition region. When the wavelength of the mode matches the depth of the hydrogen surface layer, a resonance forms and the mode is "trapped" within the hydrogen layer. Such a mode requires much less energy to move through a pulsation cycle, because most of the amplitude is in the hydrogen (or helium) surface layers which are easier to move than the core. The occurrence of trapped modes and the interval between them depends on the location of the H/He (and other) transition regions; the deeper the transition region, the smaller the overtone number of the first trapped mode, and the shorter the interval between successive trapped modes (see Brassard et al. 1992b for more details). The hydrogen layer mass of $1.5 \times 10^{-4}M_\star$ is so thick that trapping by the He/C and C/O transition regions are relatively important, especially for low overtone modes. In my models, the presence of the He/C and especially the C/O gradient is critical for the theoretical periods to duplicate the observed ones.

4 Seismology of DAV Stars

Here, I present seismological results for several DAV white dwarf stars. I use L 19–2 as an example to illustrate the procedure for seismology of DAV stars and what I can learn from them. I also present results for G 226–29, R 548, G 117–B15A, GD 165, GD 385, BPM 31594,and G 207–9. The other DAV stars do not have enough observational data to determine their mode structure; hopefully these results will stimulate the needed observations.

I use the effective temperatures and $\log g$ values listed by Koester, Allard, & Vauclair (1995), Koester & Allard (1993), Kepler & Nelan (1993), Daou et al. (1990), and Weidemann & Koester (1984) to provide constraints on the temperature and independent estimates for the stellar mass through $\log g$ values. In some cases, trigonometric parallax data is available (Harrington & Dahn 1980, McCook & Sion 1987); I use this as an independent check of the seismologically predicted distances.

4.1 L 19–2

L 19–2 has dominant periods of 113.8 s and 192.6 s (O'Donoghue & Warner 1982, 1987). The most recent temperatures for L 19–2 are about 12,200 K (Kepler & Nelan 1993; Koester et al. 1995), although Daou et al. (1990) find a hotter T_{eff} of 13,300 K. Each of the five observed modes is split into at least two components (see Table 1), offering extra constraints for mode identification. Three multiplets have splittings around 12μHz, suggesting that these three modes have the same ℓ value, and Clemens (1994, 1995) identifies tham as $\ell = 1$ modes. Frequency splittings suggest the 114 and 143 s modes are $\ell = 2$ modes, and the frequency splitting ratios $(0.6-0.7)$ are consistent with this identification, if we identify the 39.6μHz splitting as $\Delta m = 2$ instead of 1. Finally, the period ratio $192.6/113.8 = 1.69$, suggesting that these are $\ell = 1$ and 2 modes of the same $(k = 2)$ overtone. Adding these pieces together with periods predicted by my models leads to the mode identification in Table 1. A look at models with masses between 0.60 and $0.70M\odot$ shows that one at $0.66M\odot$ does the best job matching most periods. The biggest mismatches occur for the (1,1) and (2,3) modes; they are difficult for models with the fiducial composition profile to match well. Further work shows models with a slightly different composition profile will fit all of the periods present in L 19–2.

Table 1. Periods, Amplitudes, Frequency Splittings, and Predicted Splitting Coefficients and Periods for L 19–2

Period (s)	Amplitude $(\times 10^{-3}$ mfa)	Frequency Splitting $(\mu$Hz)	$1 - C_{\ell,k}$ $0.66M\odot$ model	Model Period (s)	Mode
118.52	1.8	0	0.523	126.3	(1,1)
118.68	1.1	−11.14			
192.12	0.7	+12.99			
192.61	5.8	0	0.617	192.4	(1,2)
193.09	1.0	−12.99			
348.73	0.4	+11.62			
350.15	1.0	0	0.512	350.9	(1,6)
113.27	0.6	+39.64			
113.78	2.1	0	0.918	114.6	(2,2)
143.04	0.3	+18.53			
143.42	0.6	0	0.847	133.8	(2,3)

NOTE: Observational data from O'Donoghue & Warner (1982).

Assuming the $0.66M\odot$ model is the best fitting one, I derive $T_{\text{eff}} = 12,360$ K, $\log g \approx 8.09$, a luminosity of $0.004L\odot$, a distance of 24 pc, and a seismological parallax of 0."041. The latter is less than 1σ smaller than the observed parallax of 0."044. Finally, I present $1 - C_{\ell,k}$ values for the observed modes (see Table 1); these values follow the observed frequency splitting trend and imply the rotation period of L 19–2 is just under 13 hours.

4.2 Other DAV Stars

To bolster the claim that the structure of all the DAV stars is very nearly the same, I present results for seven additional stars along with L 19–2 (see Table 2). All but BPM 31594 and G 207-9 are hot DAV stars with resolved pulsation spectra. BPM 31594 is a cooler DAV star with a resolved pulsation spectrum, while G 207-9 has only single site data available. The pulsation spectrum may not be resolved, so I consider the results for this star to be tentative at present. I am primarily interested in the group properties of these objects, so I will give only brief comments on the seismologically derived parameters, leaving detailed descriptions to a future paper.

Of these objects, I can place only limited constraints on the structure of G 226–29 and GD 385 because only one mode is present. In both cases, I find a solution that is consistent with the observational data. GD 385 has a mass near $0.6 M_\odot$ if I assume the 256 s mode is the (1,3) mode as the mode grouping suggests. A $0.6 M_\odot$ model also has a temperature consistent with the observed $T_{\rm eff}$.

In addition to L 19–2, five of these stars have multiplet splittings caused by slow rotation and/or a magnetic field. Jones et al. (1989) mention GD 385 and R 548 as magnetic field candidates, based on the presence of doublets in their power spectra. However, it is also possible that both stars could have only two (out of three possible) peaks excited to observable amplitudes. For R 548, the $(1 - C_{1,k})$ of the two observed modes give the same rotation rate. If we see R 548 almost equator-on ($i \sim 90°$), then the $m = 0$ mode would not be seen, $\Delta m = 2$, and the rotation period is 42.2 hrs. A word of caution: I am *not* saying that magnetic fields are not present in DAV stars. I only suggest the observed splittings can be explained equally well by slow rotation. All the derived rotation periods lie between 0.3 days and 4 days, suggesting that considerable angular momentum must be lost between the main sequence and white dwarf phases of stellar evolution.

To constrain the model fits, I must assume that most of the observed modes are trapped for G 207-9 and BPM 31594. BPM 31594 has only the two distinct modes present, and I can match these with the $k = 7$ and $k = 12$ modes for models between 0.65 and $0.68 M_\odot$. If I also demand the effective temperature of the best fitting model is $\sim 12,800\ K$, then $0.65 M_\odot$ models fit best. For G 207-9, a $0.60 M_\odot$ model duplicates the period distribution quite well, and the observed modes are trapped or adjacent to a trapped mode. The models also say the $k = 3$ mode is trapped, while the 557 s and 739 s modes are one overtone below the trapped modes for all models with good period matches.

Fontaine et al. (1992), Bergeron et al. (1993), and Fontaine et al. (1994) suggest hydrogen layer masses for G 226–29, GD 165, and G 117–B15A, respectively. They derive lower limits to the hydrogen layer mass of G 226–29 and GD 165 by assuming the shortest period mode is the (1,1) or (2,1) mode. For G 226–29, Fontaine et al. (1992) assume $M_\star = 0.70 M_\odot$ and derive a hydrogen layer mass of $\sim 4 \times 10^{-5} M_\star$, about 1/3 of my value. However, they interpolated between the pulsation periods of $0.6 M_\odot$ and $0.8 M_\odot$ models of Tassoul, Fontaine,

Table 2. Comparison of Observational and Seismological Data for Some DAV Stars

Object	Spectr. T_{eff} (K)	Spectr. $\log g$	Trig. parallax (mas)	Seis. T_{eff} (K)	Seis. $\log g$	Seis. Mass ($M_\odot$)	Seis. par. (mas)	Seis. Rot. Per. (hrs)
G 226–29	12,000	8.12 ± 0.05	82.7 ± 4.6	12,130	8.28	0.78	95.5	8.9
GD 165	13,300	7.96 ± 0.05	N.A.	13,150	8.07	0.66	28.7	~ 58
BPM 31594	12,800	7.80 ± 0.25	N.A.	12,590	8.07	0.65	21.8	—
L 19–2	12,200	7.96 ± 0.10	44. ± 5.	12,360	8.09	0.66	40.7	12.9
GD 385	11,800	8.05±?	N.A.	11,720	7.99	0.60	21.4	46.3 or 92.6
G 207–9	11,660	8.16±?	28.6 ± 4.4	11,860	7.99	0.60	37.8	—
G 117–B15A	12,380	7.97 ± 0.06	12.1 ± 5.3	12,300	7.90	0.55	15.9	—
R 548	12,800	7.80 ± 0.21	14. ± 2.	12,620	7.88	0.54	29.4	21.1 or 42.2

Notes: "N.A." or "?" indicates these values are unavailable.

& Winget (1990); this may be the origin of the difference. Bergeron et al. (1993) derive a hydrogen layer mass of $2 \times 10^{-4} M_\star$ for GD 165, quite close to my value. Finally, Fontaine et al. (1994) quote a hydrogen layer mass of $\sim 1.3 \times 10^{-6} M_\star$ for G 117–B15A. I derive similar hydrogen layer masses *only* if I assume the 215.2 s mode is the (1,1) mode instead of (1,2) mode, as Clemens and I claim.

5 Summary and Conclusions

These results show that seismology is possible for the DAV stars; I derive their internal structure, rotation rates, and luminosities. My results confirm and extend those of Fontaine et al. (1992) and Bergeron et al. (1993). They also lend support to Clemens's assertion that the DAV stars — and by extension, probably all DA white dwarfs — have essentially the same structure, and that the hydrogen layers must be thick, with $M_H \approx 1.5 \times 10^{-4} M_\star$. My results suggest the white dwarf luminosity function (Wood 1992) and the spectroscopic mean mass of DA white dwarfs should be re-evaluated based on models with thick hydrogen layers. Wood (1995) shows that thick hydrogen layers affect the luminosity function and increases the age estimates for the local Galactic disk. My composition profile has an 80 % oxygen core, which suggests the $^{12}C(\alpha, \gamma)^{16}O$ reaction rate is at least as great as the value given by Caughlan & Fowler (1988), depending on the amount of mixing that takes place in the core (D'Antona & Mazzitelli 1991). If this core composition is confirmed through further modeling, then I will be able to offer observational constraints on the C → O reaction rate for the first time.

If the derived hydrogen layer mass is typical for all DA white dwarfs, then the standard picture of stellar evolution from the AGB to the white dwarf stage (Iben & MacDonald 1985, 1986; D'Antona & Mazzitelli 1991) is at least qualitatively correct. This hydrogen layer is also thick enough that future models must allow for quiescent hydrogen burning near the base of the H/He transition zone (Iben & Mac Donald 1985, 1986). This could make the H/He transition zone profile sharper than predicted by diffusion theory, leading to slight changes in the theoretical periods and seismologically derived parameters. Our results are also consistent with recent X-ray and UV observations of hot DA white dwarfs having thick hydrogen layers contaminated with trace amounts of metals (Barstow et al. 1994). However, the thick hydrogen layer mass disagrees with spectral evolution theory (Fontaine & Wesemael 1987, 1991) — which requires "thin" hydrogen layer masses of $\sim 10^{-10}M_\star$ or less — applying to *all* white dwarfs. Current observational evidence suggests that at most, only a small minority of hot DA and DAO white dwarfs may have thin hydrogen layers consistent with spectral evolution theory.

As noted by Bergeron, Saffer, & Liebert (1992), the implied hydrogen layer mass of $1.5 \times 10^{-4}M_\star$ pushes the mean mass of DA white dwarfs from $0.56M_\odot$ up to $\sim 0.60M_\odot$. The mean mass of my subset of DAV stars is $0.63 \pm 0.07M_\odot$, comparable to the mean mass derived from the average $\log g$ of Daou et al. (1990) at $0.63 \pm 0.14M_\odot$, and Bergeron et al. (1992). It is also consistent with the $0.58 \pm 0.13M_\odot$ of Weidemann & Koester (1984). Thus, all of the studies of DA white dwarf masses agree in spite of the radically different DA white dwarf samples, but now the average mass is closer to $0.60M_\odot$.

Further comparisons of DA white dwarf observations to models with thick hydrogen layers should constrain their physical properties more accurately, provide a clearer understanding of their evolutionary cooling, and give us a better vantage point for determining the differences between the origin of hydrogen and helium atmosphere white dwarfs.

I am grateful to J.C. Clemens, S.D. Kawaler, S.J. Kleinman, R.E. Nather, D.E. Winget, and M.A. Wood for their encouragement and many discussions. This research was supported by a Los Alamos National Laboratory Director's Postdoctoral Fellowship.

References

Arcouragi, J.-P., & Fontaine, G. 1980, ApJ, 242, 1208
Barstow, M.A., Holberg, & Koester, D. 1994, MNRAS, 268, L35
Bergeron, P., et al. 1993, AJ, 106, 1987
Bergeron, P., Saffer, R.A., & Liebert, J. 1992, ApJ, 394, 228
Bradley, P.A. 1993, Ph. D. Thesis, Univ. Texas
Bradley, P.A. 1995, in preparation
Bradley, P.A., Winget, D.E., & Wood, M.A. 1993, ApJ, 406, 661, (BWW93)
Brassard, P., Fontaine, G., Wesemael, F., & Hansen, C.J. 1992b, ApJS, 80, 369
Brassard, P., Fontaine, G., Wesemael, F., Kawaler, S.D., & Tassoul, M. 1991, ApJ, 367, 601
Brassard, P., Fontaine, G., Wesemael, F. & Tassoul, M. 1992a, ApJS, 81, 747
Caughlan, G.R., & Fowler, W.A. 1988, Atomic and Nuclear Data Tables, 40, 334
Clemens, J.C. 1994, Ph. D. Thesis, Univ. Texas

Clemens, J.C. 1995, these proceedings

D'Antona, F., & Mazzitelli, I. 1989, ApJ, 347, 934

D'Antona, F., & Mazzitelli, I. 1991, in IAU Symp. 145, The Photospheric Abundance Connection, ed. G. Michaud & A. Tutukov (Dordrecht: Reidel), 399

Daou, D., Wesemael, F., Bergeron, P., Fontaine, G., & Holberg, J.B. 1990, 364, 242

Fontaine, G., Brassard, P., Bergeron, P., & Wesemael, F. 1992, ApJ, 399, L91

Fontaine, G., Brassard, P., & Wesemael, F. 1994, ApJ, 428, L61

Fontaine, G., & Wesemael, F. 1987, in IAU Colloq. 95, The Second Conference on Faint Blue Stars, ed. A.G.D. Philip, D.S. Hayes, & J. Liebert, (Schenectady: Davis), 319

Fontaine, G., & Wesemael, F. 1991, in IAU Symp. 145, The Photospheric Abundance Connection, ed. G. Michaud & A. Tutukov, (Dordrecht: Reidel), 421

Harrington, R.S., & Dahn, C.C. 1980, AJ, 85, 454

Iben, I.Jr., & MacDonald, J. 1985, ApJ, 296, 540

Iben, I.Jr., & MacDonald, J. 1986, ApJ, 301, 164

Jones, P.W., Pesnell, W.D., Hansen, C.J., & Kawaler, S.D. 1989, ApJ, 336, 403

Kawaler, S.D., Hansen, C.J., & Winget, D.E., 1985, ApJ, 295, 547

Kepler, S.O., & Nelan, E.P. 1993, AJ, 105, 608

Koester, D., & Allard, N. 1993, in White Dwarfs: Advances in Observations and Theory, ed. M.A. Barstow (Dordrecht: Kluwer), 237

Koester, D., Allard, N.F., & Vauclair, G. 1995, A&A, in press

Koester, D., & Finley, D. 1992, in The Atmospheres of Early Type Stars, ed. U. Heber & C.S. Jeffery, (Springer: Berlin), 314

Lamb, D.Q., & Van Horn, H.M. 1975, ApJ, 200, 306

McCook, G.P., & Sion, E.M. 1987, ApJS, 65, 603

Mazzitelli, I., & D'Antona, F. 1986, ApJ, 308, 706

O'Donoghue, D., & Warner, B. 1982, MNRAS, 200, 573

O'Donoghue, D., & Warner, B. 1987, MNRAS, 228, 949

Saio, H., & Cox, J.P. 1980, ApJ, 236, 549

Tassoul, M., Fontaine, G., & Winget, D.E. 1990, ApJS, 72, 335,

Weidemann, V., & Koester, D. 1984, A&A, 132, 195

Wood, M.A. 1992, ApJ, 386, 539

Wood, M.A. 1995, these proceedings

The Hydrogen Layer Mass of the ZZ Ceti Stars

*J. C. Clemens**

Iowa State University, Department of Physics and Astronomy,
Ames, IA 50211, U.S.A.

The ZZ Ceti stars have resisted asteroseismological analysis for two reasons: the small amplitude pulsators exhibit too few pulsation modes to define a uniquely identifiable pattern, and the large amplitude pulsators have complex power spectra which change radically on many timescales. By examining the pulsation properties of ZZ Ceti stars as a group rather than as individuals, we have found a way to overcome these difficulties.

We have combined Whole Earth Telescope observations of ZZ Ceti stars with all published single-site data in order to explore the relationships between effective temperature, pulsation period and pulsation amplitude. As expected from simple theoretical arguments, the hottest ZZ Cetis have the shortest periods and smallest amplitudes. Both quantities increase monotonically with decreasing temperature. More surprisingly, we have found striking similarities between the periods of the pulsation modes in different stars. When the the periods of all the stars are plotted on a single diagram, they do not form a random pattern, but fall into distinct bands, indicating an underlying structural similarity.

Since the structural property which has the largest effect on mode periods is the mass of the surface hydrogen layer, we interpret the similarity in pulsation periods as evidence for a very small range of hydrogen layer masses in the ZZ Ceti stars. The best identification for the common mode pattern discovered is a sequence of $l = 1$ modes. The published models which best reproduce the periods of the modes identified have hydrogen layer masses near 10^{-4} $M_\star$, the value expected from models of DA formation in which nuclear burning is the dominant process fixing the hydrogen layer mass.

These results are published in **Baltic Astronomy** and will appear in expanded form in **The Astrophysical Journal**.

---

* Hubble Fellow

Grasping at the Hot End of ZZ Ceti Variability

Alfred Gautschy and Hans-Günter Ludwig

Max-Planck-Institut für Astrophysik, Postfach 1523, D-85740 Garching, F.R.G.

Theoretical pulsation studies have shown that the temperature of the blue edge of the ZZ Ceti instability strip sensitively depends on the convective efficiency – as measured by the mixing-length parameter – which is employed in the construction of the superfical hydrogen convection zones of the white dwarf models. Here we deduce the stratification of the convective layers from 2-D hydrodynamical simulations (cf. Ludwig et al., 1994, A&A 284, 105). These simulations use opacities incorporating Ly-α satellites (cf. Allard et al., 1994, A&A submitted) and nonlocal frequency-dependent radiative transfer. The hydrodynamical envelope models are connected to chemically stratified, hydrostatic interior models in thermal equilibrium. Our current EOS accounts for ionisation, degeneracy effects, but not for Coulomb interaction. With the composite models we performed linear, nonadiabatic nonradial stability analyses (with the method described in Glatzel & Gautschy, 1992, MNRAS 256, 209) for various choices of the mass of the surface hydrogen layer q_H, of spherical degrees ℓ, and of effective temperatures. The lagrangian perturbation of the convective flux is assumed to vanish (frozen-in approximation). We obtain the following preliminary results (see Fig. 1): (i) only models with effective temperature less than 12 600 K show unstable modes; (ii) no unstable modes are found for $\ell > 1$; (iii) larger values of q_H destabilize modes with shorter periods. While being consistent with the latest observational findings our results do not agree with previous pulsation studies. Analyses to track down the origin of the disagreements are under way.

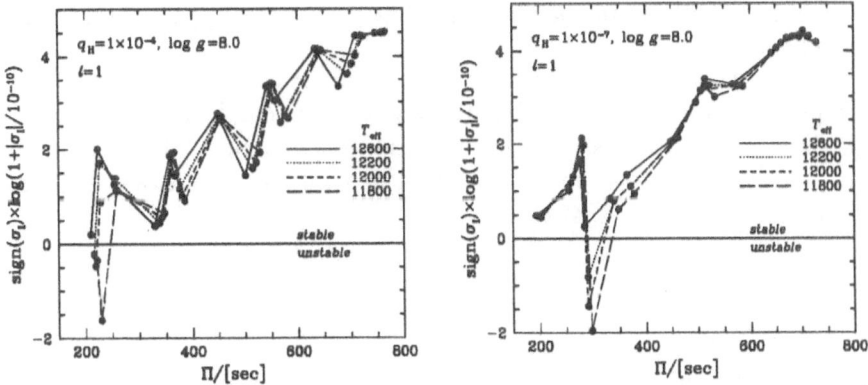

Fig. 1. Imaginary parts of the eigenfrequencies, σ_I, versus corresponding periods, Π, for eigensolutions of spherical degree $\ell = 1$ for four white dwarf models with different effective temperatures. The models have the same relative mass fraction of the hydrogen layer ($10^{-4} M_*$ left, $10^{-7} M_*$ right panel) and surface gravity $\log g = 8.0$.

WET Experiences in Central Asian Observatories

Edmundas G. Meištas

Institute of Theoretical Physics and Astronomy,Goštauto 12, LT-2600 Vilnius, Lithuania

Central Asia is located at an important longitude to fill a gap in the WET network. The Lithuanian astronomical station on Mt. Maidanak, in Uzbekistan was successfully tested for future WET campaigns in May 1992 (Meištas and Solheim 1993, in White Dwarfs: Advances in observation and theory, p.549; Solheim 1993, Baltic Astronomy, 2, 383)

But in February 1993 the observatory was nationalized by the Uzbekistan government, and almost all astronomical activities there were stopped. The future use of this observatory for WET campaigns became uncertain. We have looked for the possibility to use other Central Asian observatories for WET. Contact with the Kazakhstan Fesenkov Astronomical Institute was established and in October 1993 and in May 1994 WET observations from the Assy–Turgen observatory in the Kazakhstan took place. In 1994 Uzbekistan astronomers proposed a new agreement for the collaboration, and during Aug 1994 WET campaign we worked on Maidanak again.

In order to use telescopes in the WET campaigns at remote and not easily reachable places, a lightweight three–channel photometer has been constructed. It was built at the University of Tromsø and modified later by Vilnius University Astronomical Observatory electronic group (Kalytis et al. 1993, Baltic Astronomy, 2, 504). With their new amplifiers–discriminators it became even more reliable and lightweight.

To make observations more efficient we need to transfer observed data to our heardquarters in Texas once a day. Because there is no phone connections in both above mentioned observatories, we tested microsatellite TUBSAT for the communication and found out that it could be used for our purposes (Bruvold 1993, Baltic Astronomy, 2, 492).

The Vista for Seismological Thermometry of the DBV White Dwarfs

Darragh O'Donoghue

Dept. of Astronomy, University of Cape Town

Two new He-rich oscillators discovered in the Edinburgh-Cape Blue Object Survey are announced. One, EC15330–1403, is an AM CVn-type double degenerate binary and is described by O'Donoghue et al. (1994, MNRAS, in press). The other, EC20058–5234, is the 8th DBV to be found and will be discussed by Koen et al. (1994, MNRAS, in preparation). This star is remarkable in that the *longest* period in its power spectrum is at ~280 s, much shorter than any periods seen in the other DBVs. Moreover, EC20058-5234 has a period at 136 s, which is a close match to the period of the k = 1, ℓ = 1 g-mode in a model appropriate for GD358 calculated by Kawaler (1994, private communication). If the correlation between period and T_{eff} established for the DAVs by Clemens (1994, these Proceedings) applies to the DBVs, this star should be the hottest of all the DBVs because its period is the shortest. On these grounds, PG1351+489, which has a shorter period than GD358 should also be hotter than GD358.

The best available *classical* temperatures for DB white dwarfs from Thejll, Vennes & Shipman (1991) are not sufficiently precise to check against the *seismological* temperatures. Prospects for improvements in both methods are discussed. In particular, HST+FOS observations of GD358 can yield a S/N ratio of 100 in 30 s. As the temperature variations of GD358 during a pulsation may be several per cent, the prospects for 'real time' seismological thermometry of GD358 are encouraging.

References

Thejll P., Vennes S., & Shipman, H., 1991, ApJ 370, 355

Ionized Helium and Carbon in the DBV GD358

J. L. Provencal and H. L. Shipman

Department of Physics and Astronomy, University of Delaware, Newark, DE 19716

The prototype helium variable white dwarf (DBV) GD358 is a multiperiodic, g-mode pulsator, with dominant periods ranging from about 770 to 400 s. The Whole Earth Telescope has resolved the light curve into a series of g-modes nearly equally spaced in period, suggesting the variations originate in a thin helium layer ($M_{He} \simeq 10^{-6} M_*$) over a carbon/oxygen core (Bradley & Winget 1994). In 1988, Sion et al. reported the detection of CII (1335.75Å) and HeII (1640.43Å) in GD358's IUE spectra. The authors theorized that carbon's presence is consistent with dredge up from the underlying core by the helium convection zone. GD358 is therefore a unique object: pulsation analysis has given us an internal structure which we can use to calibrate convection.

The dramatic improvement in spectral resolution and signal to noise ratio provided by HST enables us to isolate faint features, and measure strengths and velocities with great accuracy, allowing us to separate circumstellar, photospheric, and interstellar features. Intrigued by the CII and HeII in GD358's spectra, we obtained four GHRS spectra of GD358, centered on 1215, 1310, 1335 and 1640 Å. Each spectrum covers about 35Å, with a resolution of about 0.07Å per diode and has an integration time of 897.6 seconds. We reduced the data using the IRAF routines STSDAS and NOAO (onedspec and splot) running on a Sparc LX.

The HeII and CII features are clearly present in our spectra. The CII doublet, at 1334 and 1335Å, has equivalent widths and central velocities of 158 mÅ and 8 km s^{-1}, and 220 mÅ and 11 km s^{-1}, in agreement with the numbers quoted by Sion et al. (1988). The HeII (1640) line is fairly symmetric, with an equivalent width of 107 mÅ and a central velocity close to 0 km s^{-1}. Comparison with models presented by Thejll indicate a best fit with a temperature of 28,000 K.

The Lyman α + HeII (1215) region reveals a subtle hint of absorption wings flanking the geocoronal emission, and will provide us with an improved limit on GD358's hydrogen abundance. Our first, conservative, estimates, using older models, and assuming $T_{eff} = 27500$K, is $\log(n(H)/n(He)) < 10^{-7}$. Work is in progress to derive an accurate abundance for both hydrogen and carbon.

Stabilized Restoration of Pulsating White Dwarf Spectra

B. Serre, S. Roques, B. Pfeiffer, N. Dolez, G. Vauclair, P. Maréchal, A. Lannes.

Observatoire Midi-Pyrénées — LAT/CERFACS & GdR 134 CNRS
14, Avenue Edouard Belin — 31400 Toulouse — FRANCE

The temporal spectra obtained from incomplete records of luminosity variations of white dwarfs can be reconstructed from a deterministic procedure related to band-limited interpolation (Lannes et al. 1987).

In such a problem, the stability of the underlying deconvolution procedure is governed in particular by the choice of the frequency supports over which the restoration has to be achieved. Thus, the necessity of getting information about the energy together with the lifetime of a given peak in the Fourier spectrum leads us naturally to time-frequency analysis.

In this context, the matching pursuit algorithm (Mallat and Zang 1993) plays a descisive role. The associated constructive process allows us to detect and characterize the time-frequency components of the signal one by one, from the highest energy one to the lowest.

In the final stage, the spectrum of the star can be reconstructed step by step, provided that it can be broken up, whereas a global reconstruction would be unstable at the same target resolution.

The method has been applied to 7 hours of data on the DAV GD154. The final resolution corresponds to a 12 hour observation synthesis. Finally, the results are successfully compared with raw data obtained during a WET (Nather et al. 1990) campain.

The reader is referred to Serre et al. (1994) for more details.

References

A. Lannes, S. Roques and M.J. Casanove, J. Opt. Soc. Am. A, Vol. 4, No. 1, pp. 189-99, 1987.

S. Mallat and Z. Zhang, in Progress in Wavelet Analysis and Applications, edited by Meyer & Roques, Frontières, Gif-sur-Yvette, pp. 155-64, 1993.

R.E. Nather, D.E. Winget, J.C. Clemens, C.J. Hansen, and B.P. Hine, ApJ,361, pp. 309-317, 1990.

B. Serre, P. Maréchal, S. Roques, B. Pfeiffer, G. Vauclair and A. Lannes, IEEE Int. Symp. on Time-Frequency and Time-Scale Analysis, Philadelphia, pp. 349-352, 1994.

Part VII
EUV and X-Ray Observations

Extreme Ultraviolet Spectroscopy of White Dwarfs

*M.A. Barstow[1],   J.B. Holberg[2],   D. Koester[3],   J.A. Nousek[4]   &
K. Werner[3]*

[1] Department of Physics and Astronomy, University of Leicester, University Road, Leicester LE1 7RH, UK
[2] Lunar and Planetary Laboratory, University of Arizona, Tucson AZ, USA
[3] Institut für Astronomie und Astrophysik der Universität, D-24098 Kiel, Germany
[4] Department of Astronomy and Astrophysics, Pennsylvania State University, University Park, PA, USA

Abstract: It has long been predicted that Extreme Ultraviolet (EUV) observations of white dwarfs would be one of the principal keys to understanding the composition and structure of white dwarf atmospheres. Indeed, the ROSAT X-ray and EUV sky survey has revealed that most H-rich white dwarfs hotter than 40,000K contain significant quantities of heavy elements in their atmospheres. However, the ROSAT data are unable to say much about the role of helium in the hottest white dwarfs, which remains an important question. Spectroscopic observations with the Extreme Ultraviolet Explorer allow a direct and sensitive search for HeII absorption to be performed. Furthermore, they potentially reveal the detailed nature of the sources of EUV opacity. In most cases examined no trace of He is found but HZ43 and G191−B2B show tantalising evidence for the presence of He, which would have important implications for our views of white dwarf evolution.

1 Introduction

One of the most significant problems in the study of hot white dwarf evolution has been the existence of two distinct groups having either H or He dominated atmospheres and the possible relationships between them and their proposed progenitors, the central stars of planetary nebulae (CPN). While the very hottest H-rich DA white dwarfs outnumber He-rich DOs by a factor 7 (Fleming Liebert & Green 1986), the relative number of H- and He-rich CPN is about 3:1. Secondly, the absence of He-rich stars in the temperature range 30,000–45,000K would suggest that the two groups are not entirely distinct and that at least some He-rich objects must evolve into DA white dwarfs and vice versa. Several competing physical processes can affect the composition of a white dwarf atmosphere. He and heavier elements tend to sink out under the influence of the strong gravitational field, leaving a layer of H at the surface, but this process can be counteracted by radiation pressure. Convective mixing may also play a significant role.

A thorough understanding of the physical mechanisms and their influence on white dwarf atmospheres must be founded on a detailed knowledge of the actual composition and structure of white dwarf envelopes for a range of temperatures and spectral types. For the brightest stars, IUE echelle spectra have given us important insight through the detection of C, N, O, Si, Fe and Ni (eg. Vennes et al. 1992, Holberg et al. 1993 & 1994, Werner & Dreizler 1994). Similar studies have also been carried out with the HST GHRS instrument (Vidal-Madjar et al. 1994). However, these features are only visible in those objects with the most extreme abundances. Furthermore, the limits placed on the presence of He (He/H$< 10^{-3}$) by UV spectroscopy are too crude to be of physical significance. It has been realised for some time that extreme ultraviolet (EUV) spectroscopy provides the most sensitive search for the presence of photospheric He in DA white dwarfs. In addition, the effects of heavy element absorption are expected to be visible at abundances lower than in far UV observations, at least through their influence on the shape of the continuum flux if not by direct detection of lines.

In this paper, we review the development of EUV spectroscopy as a tool to study white dwarf atmospheres. We also consider the complementary role of broad band EUV photometry and observations in other wavebands in aiding our understanding of white dwarfs. Much of this work underpins the Extreme Ultraviolet Explorer (EUVE) spectroscopy programme, which is now yielding exciting new results.

2 Early EUV spectroscopy

The 1980s saw the first use of EUV spectroscopy in the study of white dwarf atmospheres, concentrating on He in particular as the dominant opacity source. Malina, Bowyer and Basri (1982) reported the detection of an HeII absorption edge at 228Å in a spectrum of HZ43 recorded by a sounding rocket borne spectrometer. The strength of the feature corresponded to a helium/hydrogen abundance ratio of $1 - 6 \times 10^{-5}$. Subsequently, more sensitive observations made with the EXOSAT transmission grating spectrometer, failed to confirm this report (Paerels et al. 1986a). With a peak effective area of about 0.5cm^2 and a spectral resolution of 6Å, the EXOSAT data were able to place an upper limit (90% confidence) on the He/H abundance ratio of 3×10^{-6}, based on the non-detection of an HeII 228Å edge. For technical reasons, associated with the deployment of the transmission gratings, only two other white dwarf spectra were obtained by EXOSAT. Helium was not detected in either of these (Feige 24 and Sirius B), an upper limit of 1×10^{-5} being obtained for the He abundance in Sirius B (Paerels et al. 1988). In just treating the He 228Å absorption as a simple edge rather than a series of lines, the abundance limits quoted might well be somewhat misleading. On the other hand, using the EXOSAT spectrum as the sole measure of temperature and gravity increases the number of unconstrained parameters in the study. Reanalysing the spectrum of HZ43 with LTE synthetic spectra which are fully line-blanketed and imposing T and log g contraints from optical data (Napiwotzki et al. 1993), we obtain an upper limit to the He abundance of 2×10^{-6}, just slightly improved (but better founded) compared to the earlier result (figure 1).

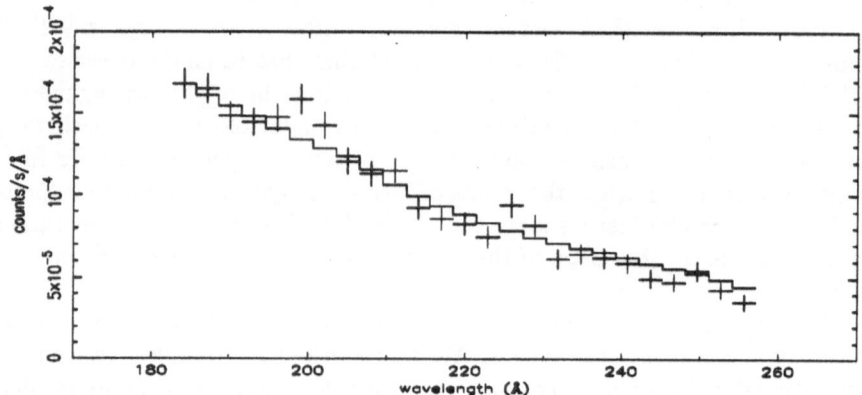

Fig. 1. EXOSAT TGS spectrum of HZ43 compared to the best fit model with T=47000K and log g=7.88 with He/H=2 × 10^{-6}, corresponding to the 90% upper limit discussed in the text

In contrast to HZ43 and Sirius B, the EXOSAT spectrum of Feige 24 could not be explained by a simple pure H or H+He model atmosphere (Paerels et al. 1986b). A plausible explanation of this result has been provided by Vennes et al. (1989), who find that introducing small traces of large numbers of heavy elements such as C, N, O, Ne, Si, Ar and Ca into a homogeneous model can qualitatively reproduce the spectrum. However, the precise abundance and composition (even which species were really present) could not be uniquely determined from the data. This explanation of the Feige 24 spectrum seems reasonable but it does require that, although only a few thousand degrees hotter than HZ43, is has a very different composition. Further evidence for the presence of heavy elements in at least a few DA is also found in IUE spectra. For example, C, N, Si and Fe have been observed in both Feige 24 and G191−B2B (eg. Vennes et al. 1992). In addition, a rocket EUV spectrum of G191−B2B (Wilkinson, Green & Cash, 1992) show features which the authors identify as OIII and FeIV absorption but again, no He.

3 EUV/X-ray photometry, IUE and optical spectroscopy

The combination of EXOSAT spectroscopic and photometric results (eg. Paerels & Heise, 1989; Koester 1989) led to rather uncertain conclusions concerning the general population of DA white dwarfs. It was not clear whether He, if present, was homogeneously distributed or in the more physically realistic stratified configuration (see eg. Vennes et al. 1988). Nor was it possible to determine whether

Feige 24 or HZ43 were most representative of DA atmospheres. The ROSAT EUV and X-ray all sky survey has been able to address the general properties of the population (Barstow et al. 1993). Although providing broadband data, like EXOSAT, the better spectral coverage and narrower bandpasses allowed tighter constraints to be placed on the possible atmospheric composition and structure of each star in the sample. This work demonstrated that *most* of the hottest DA white dwarfs (T> 40,000K) must have heavy elements in their atmospheres, whereas stars below 40,000K mainly seem to have little or no opacity in addition to that provided by hydrogen. For the hottest stars it is clear that He alone, in either configuration, is insufficient to account for the total opacity and heavier elements must be involved. However, the ROSAT data do to not allow us to decide whether He makes any contribution or not.

In addition to the above work, the ROSAT survey led to the discovery of a number of new white dwarfs. Several of these were sufficiently bright to be studied by the IUE echelle. RE2214−492 and RE0623−377 in particular were found to be very similar to Feige 24 and G191−B2B but contain even greater abundances of C, N, O and Fe (Holberg et al. 1993). More detailed work on these stars has also led to the detection of Ni ions (Holberg et al. 1994, Werner & Dreizler 1994).

While the role of He in DA white dwarf atmospheres has remained unclear, important results have emerged from an optical spectroscopic study of DAO white dwarfs by Bergeron et al. (1994). DAO stars, hybrid objects having a predominantly H envelope with spectroscopically detectable traces of He, have often been considered to represent a natural intermediate stage of evolution from DO to DA white dwarfs, where the proposed upward migration of residual H in the envelope is not yet complete. This hypothesis makes the important prediction that the atmospheric structure of DAO and DA stars should be stratified. However, it is clear from the work of Bergeron et al. that, at least in the optical line forming region, the majority of DAO stars have homogeneous atmospheres. As radiative levitation is unable to support the observed quantities of He (Vennes et al. 1988) an alternative, but as yet unknown, mechanism must be found to explain this result. Furthermore, we must remain open-minded about the structure of DA atmospheres, since a similar effect may exist in these stars.

Bergeron et al. also make a potentially important point concerning the effect of He on DA white dwarf atmospheres, finding that even small traces $(He/H> 10^{-5})$, at levels spectroscopically undetectable in the visible or UV bands, affect the Balmer line profiles. Consequently, the temperature measurements of the hottest white dwarfs may be erroneously high if trace He is present, but not accounted for in the models because it is not seen and, therefore, its abundance unknown. However, we note that Napiwotzki (these proceedings) finds that NLTE models do not show the dependency of the Balmer lines on the He abundance. Such a potential problem is not necessarily restricted to He. It has been demonstrated that similar effects may be caused by the presence of heavier elements (Lanz & Hubeny 1994, Werner & Dreizler 1993). Nevertheless, the importance of He abundance measurements in the EUV is increased, as they can potentially achieve a level of sensitivity where the possible effects on the Balmer lines become insignificant.

4 Spectroscopy of DA white dwarfs with EUVE

The launch of NASA's EUVE mission has provided us with a renewed opportunity of searching for He in DA stars and examining other sources of opacity. With a peak effective area about 20 times ($\approx$ 1cm^2) that of EXOSAT and a superior spectral resolution ($\approx 0.5 - 2.0$Å), EUVE allows us to obtain spectra throughout most of the EUV band, from $\approx 60 - 700$Å.

We have carried out an extensive search for photospheric He features in a number of hot DA white dwarfs. In these studies both a stratified H upon He configuration and, in the light of the work of Bergeron et al. (1994), a homogeneous mixture are considered. From an analysis point of view, the white dwarfs in question (drawn both from the EUVE archive and our own guest observer programme) divide into two distinct groups – those with and those without significant heavy element opacity. EUV spectra of the latter category can be described completely by a synthetic spectrum calculated for an H+He model atmosphere, together with a suitable model of the opacity of the interstellar medium (Rumph, Bowyer & Vennes 1994). If heavy elements are present the stars can only be properly studied using complex line-blanketed models with a comprehensive selection of opacity sources. However, it is possible to study just the region of the EUV spectrum near the HeII Lyman series, reproducing the general shape of the spectrum with an H+He model plus an exponential roll-off towards short wavelengths. A complete description of the data reduction and spectral fitting techniques used has been published elsewhere (Barstow, Holberg & Koester 1994a & 1994b).

Data obtained during the first year of EUVE observations are limited in signal-to-noise by a detector fixed pattern efficiency variation of $\approx \pm 20\%$. We find no evidence for any photospheric He in any of several DA white dwarfs studied (table 1). It is important to note that, in the homogeneous case the abundance limits are well below the level at which the He will significantly effect the Balmer line temperature determinations.

Table 1. Measured H layer mass and He abundance 90% limits in DA white dwarfs

Name	T($\pm 1\sigma$)	log g($\pm 1\sigma$)	H mass ($M_\odot$)	He/H
Feige 24	59,800(3,400)	7.5(0.51)	2.4×10^{-13}	3.0×10^{-6}
G191−B2B	57,314(750)	7.50(0.07)	2.6×10^{-13}	
GD71	32,045(148)	7.72(0.05)	1.0×10^{-12}	1.1×10^{-6}
GD246	54,987(931)	7.71(0.09)	9.7×10^{-14}	5.0×10^{-6}
HR8210	34,540(100)	8.95(0.06)	1.0×10^{-14}	
HD15638	46,000(+1,800/-850)	8.5(+0.1/-0.2)	4.1×10^{-14}	

Towards the end of the first year of EUVE spectroscopic observations, a technical remedy for the fixed pattern efficiency variation was implemented. The 'dither mode' divides each observation into a series of separate pointings to average out the fixed pattern, providing improved s/n by eliminating most of the spurious features. Figure 2 shows the complete spectrum of HZ43, recorded in this way. Comparing this with a smooth model spectrum indicates, from the excess scatter on the data points, that the residual efficiency variation has improved to an amplitude of $\pm 5\%$. Closer inspection of the HZ43 data, near the HeII Lyman series (figure 3) shows a weak, but distinct, absorption dip at the position of HeII 304Å, with an equivalent width of 200 ± 100mÅ. However, as there are other 'features' in the spectrum which are only a little weaker, that at 304Å may not be real. Even so, it gives a good indication of the sensitivity of the observation to the presence of photospheric He. If attributed to an interstellar component, the implied HeII column density would be $\approx 5 \times 10^{17}$cm^{-2}. This would give rise to a large absorption edge, as illustrated in figure 3, which is not seen.

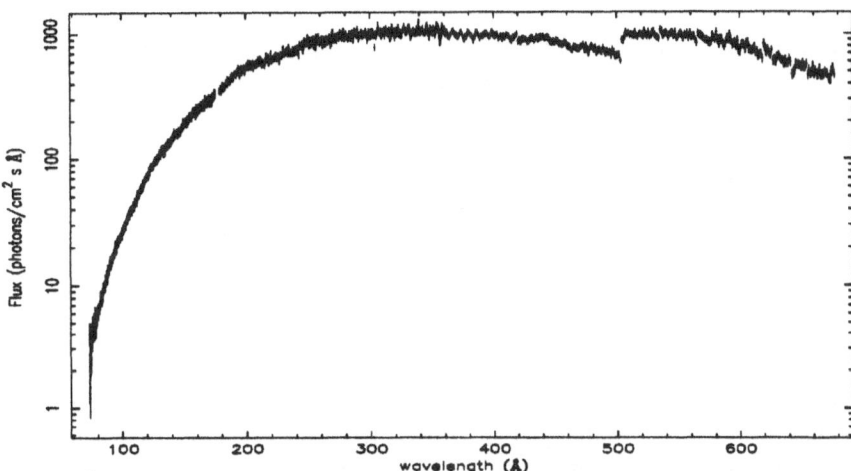

Fig. 2. Complete EUVE spectrum of HZ43 after deconvolution with the instrument response. The best fit model corresponds to T=50560K, log g=7.9, and log He/H=-5.9.

In fact, the best fit homogeneous model to the complete spectrum does require a small but finite amount of He, with an abundance of 1.3×10^{-6}, when the allowed temperature and gravity are constrained by the results of Napiwotzki et al. (1993). Forcing the He abundance down to 1×10^{-7} or lower leads to an increase in the value of χ^2 which is significant at the 95% confidence level. However, since the shape of the continuum flux is quite sensitive to possible systematic effects in the models and instrument calibration, we are cautious about

interpreting this result as a real detection of He in HZ43. Concentrating purely
on the 180–260Å region covered in the EXOSAT analysis and on the limits placed
on the strength of the HeII lines gives a much lower value for the He abundance

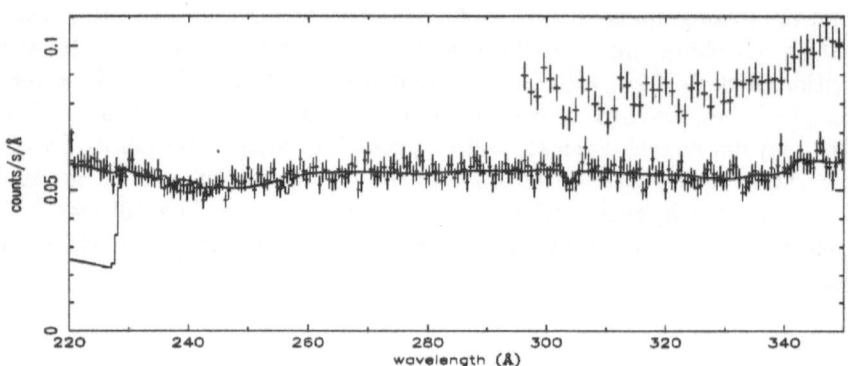

Fig. 3. Count spectrum of HZ43 between 220 and 310Å, showing a model with 10^{-7}
of photospheric He and demonstrating the size of interstellar edge which would give
the same 304Å line strength.

of $\approx 2 \times 10^{-7}$ or less. Whichever limit or value we adopt, it is at odds with the
calculations of radiative levitation theory (Vennes et al. 1988), which predict an
He abundance of at least 3×10^{-6} for a star with the mass and temperature of
HZ43.

In contrast to HZ43, the spectrum of G191−B2B (figure 4) shows a large
number of features, most of which can probably be attributed to the heavy elements which we already know to be present in the photosphere of this star (eg.
Vennes et al. 1992, Holberg et al. 1994). However, we note that some of the
absorption features correspond to the 304Å, 256Å, 237Å and 234Å lines of HeII,
only the 243Å line appears to be absent from the series. This is illustrated in figure 4, where a homogeneous H+He model (T=57,300; log g=7.5; log He/H= −5)
incorporating an exponential roll-off is compared with the observed spectrum,
highlighting the line positions. There are two very strong features adjacent to
the position of 243Å and it is possible that the line is comparatively too weak
to be observed at the resolution of EUVE. Indeed, synthetic spectral models
incorporating heavy elements and He demonstrate that this will happen. At the
moment, it is necessary to add a strong note of caution to this interpretation.
Although there do not appear to be any other strong lines coincident with the
HeII wavelengths, G191−B2B has a complex EUV spectrum which we are not
yet able to model in detail and there may be important lines which are omitted
from our calculations. This is amply illustrated by the fact that incorporating

all the elements detected by UV observations, we still fail to reproduce even the approximate shape of the EUV energy distribution. Furthermore, the large number of features observed, many of which will be blends of several lines at the resolution of EUVE, may give a false continuum. Coupling this with the possible effect of the heavy elements on the HeII level populations, compared to simple H+He models, the estimated He abundance, if He is really present, can only be regarded as very crude. However, if real, an abundance of 10^{-5} could affect the Balmer line profiles in this star, requiring a re-evaluation of its temperature.

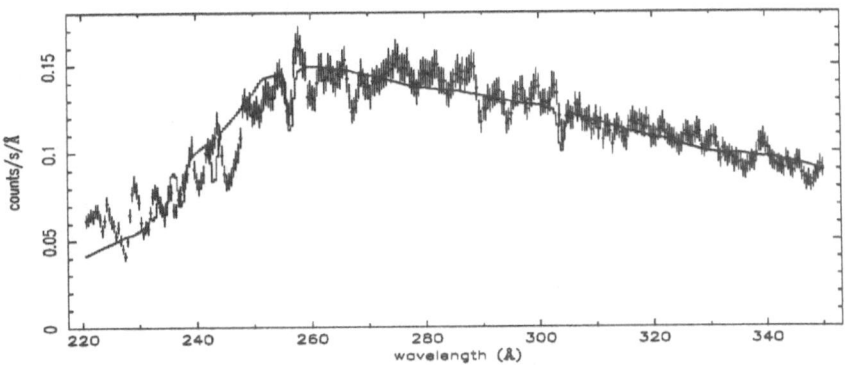

Fig. 4. Count spectrum of G191−B2B between 220 and 310Å. The best fit model for G191−B2B includes 1×10^{-5} of He.

5 EUVE studies of H deficient stars

One of the most interesting EUV sources discovered by the ROSAT sky survey is the DO white dwarf RE0503−289 (Barstow et al. 1994). The only DO white dwarf detected in the EUV surveys it lies along a very low column density line of sight, making it an ideal target for the EUVE spectrometers. With a temperature of 70,000K and log g ≈ 7 the dominant component in the atmosphere in addition to He is C, with an abundance ≈ 1% . The IUE high dispersion spectrum shows CIII, CIV, NV, OV and SiV, in addition to He, but little evidence for any other species. There is also good evidence for episodic mass loss, from variations in the line profiles (Barstow & Sion, 1994) The EUV spectrum (figure 5) is dominated by the HeII Lyman absorption series, which effectively reduces the flux to zero at wavelengths below 228Å. This illustrates why this is the only DO white dwarf detected in the EUV sky surveys, as a very low column density is needed to see out to 228Å. One ion not seen in the IUE data, but clearly detected in the EUV is NIV. In figure 5 we show, for comparison with the

observed spectum, a synthetic spectrum calculated for a model atmosphere with
T=70,000K log g=7.0, C/He=1% and N/He=0.01% . The general shape of the
observational data is reproduced by the model but the NIV lines are clearly
too strong, implying an abundance substantially lower than 0.1%. The absolute
level of the model spectrum is scaled to match the EUVE data. However, this
hides the most important discrepancy between the two, since the absolute flux
predicted by the model atmosphere in the wavelength range 228–400Å is nearly
an order of magnitude higher than observed. Hence, there must be significant
sources of EUV opacity present in the atmosphere of the star which are not
included in the model.

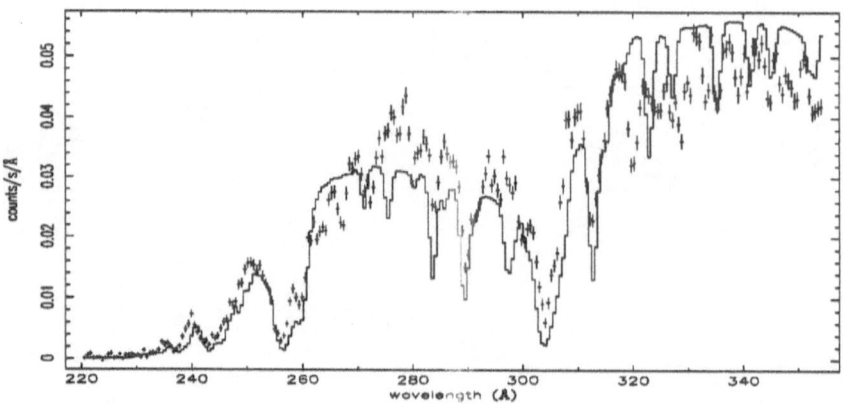

Fig. 5. EUVE spectrum of RE0503−289 compared to an HeCN model as discussed the
text.

H1504+65 is a peculiar pre-white dwarf object, which appears to be completely devoid of H or He (Nousek et al. 1986). A recent optical analysis, using
NLTE models, gives a temperature of $170,000 \pm 20,000$K and log g= 8 ± 0.5
(Werner, 1991). The atmosphere is solely made up of C and O by equal parts
with less than 1% He. It appears to be an extreme PG1159 star, a pre-white
dwarf which has lost its entire H and He envelopes as a consequence of a late
He shell flash. As the naked core of a former AGB red giant this gives us, in
principle, a unique opportunity to study the thermonuclear $^{12}C(\alpha, \gamma)^{16}O$ reaction rate. H1504+65 is the second brightest EUV source in the ROSAT WFC
survey but, having a relatively high interstellar column density of $6 \times 10^{19} \text{cm}^{-2}$,
is only visible at the shorter EUV wavelengths. H1504+65 is only detected in
the EUVE short wavelength spectrometer (figure 6). The whole spectrum covers the range 80–180Å and is dominated by absorption from OVI. A 180,000K
model spectrum matches the positions of the lines accurately but fails to match
the data in a number of respects. As for RE0503−289 the predicted level of the

continuum flux is about an order of magnitude higher than that observed. A number of prominent absorption lines are not accounted for by the model and, at the time of writing, these remain unidentified. There is clearly some source of opacity which decreases the flux between 100Å and 115Å more rapidly than the OVI line series alone. Furthermore, there is a signficant tail of continuum emission below the OVI edge, which is not predicted by the model. Neither C nor N have any important transitions which contribute to the opacity in this region of the EUV but Ne does, although this has not been detected spectroscopically at other wavelengths. The detection of Ne would be an indication that H1504+65 may have burned carbon in the past.

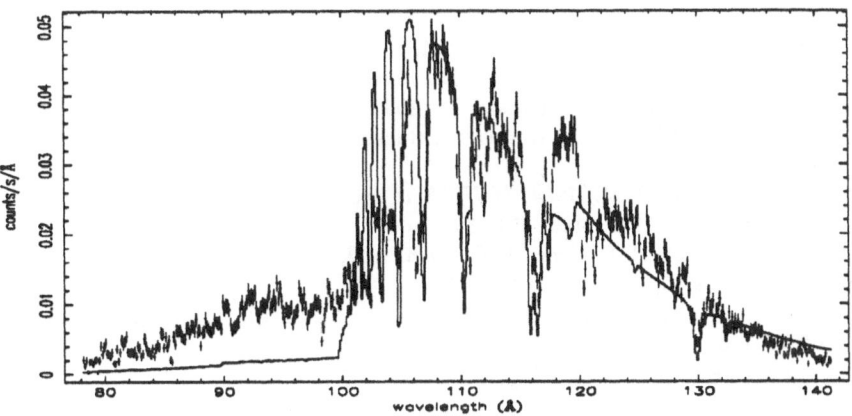

Fig. 6. EUVE spectrum of H1504+65 with a 180,000K C/O model for comparison.

6 Conclusions

It is clear that spectroscopy with EUVE is already making significant contributions to our understanding of white dwarf atmospheres, based on a firm foundation from EXOSAT, ROSAT, IUE and ground-based results. In particular, important progress has been made in understanding the role of He in DA stars. Limits placed on the abundance of He in white dwarf spanning a temperature range from 30,000K to 50,000K are 2-3 orders of magnitude below the values typical of DAO white dwarfs, supporting the growing view that DAOs do not form a direct link between DO and DA stars. However, this does not rule out the possibility that DOs still evolve into DAs by upward migration of residual H, but with the process being too rapid for a significant population of intermediate objects to build up. With the exception of (possibly) G191−B2B, the allowed abundances of He present are all below a level where the Balmer line temperature measurement becomes problematic. Therefore, it seems likely, that there is no general problem with the DA temperature scale as established so far as the effects of He are concerned. However, the contribution of heavier elements has not yet been addressed in detail.

Although DA white dwarfs are by far the largest population which can be studied by EUVE, the few H (and He) deficient stars, while rare, are very important. Since only two such objects have been discussed here it is difficult to draw any general conclusions. What is clear is that even the best state-of-the-art NLTE models are unable to completely explain either of the spectra presented here. Further work in this area is required before the EUVE data will yield the promised results.

References

Barstow M.A., et al., 1993, *MNRAS*, **264**, 16.

Barstow M.A., Holberg J.B., Koester D., 1994, *MNRAS*, **268**, L35.

Barstow M.A., Holberg J.B., Koester D., 1994, *MNRAS*, **270**, 516.

Barstow M.A., Holberg J.B., Werner K., Buckley D.A.H., Stobie R.S., 1994, *MNRAS*, **267**, 653.

Barstow M.A., Sion E.M., 1994, *MNRAS*, in press.

Bergeron P., et al., 1994, *ApJ*, **432**, 305

Fleming T.A., Liebert J., Green R.F., 1986, *ApJ*, **308**, 176.

Holberg J.B., et al., 1993, *ApJ*, **416**, 806.

Holberg J.B., Hubeny I., Barstow M.A., Lanz T., Sion E.M., Tweedy R.W., 1994, *ApJ*, **425**, L105.

Koester D., 1989, *ApJ*, **342**, 999.

Lanz T., Hubeny I., 1994, *ApJ*, in press.

Malina R.F., Bowyer, S., Basri G., 1982, *ApJ*, **262**, 717.

Napiwotzki R., Barstow M.A., Fleming T.A., Holweger, H., Jordan S., Werner K., 1993, *A& A*, **278**, 478.

Nousek J.A., Shipman H.L., Holberg J.B., Liebert J., Pravdo S.H., White N.E., Giommi P., 1986, *ApJ*, **309**, 230.

Paerels F.B.S., Bleeker J.A.M., Brinkman A.C., Gronenschild E.H.B.M., Heise J., 1986a, *ApJ*, **308**, 190.

Paerels F.B.S., Bleeker J.A.M., Brinkman A.C., Heise J., 1986b, *ApJ*, **309**, L33.

Paerels F.B.S., Bleeker J.A.M., Brinkman A.C., Gronenschild E.H.B.M., Heise J., 1988, *ApJ*, **329**, 849.

Paerels F.B.S., Heise J., 1989, *Ap.J.*, **339**, 1000.

Rumph T., Bowyer S., Vennes S., 1994, *AJ*, in press.

Vennes S., Chayer P., Fontaine G., Wesemael F., 1989, *ApJ*, **390**, 590.

Vennes S., Chayer P., Thorstensen J.R., Bowyer S., Shipman H.L., 1992, *ApJ*, **392**, L27.

Vennes S., Pelletier C., Fontaine G., Wesemael F., 1988, *ApJ*, **331**, 876.

Vidal-Madjar A., et al. 1994, *A& A*, **287**, 174.

Werner K., 1991, *A& A*, **251**, 147.

Werner K., Dreizler S., 1993, *Acta Astronomica*, **43**, 321.

Werner K., Dreizler S., 1994, *A& A*, **286**, L31.

Wilkinson, E., Green, J.C., Cash, W., 1992, *ApJ*, **397**, L51.

White Dwarfs in Close Binaries in the Extreme Ultraviolet Explorer All-Sky Survey

Stéphane Vennes[1] and John R. Thorstensen[2]

[1] Center for EUV Astrophysics, University of California
2150 Kittredge Street, Berkeley, CA 94720, USA
[2] Department of Physics and Astronomy, Dartmouth College
Hanover, NH 03755, USA

Abstract: We examine the orbital and physical properties of the *EUVE* sample of white dwarfs found in binary systems with short orbital periods ($\sim 1^{d}$). These systems, consisting of a hot white dwarf and a main sequence star, are recent survivors of a common envelope phase responsible for dramatic momentum losses and orbital shrinkage. The white dwarfs in close binaries also constitute a surprisingly large fraction of the total sample of white dwarfs discovered in the *EUVE* all-sky survey. Most secondary stars are M-dwarfs, with an average mass of $> 0.5~M_\odot$, and comparisons with some simulated samples of close binaries suggest that many more systems with very low mass secondary stars ($< 0.3~M_\odot$) remain to be discovered.

1 Introduction

A large number of hot white dwarfs were detected during the ROSAT WFC (Pounds et al. 1993) and Extreme Ultraviolet Explorer (*EUVE*) all-sky surveys (Bowyer et al. 1994). Follow up optical observations show that many white dwarfs are in fact members of binary systems; Vennes & Thorstensen (1993) have established that several binaries underwent dramatic orbital evolution and present spectroscopic characteristics similar to the well known $4.^{d}2$ binary Feige 24 (Thorstensen et al. 1978). Evolutionary models predict a number of sample characteristics, for example the primary and secondary masses and orbital period distributions (see de Kool & Ritter 1993, Yungelson et al. 1993), that can be tried in the context of the EUV-selected sample of close binaries. In the following section we examine the binary properties and in §3 we study the photospheric properties of the white dwarfs in close binaries.

2 The Orbital Properties

We study the phase variation of white dwarf photospheric absorption lines as well as red dwarf Balmer emission resulting from reprocessing of EUV radiation from the white dwarf. Optical spectroscopy obtained at various observatories (*IUE*, CTIO, KPNO, Michigan-Dartmouth-MIT, and Lick) has allowed to determine the orbital parameters of four close binaries (Table 1 and Figures 1 & 2): Feige 24 (Vennes & Thorstensen 1994), EUVE J0720-317 (Vennes & Thorstensen 1993), EUVE J1016-053 (Thorstensen, Vennes, & Bowyer 1995) and EUVE J2013+400 (Thorstensen, Vennes, & Shambrook 1994, and Vennes & Thorstensen 1995). Adding to this sample the binary V 471 Tau (Young & Lanning 1975) we find an average secondary mass of $M_2 = 0.66M_\odot$. This mass estimate is consistent with the fact that secondary spectral types are also found in the range dM1 to dM4 in four systems.

Table 1. Orbital Parameters for a Sample of EUV-Selected Close Binaries

Name	Types	P (d)	$f_{wD}(M_\odot)$	M_1/M_2
Feige 24	DA+dM1-2	4.2316	0.19	1.59
V471 Tau	DA+dK2	0.5212	0.175	0.97
0720-317	DAO+dM0-2	1.32	0.12	0.92
1016-053	DAO+dM1-3	0.7893	0.28	1.8
2013+400	DAO+dM1-4	0.7055	0.063	1.3

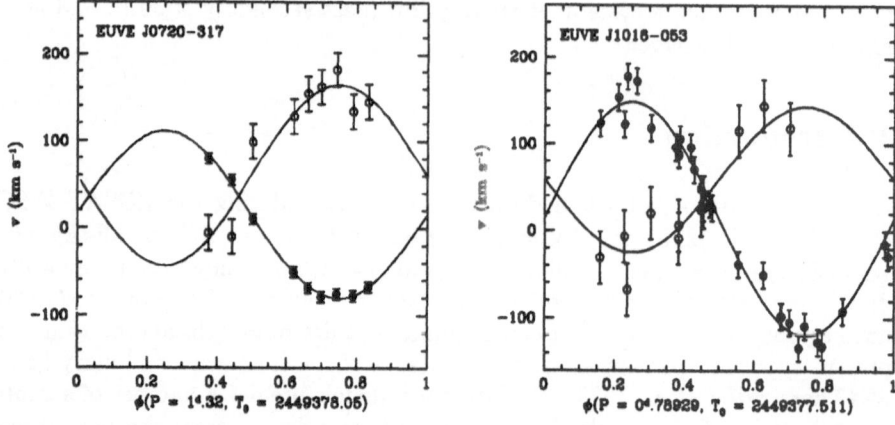

Fig. 1. Orbital velocities of both components of the binaries EUVE J0720-317 and EUVE J1016-053: (*filled circle*) the dM star and (*open circle*) the DAO white dwarf.

The absence of low mass secondaries can, in a large measure, be attributed to a bias in the selection of the systems in the *EUVE* all-sky survey. The Balmer emission strength is a direct function of the intensity of the in-coming EUV radiation on the red dwarf upper atmosphere. Large orbital separation or low intercepting surface would both conspire in reducing the emission strength. Late M-dwarfs are less massive and accordingly have smaller radius than the earlier M-dwarfs found in the *EUVE* sample of close binaries. Kirkpatrick & McCarthy's

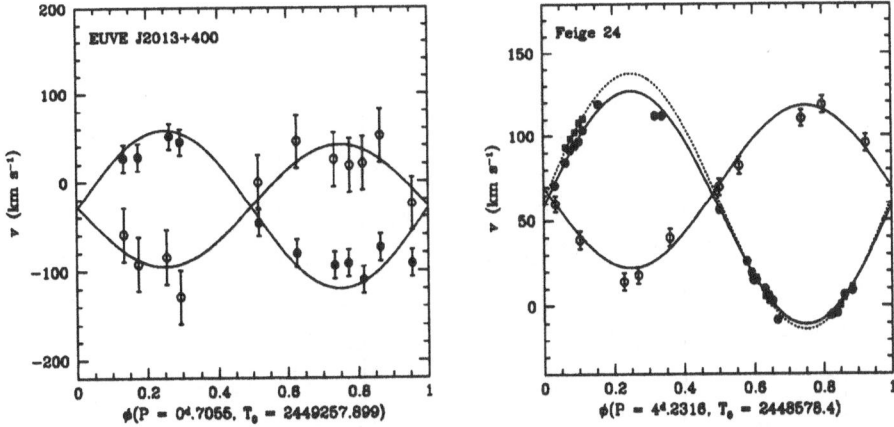

Fig. 2. Same as Figure 1 but for the binaries Feige 24 (EUVE J0235-037) and EUVE J2013+400. High resolution optical and ultraviolet spectroscopy of the binary Feige 24 provide orbital parameter measurements of unprecedented accuracy. The *true* orbital velocity of the dM star in Feige 24 (*filled square*) was also measured from photospheric absorption (mainly TiO bands). The gravitational redshift of the white dwarf in EUVE J2013+400 appears small but cannot be estimated accurately from available data.

(1994) mass-luminosity relation implies that $0.1-0.2M_\odot$ dwarfs may offer only a tenth or a hundredth of the emitting surface of our selected M-dwarf companions. Infrared photometry or high-resolution Hα spectroscopy are possibly the only means by which additional WD+dM systems will be uncovered.

There is a possibility, however, that such low mass secondaries may not exist at all. de Kool & Ritter (1993) have explored the effect of the assumed initial masses: they devised two drastically different distributions of initial mass ratios, one strongly peaked at systems of equal masses (the dN $\propto$ dq distribution) and the other described as a random associations of stars from the same IMF. In the former case de Kool & Ritter predict a paucity of low mass dwarfs and they infer a secondary mass distribution which peaks near $0.4M_\odot$ in agreement with the distribution observed in the EUV-selected sample. We cannot at this time conclude that masses in binaries are strongly correlated until we have obtained a definitive assessment of the population of close binaries in the *EUVE* survey.

Studies of the binary orbital parameters and of the white dwarf spectroscopic properties constrain the white dwarf mass and radius (Figure 3). Feige 24 appears to have a distinctively large radius attributed to a massive hydrogen envelope (Vennes & Thorstensen 1994). Our measurement of the gravitational redshift of EUVE J1016-053 is slightly inconsistent with Bergeron et al.'s (1994) surface gravity. A new ultraviolet velocity study of EUVE J1016-053 and other systems with the *Hubble Space Telescope* is now necessary.

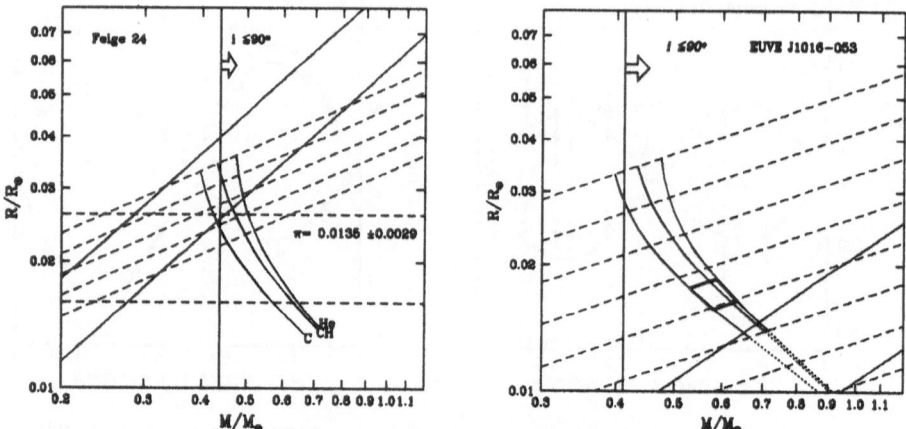

Fig. 3. Constraints placed on the mass and radius of the white dwarfs in Feige 24 and EUVE J1016-053 by theoretical mass-radius relations (Vennes, Fontaine, & Brassard 1995), spectroscopic estimates of the surface gravity and gravitational redshift measurements (*full line*). The mass-radius relations were computed for helium interiors (labelled *He*) as well as carbon interiors with and without an envelope of hydrogen (labelled *CH* and *C* respectively). The minimum mass of both white dwarfs is also derived from the mass function and mass ratio of the systems ($i \leq 90°$). (*left*) The radius of Feige 24 is also constrained by a parallax measurement and a photometric distance modulus. Lines of constant gravity (*dash line*) are drawn from $\log(g) = 7.0$ (*top*) to 7.4 (*bottom*). (*right*) The mass and radius of EUVE J1016-053 are further constrained by a measurement of the surface gravity (*thick line*; Bergeron et al. 1994). Lines of constant gravity are drawn from 7.0 (*top*) to 8.4 (*bottom*).

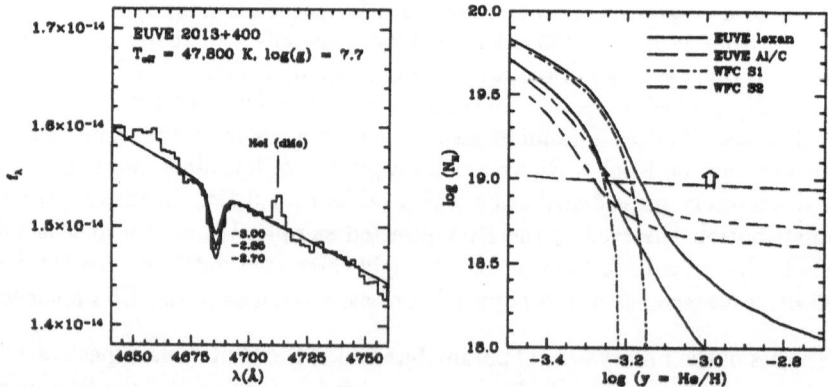

Fig. 4. Photospheric abundance of helium from EUV count rate measurements (*right panel*) and the He II $\lambda4686$ line profile (*left panel*). The EUV count rates are analyzed in the $\log(y) - \log(n_H)$ plane for a given temperature and surface gravity. The optical spectrum, f_λ vs λ, is shown with three models with helium abundance of $He/H = 1.0, 1.4, 2.0 \times 10^{-3}$.

3 Accretion and Diffusion in White Dwarf Atmospheres

The photospheric layers of white dwarfs in close binaries are possibly showing
evidence of accretion from the close companion. Spectroscopic detection of helium in all these objects (Tweedy et al. 1993; Bergeron et al. 1994; Vennes &
Thorstensen 1993, 1994) and *not* in any other so-called isolated white dwarfs
indicates the occurrence of a unique phenomenon. Detailed accretion-diffusion
models suggest strong time-dependent inhomogeneity in the photospheric layers of these peculiar white dwarfs. Current EUV/optical abundance analysis of
the white dwarfs in EUVE J1016-053, EUVE J0720-317 and EUVE J2013+400
(Figure 4) show a depth-dependent helium abundance expected under steady
state at low rate of accretion (Dupuis & Vennes 1994, private communication).
A well constrained model would therefore provide estimates of the red dwarf
mass loss in conjunction with an estimate of the accretion cross-section of the
white dwarf; it should also provide critical tests of diffusion time scales.

4 Conclusion

The sample of close binaries discovered in EUV all-sky surveys provides stringent
tests of theory of close binary evolution. Because the sample is defined by few
simple selection criteria and because it is enclosed in a larger and also well
defined sample of EUV-selected white dwarf stars it is possible to conduct a
population study and measure the space density and birthrate of this important
class of evolved objects. This work is supported by NASA contract NAS5-30180
and NASA grant NAG5-2405. We thank S. Bowyer for his advice.

References

Bergeron, P., Wesemael, F., Beauchamp, A., Wood, M.A., Lamontagne, R., Fontaine,
    G., & Liebert, J. 1994, ApJ 432, 305
Bowyer, S., Lieu, R., Lampton, M., Lewis, J., Wu, X., Drake, J.J., & Malina, R.F.
    1994, ApJS 93, 569
de Kool, M., & Ritter, H. 1993, A&A 267, 397
Kirkpatrick, J.D., & McCarthy, Jr., D.W. 1994, AJ 107, 333
Pounds, K.A., et al. 1993, MNRAS 260, 77
Thorstensen, J. R., Charles, P. A., Margon, B., & Bowyer, S. 1978, ApJ 223, 260
Thorstensen, J.R., Vennes, S., & Shambrook, A. 1994, AJ 108, 1924
Thorstensen, J.R., Vennes, S., & Bowyer, S. 1995, ApJ, submitted
Tweedy, R., Holberg, J.B., Barstow, M.A., Bergeron, P., Grauer, A.D., Liebert, J., &
    Fleming, T.A. 1993, AJ 105, 1938
Vennes, S., Fontaine, G., & Brassard, P. 1995, A&A, in press
Vennes, S., & Thorstensen, J. R. 1994a, ApJ 433, L29
Vennes, S., & Thorstensen, J. R. 1994b, AJ 108, 1881
Vennes, S., & Thorstensen, J. R. 1995, in preparation
Young, A., & Lanning, H. 1975, PASP 87, 461
Yungelson, L.R., Tutukov, A.V., & Livio, M. 1993, ApJ 418, 794

Constraints on DAO White Dwarf Composition from the ROSAT EUV Survey

M.R. Burleigh & M.A. Barstow

Department of Physics and Astronomy, University of Leicester, University Road, Leicester LE1 7RH, UK

1 Introduction

DAO white dwarf stars have hybrid spectra showing both hydrogen and helium lines, in particular a weak HeII feature at 4686Å. They have effective temperatures in the range 50,000K-70,000K, and typical abundances N(He)/N(H) 10^{-2}.

It has been suggested that DAOs might be intermediate objects on a single channel white dwarf cooling sequence between helium rich Planetary Nebula Nucleii (PNN) and PG1159 stars, through DO white dwarfs to hydrogen rich DAs. The idea was that DAs arise from DOs by gravitational settling of helium and the upward diffusion of hydrogen in the atmosphere to form an overlying hydrogen layer. The absence of any known DAs above 70,000K and the gap in the helium cooling sequence between 30,000K-45,000K supported this idea. DAOs were thus interpreted as white dwarfs undergoing metamorphosis from helium rich types to hydrogen dominated stars. If this interpretation was correct, they should have chemically stratified atmospheres.

However, recent work has begun to undermine this theory. Bergeron et al. (1994) studied a sample of 14 DAO white dwarfs and concluded that the majority of DAOs have homogeneously mixed rather than stratified atmospheres. Furthermore, the spectroscopically determined masses indicate that the DAOs are a product of Extended Horizontal Branch (EHB) rather than Asymptotic Giant Branch (AGB) evolution. This implies that DAOs are probably not a halfway stage between hot DOs and DA white dwarfs. The discovery of a hot (90,000K) DA white dwarf, RE1738+665 (Barstow et al. 1994), supports this: RE1738 can be viewed as a "missing link" between hydrogen-rich PNN and DA stars, without having to invoke evolution through the helium channel first.

Only three DAO white dwarfs were detected in the ROSAT Wide Field Camera (WFC) survey (Pounds et al. 1993): RE0720−318 (Barstow et al. 1994), RE1016−053, and RE2013+400 (Barstow et al. 1994). All are in binary systems with M dwarf companions.

Originally ROSAT was expected to detect between 1,000-2,000 White Dwarfs, assuming pure hydrogen atmospheres like the prototype Extreme Ultra-Violet (EUV) white dwarf source HZ43. In the end only around 120 were found. Barstow et al. (1993) concluded that trace amounts of elements heavier than helium in the atmospheres of hot DA white dwarfs prevent many of them being detected by blocking the EUV flux, although helium may still contribute to the opacity. Similarly, Bergeron et al. hypothesise that heavy elements in the atmospheres of DAO white dwarfs can account for their non-detection by ROSAT by providing most of the soft Xray and EUV opacity. The aim of this study is to test their hypothesis using data from the ROSAT WFC survey, and to try to explain why the only DAOs actually detected by ROSAT all lie in binaries.

2 Sample

Table 1 shows the DAO stars in the sample, with temperature, log gravity and chemical abundance log He/H taken from Bergeron et al. Only one star, PG1305, can be interpreted with a stratified atmosphere. PG1210 appears on both lists because it is not well fitted by either kind of model. The parameters of RE0720 were taken from Barstow et. al. (1994).

Table 1. DAO Sample Stars and Atmospheric Parameters

A) Homogeneous models Name	Teff (K)	log g	log He/H	Name	Teff (K)	log g	log He/H
LS V +46 21	77300	7.31	-2.06	PG0134+181	56400	7.40	-2.98
TON 320	68800	7.68	-2.74	RE1016-053	56400	7.74	-3.26
TON 353	66100	7.11	-2.88	RE0720-318	56360	7.64	-3.56
LB2	65700	7.67	-2.51	PG1413+015	48100	7.69	-2.57
GD 561	65300	6.71	-2.40	RE2013+400	47800	7.69	-2.62
PG0834+501	60400	7.11	-2.18	PG1210+533	44800	7.89	-2.08
Feige 55	58300	7.15	-2.92				
B) Stratified models Name	Teff (K)	log g	log q_H	Name	Teff (K)	log g	log q_H
PG1210+533	46600	7.91	-15.98	PG1305-017	44400	7.76	-15.98

3 Method

Images were selected from the ROSAT WFC survey and the background examined at the positions of each star. The images have been reprocessed to improve signal to noise, and with improved attitude and background rejection. The number of counts in a source box of radius three arcmin was measured, together with the background counts in an annulus around the source box. A Baysian statistical technique was used to determine the three sigma upper limit on the source counts at the position of each DAO white dwarf in the sample.

The parameters in Table 1 were applied to homogeneous and stratified models, supplied by Detlev Koester, to predict the count rate for each star as a function of interstellar hydrogen column density. The V magnitude was used to normalise each model spectra. The normalisation parameter, a measure of $(R/D)^2$, was used with Matt Wood's evolutionary code (which gives mass and radius for a given T and log g) to estimate the distance to each star. Hence the column density to each star could be obtained using the published work of Paresce and Frisch and York.

4 Results

Comparing the predicted count rates with observed count rates and upper limits allows us to draw several conclusions.

The non-detection of four stars (PG0834, PG0134, PG1413, PG1210), all $T \leq 60,000K$, can be accounted for purely in terms of absorption by interstellar hydrogen gas. For example, Figure 1 shows the predicted count rates for the S1 and S2 filters for PG1413 as a function of column density. The straight horizontal lines are the upper limits on the count rates for the two filters. From the Frisch and York maps the estimated column density to this star (at a distance of 490 parsecs) is greater than 10^{20}. Quite clearly the predicted count rates (1.5×10^{-4} counts/sec in S1 and 3×10^{-6} counts/sec in S2) are much lower than the upper limits (8×10^{-3} counts/sec in S1 and 12.1×10^{-3} counts/sec in S2). Thus this star's non-detection is simply accounted for by absorption.

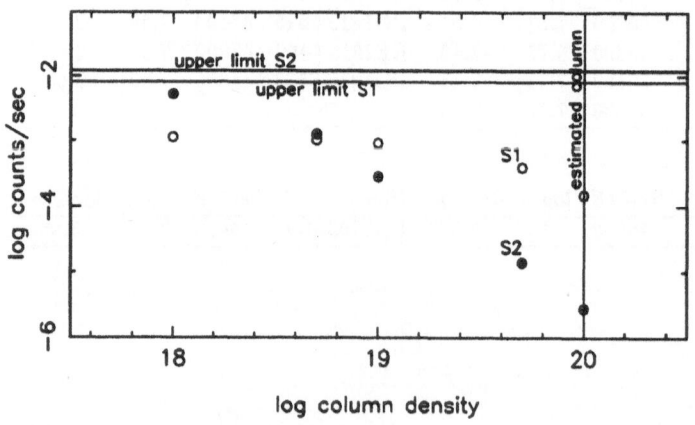

Fig. 1. PG1413: Model predicted count rates for the S1 and S2 filters plotted as a function of interstellar column density

The non-detection of the stratified star PG1305 can be easily explained because it has a very thin hydrogen layer; the EUV flux is blocked by the underlying helium. However, the non-detection of six stars (LS V +4621, Ton 320, Ton 353, LB2, GD561, Feige 55), all $T \geq 60,000K$, cannot be accounted by column density alone, or by their helium content. A good example is LB2. Figure 2 shows the predicted count rates for this star in the S1 and S2 filters as a function of column density. The straight horizontal lines are the upper limits on the count rates for the two filters. The Frisch and York maps give an estimated column density to this star (at a distance of approximately 300 parsecs) of less than 5×10^{18}. Paresce gives the column density as 2×10^{19} or less. Either way it is clear from the diagram that column density cannot account for the non-detection of this star since the predicted count rates are far in excess of the upper limits. Therefore we conclude that the column density to these six stars, or their helium content, cannot account for their non-detection. Instead, we suggest that heavy elements, in addition to helium, are providing the opacity to block the EUV flux.

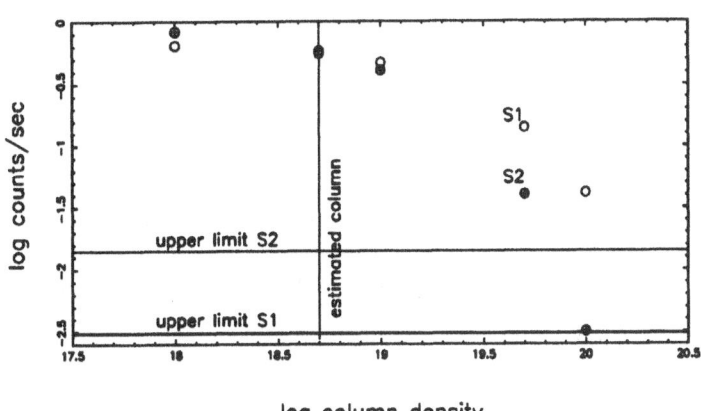

log column density

Fig. 2. LB2: Model predicted count rates for the S1 and S2 filters plotted as a function of interstellar column density

5   Discussion

We find that although the non-detection of some DAOs can be accounted for by interstellar hydrogen absorption, for DAOs $T \geq 60,000K$ heavy elements in addition to helium must be blocking the emergent EUV flux and preventing their detection. We would expect these hotter DAOs to have greater abundances of heavy elements as a result of the higher efficiency of radiative levitation, providing the additional necessary opacity source.

There are, however, several hydrogen-rich DA white dwarfs (e.g. RE1738) detected by ROSAT at temperatures greater than 60,000K. It may be that the DAOs in this temperature range are not detected either because of the mixture of helium and heavier elements, or because the presence of helium and a different atmospheric structure allows more heavy elements to be supported.

On the evidence of this work there is no reason to suppose that the binaries are in any way intrinsically different from the rest of the sample. They all lie relatively close by in low column directions, so we would not expect much interstellar absorption.

References

Barstow M.A., et al., 1993, *MNRAS*, **264**, 16.
Barstow M.A., O'Donoghue D., Kilkenny D., Burleigh M.R., Fleming T.A., 1994, *MNRAS, in press*
Barstow M.A., et al., 1994, *MNRAS*, **271**, 175
Barstow M.A., et al., 1994, *MNRAS, in press*
Bergeron P., et al., 1994, *ApJ*, **432**, 305
Frisch P., York D., 1983, *ApJ*, **271**, L59
Paresce F., 1984, *AJ*, **89**, 1022
Pounds K.A., et al., 1993, *MNRAS*, **260**, 77

Spectroscopy of the Lyman Continuum of Hot DAs Using the Extreme Ultraviolet Explorer

Jean Dupuis and Stéphane Vennes

Center for EUV Astrophysics, 2150 Kittredge Street University of California, Berkeley, CA 94720, USA

Abstract: We report observations by the Extreme Ultraviolet Explorer (*EUVE*) of the hot DA stars Feige 24, G191-B2B, HZ 43, GD 71, and GD 153 which are among the brightest white dwarfs observed by EUVE. Spectral features such as the HeI photoionization edge at 504 Å and the autoionization feature at 206 Å as well as, the HeII photoionization edge at 228 Å are fingerprints of the ISM typically detected in the EUVE spectra of hot DA. Effective temperatures both obtained from the fitting of pure hydrogen and metal-blanketed models are derived along with the ISM column densities of HI and HeI.

1 Introduction

A large number of hot white dwarfs were detected during the Extreme Ultraviolet Explorer (*EUVE*) all-sky survey (Bowyer et al. 1994). These stars are among the brightest EUV sources and span a range of distances from a few pc to a few hundred pc making them ideal probes of the distribution of hydrogen and helium in the local interstellar medium (LISM). Some white dwarfs were revealed as strong Lyman continuum sources by their detections in the two longest wavelengths bandpasses surveyed by *EUVE*, i.e. in the Dagwood (345-605 Å) and in the Tin (500-740 Å) bandpasses. An analysis of the *EUVE* photometry data for these stars has already imposed valuable constraints on the extent of the Local Fluff in which the solar system is thought to be embedded and how devoid the LISM is of neutral gas in certain directions up to a distance of 90 pc (Vennes et al. 1994). In spite of the fact that the *EUVE* survey data lead to useful estimates of HI column densities and to a least extent of HeI and HeII column densities, the medium resolution spectroscopy (approximatively 200) in the extreme ultraviolet(70-760 Å) offer by *EUVE* allows for a far more precise determination of the previous column densities. An early example in the *EUVE* history of its usefulness for ISM study was the case of GD 246 (Vennes et al. 1993) whose EUV spectrum exhibited the HeI autoionization transition at 206 Å(see Rumph, Bowyer, Vennes 1994) and the HeII photoionization edge at 228 Å. From this observation, a degree of ionization of 25 percent was derived for helium and an upper limit of 27 percent was set on the level of ionization of hydrogen. A knowledge of the ionization state of the main constituents of the ISM (H and He) is of fundamental importance in the understanding of the interaction between diffuse clouds and the hot phase of the ISM. In order to gain more insights on this topic, we have obtained EUV spectroscopy for a sample of hot white dwarfs sampling low column density lines of sight and therefore more likely to probe the local cloud. Our sample comprises low column densities white dwarfs (see Fig. 1) with relatively high level of contamination by heavy elements (Feige 24 and G191-B2B) and others consistent with pure hydrogen composition (HZ 43, GD 71, and GD 153).

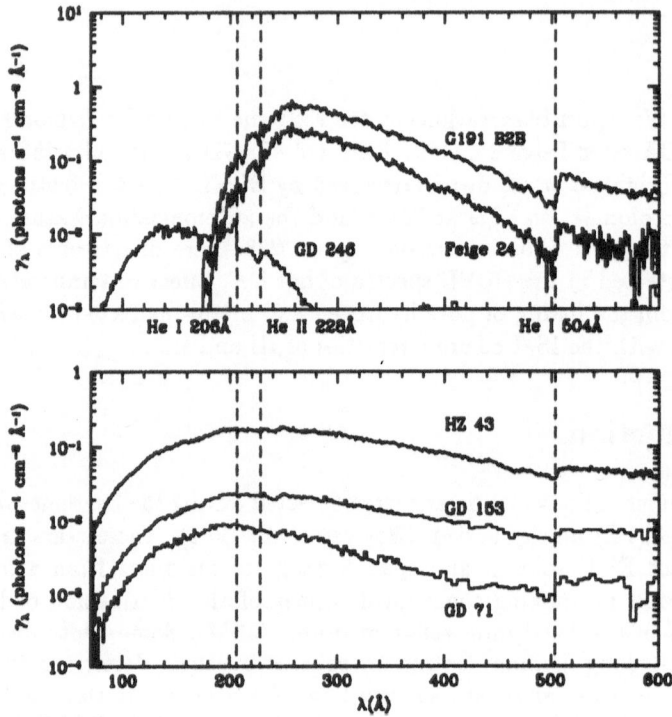

Fig. 1. *EUVE* Spectra of Hot DA Stars.

2 Results: Effective Temperatures and Column Densities of HI and HeI

In addition to information on the LISM, the spectroscopy of the Lyman continuum of hot DA can also provide a measurement of the fundamental atmospheric parameters. In fact, one does not go without the other. We have derived effective temperatures by fitting the continuum from 400 Å to 550 Å and obtained results in agreement with results based on fits of balmer lines, lyman alpha and continuum in the ultraviolet. The analysis based on pure hydrogen models, as expected, have produced results in excellent agreement with previous determinations for HZ 43, GD 71, and GD 153. In the case of Feige 24 and G191-B2B, these models are certainly inappropriate below 400 Å where the continuum is affected by various heavy element absorption features. To quantify the effect of blanketing by the combination of metals radiatively supported in the photosphere, synthetic spectra computed from LTE metal-blanketed model atmospheres were used to reproduce the Lyman continuum. We made use of a grid of models computed by Vennes, Pradhan, and Thejjl (1995) including bound-bound and bound-free transitions of C, N, O, and Fe using new radiative opacities from the Opacity Project (Seaton et al. 1994). This is part of a larger effort aimed at understanding the complex EUV spectra of hot white dwarfs, especially those with high metallicity. Of immediate concern to us is how much the level of the continuum increases because of metal blanketing with respect to a pure hydrogen model.

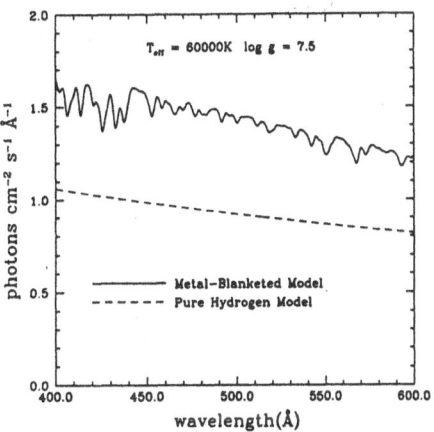

Fig. 2. Effect of blanketing by metals on the Lyman continuum of hot DA. The composition is C/H = 4×10^{-7}, N/H = 5×10^{-6}, O/H = 10^{-6}, and Fe/H = 10^{-5}.

We show in fig. 2 a comparison between a pure hydrogen and metal-blanketed models for an effective temperature of 60,000K and a gravity of $10^{7.5}$ cm s^{-2} and Fe/H of 10^{-5}. Indeed the level of the metal-blanketed model is higher by about 50 percent, for a model with Fe/H of 10^{-6} the level of the continuum approximately matches that of pure hydrogen limit. A remarkable fact for G191-B2B is that the effective temperatures derived from fitting both pure hydrogen and metal-blanketed models are significantly smaller than the optical temperature. In case of Feige 24, we obtain a higher effective temperature instead but its optical temperature has to be considered with caution because of the contamination by the late-type star companion. This is not unlike the effect noted by Bergeron et al. (1994) of the lowering of the fitted effective temperature when blanketing by a low level of helium abundance (not detectable in the optical) is included. We believe that our results point out a potential effect of opacity of metals on the Balmer line profiles and that G191-B2B is significantly cooler than previously thought from fitting pure hydrogen models. This could be related to the discrepancies found between predictions from radiative forces computation and the measured abundances in the two previous stars(Chayer, Fontaine, and Wesemeal 1995). We are currently investigating this in more detail.

Table 1. Determination of effective temperatures, HI and HeI column densities from *EUVE* spectroscopy

Name	T_{eff} (K)	NHI $(10^{17}\mathrm{cm}^{-2})$	NHeI $(10^{17}\mathrm{cm}^{-2})$	NHeI/NHI
G191-B2B	57186±400(pure H)	18.6±0.3	1.46±0.08	0.078±0.005
	53670±450(metal)	20.7±0.4	1.50±0.08	0.072±0.004
Feige 24	62055±810(pure H)	28.0±0.5	1.54±0.14	0.055±0.005
	58280±870(metal)	30.0±0.7	1.63±0.11	0.054±0.004
HZ 43	51156±430	8.9±0.4	0.55±0.03	0.062±0.004
GD 153	39700±280	9.8±0.7	0.66±0.07	0.067±0.008
GD 71	32362±380	6.4±1.7	0.52±0.08	0.081±0.024

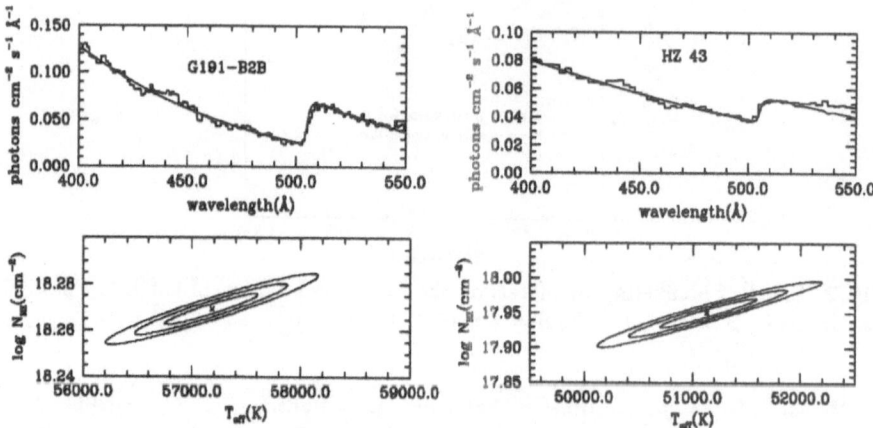

Fig. 3. Fitting of the Lyman Continuum for G191-B2B and Feige 24.

In figure 3, we show examples of the fit of the Lyman continuum for G191-B2B and HZ 43. These fits are based on pure hydrogen models described in Vennes and Fontaine (1992). In summary, the HeI edge is first fitted alone and the convergent lines of the ground-state series of HeI in the ISM is included to better reproduce the shape of the edge. Fixing the HeI column density, we then fit the continuum between from 400Å to 550Å for the effective temperature and the hydrogen column density. The results are summarized in table 1. A more detailed description of the analysis will be described in Dupuis et al. (1994). Figure 4 shows the result of table 1 and demonstrate that the measured HeI to

HI ratios are significantly smaller than 0.1 (see Kimble et al. 1993) for each of the lines of sight. Ionization of hydrogen and helium in the LISM will drive the ratio of HeI to HI away from the cosmic ratio and our results can be interpreted in terms of preferential ionization of helium. This nicely fits with our interpretation of the GD 246 EUV spectra in terms of helium ionization. Direct measurements of the HeII edge(228Å) are more difficult because of their low column densities and possible confusion with photospheric features(specially for Feige 24 and G191–B2B).

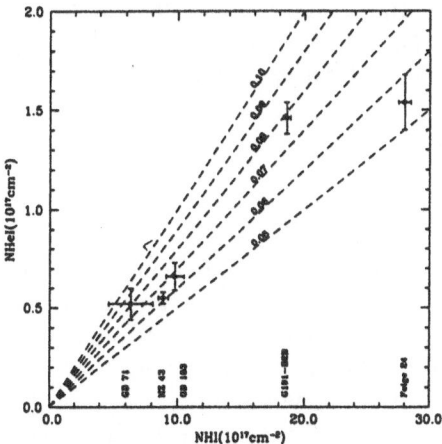

Fig. 4. Ratio of HeI/HI in the LISM in the direction of the sample's white dwarfs. The staight lines indicate constant HeI/HI ratios.

Acknowlegdement. This research is supported by NASA contracts NAS5-30180, NAG5-2405, and NAG5-2620.

References

Bergeron, P., Wesemael, F., Beauchamp, A., Wood, M.A., Lamontagne, R., Fontaine, G., and Liebert, J. 1994, ApJ, 432, 305.

Bowyer, S., Lieu, R., Lampton, M., Lewis, J., Wu, X., Drake, J.J., and Malina, R.F. 1994, ApJS, 93, 569.

Chayer, P., Fontaine, G., and Wesemael, F. 1995, ApJ, in press.

Dupuis, J., Vennes, S., Bowyer, S., Pradhan, A.K, and Thejll, P. 1994, in preparation.

Green, J., Jelinsky, P., and Bowyer, S. 1990, ApJ, 359, 499.

Kimble, R.A. et al. 1993, ApJ, 404, 663.

Rumph, T., Bowyer, S., and Vennes, S. 1994, AJ, 107, 2108.

Seaton, M.J., Yan, Y., Mihalas, D., and Pradhan, A.K. 1994, MNRAS, 266, 805.

Vennes, S. and Fontaine, G. 1992, ApJ, 401, 288.

Vennes, S., Dupuis, J., Rumph, T., Drake, J., Bowyer, S., Chayer, P., and Fontaine, G. 1993, ApJ, 410, L119.

Vennes, S., Dupuis, J., Bowyer, S., Fontaine, G., Wiercigroch, A., Jelinsky, P., Wesemael, F., and Malina, R.F. 1994, ApJ, 421, L35.

Vennes, S., Pradhan, A.K., and Thejll, P. 1995, in preparation.

An EUV Selected Sample of DA White Dwarfs

M.C. Marsh[1], M.A. Barstow[1], J.B. Holberg[2], D. O'Donoghue[3],
D.A. Buckley[3], T.A. Fleming[4,5], D. Koester[6], M.R. Burleigh[1]

[1] Department of Physics and Astronomy, University of Leicester, UK
[2] Lunar and Planetary Laboratory, University of Arizona, Tuscon, USA
[3] Department of Astronomy, University of Capetown, SA.
[4] Steward Observatory, University of Arizona, Tuscon, USA.
[5] Max Planck Institut fur Extraterrestrische Physik, Garching, Germany
[6] Institut für Astronomie und Astrophysik der Universität, D-24098 Kiel, Germany

1 Introduction

In order to answer some of the problems facing white dwarf evolution, such as
how distinct are the DO and DA evolutionary paths, one needs to know the
photospheric composition of a large number of objects and how this varies under different physical conditions. Several groups have shown that trace heavy
elements can be supported in the atmosphere against gravitational settling by
radiation pressure. These 'opacity' sources have their greatest effect on the emergent flux at EUV wavelengths. Although we now have instruments such as the
EUVE spectrometers which can actually identify photospheric species directly,
most objects are too faint to be observed with realistic exposure times. We do
however have at our disposal the ROSAT all-sky survey (Pounds et al. 1993;
Trümper 1992) which has X-ray/EUV photometry for approximately 120 DA
white dwarfs. Although this photometry is limited in the detail we can obtain
about the nature of the trace elements, it does provide us with a view of the
overall pattern of photospheric opacity across the hot DA temperature range.

2 The EUV Sample

In order to utilize the EUV/X-ray information one needs an accurate temperature and gravity for each star, obtained from fitting the Balmer line profiles of
the optical spectra (eg. Bergeron et al. 1992). An accurate V magnitude is also
required in order to normalise the fluxes for distance and radius. Earlier work
on the ROSAT survey sample (Barstow et al. 1993, hereafter B93) examined 30
DA white dwarfs selected on the basis that they were well studied objects with
known T_{eff}, logg and m_v. We now have such optical data for 79 DA's in the
ROSAT survey, and will shortly extend the list to 110 objects. A large proportion of these new stars are at the hotter end of the DA cooling sequence which
is where the effects of opacity are greatest. This places us in an excellent position to examine the dependence of opacity and, therefore, inferred trace element
abundances on physical parameters such as temperature and gravity. Also as this

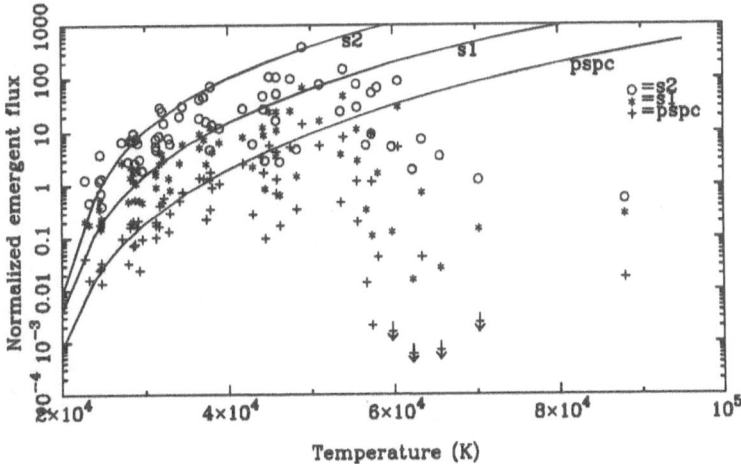

Fig. 1. Normalised emergent EUV/X-ray fluxes as a function of temperature for S2, S1 & PSPC survey bands as indicated. Lines correspond to predicted fluxes for a pure H atmosphere and HI column of $2.5 \times 10^{18} \mathrm{cm}^2$ for each filter/instrument combination normalised to HZ43. Arrows depict upper limits.

is a more comprehensive sample, any objects such as HZ43 (which appears to be unique among very hot DA's in being pure H) should now become apparent.

ROSAT yields little information about the exact nature of the absorbing material (see B93). Dividing the observed count rate by the nominal effective area of each instrument and normalising by the V magnitude gives an estimate of the relative emergent flux for each star (figure 1). This confirms the results of B93, that almost all stars below 40,000K have nearly pure H atmospheres with those above 40,000K having significant additional opacity. However, the increased sample size provides more detail, particularly in the 40,000 - 50,000K range which was sparsely populated in the B93 work.

3 Temperature Dependence

In order to show more clearly how the levels of opacity vary with temperature, figure 2 shows the ratios of the PSPC fluxes (which are least sensitive to column) to the predicted fluxes for the pure H model, now including uncertainties in temperature. The increase in opacity appears to be modest in the 40,000 - 50,000K range but steepens dramatically above 55,000K with several stars in this sample only having upper limits in the PSPC. Some of these objects, RE0623-37, RE2214-49 and Feige24 have been observed with IUE (eg. Holberg et al. 1993; Vennes et al. 1992) and show many heavy element features including Fe. Chayer et al. (1991) predict that these Fe species, which have a large opacity in the EUV, can only be radiatively levitated against gravity above 50000K. From this

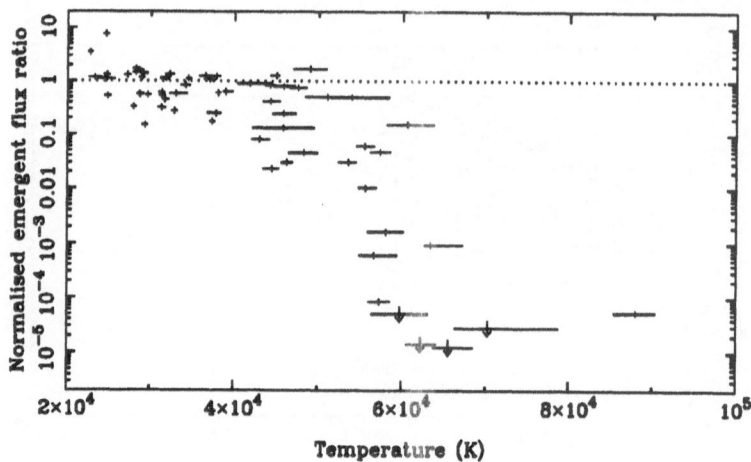

Fig. 2. Ratio of PSPC emergent fluxes to those of a pure H model with a HI column of $2.5 \times 10^{18} \, \mathrm{cm}^2$ as a function of temperature. Arrows depict upper limits.

we may infer that the large decrease in flux above 55000K is infact due to the prescence of Fe in the photospheres.

Another interesting feature of figure 2 is the dispersion in the levels of opacity seen at a given T_{eff}. This is very small up to 40000K, beginning to increase above 40000K and then dramatically above 55000K. This suggests that the abundances of the absorbing material vary quite considerably from star to star.

4 Gravity dependance

A possible explanation for the spread in opacity at constant T_{eff} could be the effect of gravity, with higher gravities competing more effectively against radiation pressure, reducing the abundances of photospheric absorbing material. Figure 3 shows the flux ratios against measured surface gravity for stars hotter than 40000K. There is a weak correlation between the level of opacity and gravity, objects with the lowest gravities (RE2214 etc) having the highest opacities. However, this may simply be a result of normal evolution where gravity increases as the white dwarfs cool. Theoretical work also suggests that the sensitivity of opacity to gravity is too weak to explain the absence of trace elements in some stars (eg. Vauclair et al. 1989).

5 Conclusions

Through the use of the ROSAT EUV/X-ray all.sky survey we have been able to examine the photospheric composition of DA stars across most of the hot DA range. We confirm previous findings that the cooler stars ($< 40000K$) have

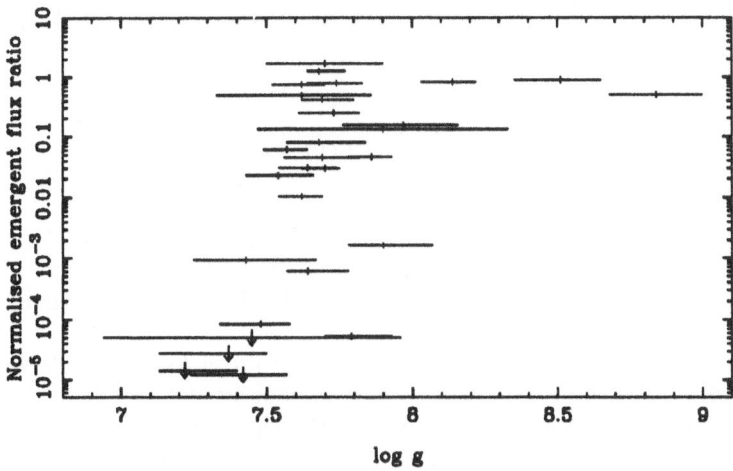

Fig. 3. Ratio of PSPC emergent fluxes to those of a pure H star with a HI column of $2.5 \times 10^{18} \mathrm{cm}^2$ as a function of log gravity for stars hotter than 40000K. Arrows depict upper limits.

more or less pure H atmospheres, while stars hotter than 40000K contain varying amounts of trace elements. Above 55000K the average opacity increases significantly suggesting the appearance of an additional absorption component, probably Fe. In the hotter T_{eff} regions there is a large dispersion in the levels of opacity showing that when DA's arrive on the cooling sequence the abundances of heavy elements in the photospheres must differ considerably.

Indeed, with this more complete sample we also see that there are several stars with nearly pure H atmospheres in the 40,000 - 50,000K region. It seems difficult to explain such a large range of extremes without invoking other physical effects such as varying layer masses and/or mass-loss.

References

Barstow M.A., et al., 1993, *MNRAS*, **264**, 16

Bergeron P., Saffer R.A., Liebert J., 1992, *Ap.J.*, **394**, 228

Chayer P., Fontaine G., Wesemael F., 1991, in *White Dwarfs*, eds. G. Vauclair and E. Sion., Kluwer, p.249

Holberg J.B., et al., 1993, *Ap.J.*, **416**, 806

Pounds K.A., et al., 1993, *MNRAS*, **260**, 77

Trümper J., 1992, *QJRAS*, **33**, 165

Vauclair G., 1989, in *White Dwarfs*, ed. G. Wegner, Springer-Verlag, p.176

Vennes S., et al., 1992, *Ap.J.*, **392**, L27

The EUVE Spectrum of the Hot DA White Dwarf PG 1234+482

Stefan Jordan[1], Detlev Koester[1], David Finley[2], Klaus Werner[1], and Stefan Dreizler[3]

[1] Institut für Astronomie und Astrophysik der Universität, D-24098 Kiel, Germany
[2] Center for EUV Astrophysics, Berkeley, CA 94720, USA
[3] Dr.-Remeis-Sternwarte, Universität Erlangen-Nürnberg, D-96049 Bamberg, Germany

Abstract: We have analyzed the EUVE spectrum of PG 1234+482 with fully blanketed model atmospheres taking into account several million lines of heavy elements. Most of the spectral features in the short (SW) and medium (MW) wavelength spectrum can be identified with lines of iron, nickel and calcium ions.

1 Introduction

Observations in the EUV and soft X-ray region of the electromagnetic spectrum have revealed that white dwarfs of spectral type DA, showing only Balmer lines of hydrogen in the optical, can possess significant amounts of heavier elements. For several objects the EINSTEIN (Kahn et al. 1984, Petre et al. 1986) and EXOSAT (Jordan et al. 1987; Paerels & Heise 1989) satellites measured a flux smaller than predicted from pure hydrogen atmospheres, clearly indicating the presence of material absorbing at wavelengths below about 300 Å. These data already indicated a trend that the hotter objects have larger amounts of absorbers in their atmospheres, with the exception of HZ 43, where the observations are compatible with an outer layer of pure hydrogen (e.g. Napiwotzki et al. 1993).

Approximately 120 DA white dwarfs have been detected during the ROSAT all-sky survey (Pounds et al. 1993). Barstow et al. (1993) and Marsh et al. (these proceedings) have analyzed 79 DAs with optically determined $T_{\rm eff}$ and $\log g$. Jordan et al. (1994), Wolff et al. (1994, and these proceedings) compared optically determined temperatures with those derived from ROSAT pointed observations of approximately 50 DAs. Both groups concluded that stars with $T_{\rm eff} \lesssim 38,000\,\mathrm{K}$ have nearly pure hydrogen atmospheres while at higher temperatures most objects contain additional opacity. Unfortunately, the limited energy resolution of these detectors did not allow a unique determination of the chemical composition. Therefore the analysis of the X-ray data was mostly restricted to hydrogen atmospheres with small traces of helium. The need to invoke elements heavier than H and He had first been demonstrated by Vennes et al. (1989). In their analysis of one of the very few EUV spectra obtained by EXOSAT, they found that a mixture of several heavy elements is necessary to explain the spectrum of Feige 24. Other direct evidence for such absorbers was the detection of ions of heavier elements (Si, Fe, Ni, C, N, O) in IUE high resolution and HST GHRS spectra of the bright DA star G191-B2B (Vennes et al. 1992; Holberg et al. 1993, 1994; Werner & Dreizler 1994; Vidal-Madjar et al. 1994).

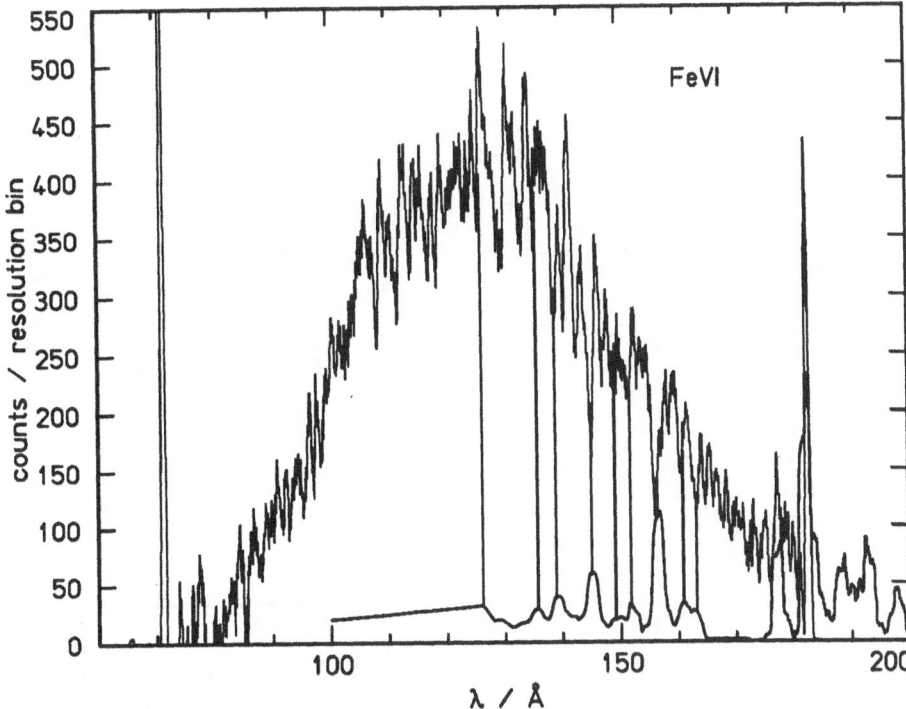

Fig. 1. SW spectrum of PG1234+234 compared to the FeVI opacity (arbitrarily scaled) at $\tau \approx 1$ in a $T_{\text{eff}} = 50,000$ K atmosphere. The spectral resolution is about 300, the data were smoothed with a boxcar of 1 Å

2 Observations and model atmosphere analysis

PG 1234+482, after being classified as a sdB (Green et al. 1986), was found to be a hot DA (Jordan et al. 1991). From the optical spectrum we determined an effective temperature of $55,600 \pm 1,500$ K (with a pure H atmosphere), which is compatible with the IUE low resolution spectrum. A pointed observation with the ROSAT satellite could not be reproduced by assuming model atmospheres containing only hydrogen and helium (Jordan 1993). With the limited energy resolution of ROSAT it was, however, not possible to determine the metal abundance unambiguously. With its much higher spectral resolution the EUVE satellite is able to detect spectral lines in the extreme UV.

The exposure time of the EUVE spectrum of PG 1234+482 was 99 ksec. Significant flux could be detected with the SW and MW spectrographs.

For the interpretation of EUV spectra Koester's model atmosphere code was expanded to include the line blanketing by many millions of metal lines taken from the Kurucz (1991) tables. For this study typically 10^7 spectral lines of C, N, O, Si, Ca, Fe, and Ni were considered with an opacity sampling algorithm. For the photoionization cross sections Opacity Project data (Seaton et al. 1992) were used with the exception of nickel, for which the hydrogenic approximation had to

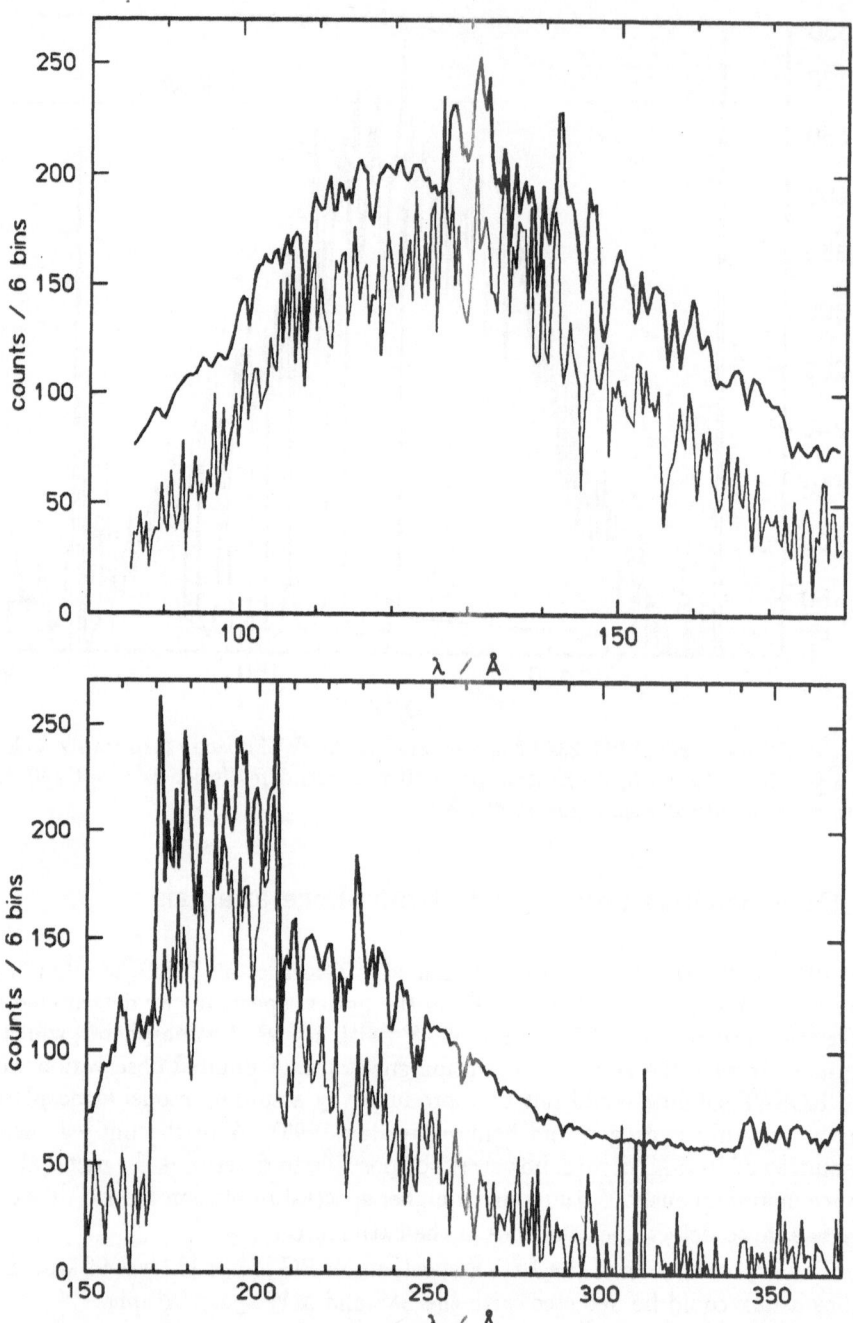

Fig. 2. The SW (top) and MW (bottom) spectrum of PG 1234+482 is compared to a synthetic spectrum for $T_{\mathrm{eff}} = 55,600\,\mathrm{K}$, $\log g = 7.7$, Fe/H=$2 \cdot 10^{-7}$, Ni/H=$2 \cdot 10^{-8}$, Ca/H=$2 \cdot 10^{-8}$, C/H=$1 \cdot 10^{-7}$, N/H=$1.5 \cdot 10^{-6}$, O/H=$3 \cdot 10^{-8}$. For the fit of the SW spectrum the flux of the model spectrum has been artificially reduced by a second order polynomial (about 50% at $\lambda < 120\,\text{Å}$, and 40% at $\lambda > 140\,\text{Å}$). No flux reduction was applied for the MW spectrum. For clarity the model spectra are offset by 50 counts

be applied. For a comparison with the observed EUVE spectra the model flux is folded with the detector response matrix of the spectrograph taking into account order overlapping. Moreover, the spectra are normalized at the V magnitude of $14^{\mathrm{m}}38$; the interstellar absorption is calculated according to Rumph et al. (1994) and Morrison & MacCammon(1983).

In a first step we calculated opacities for various elements for a temperature and pressure representative for the line forming region in the atmosphere. This allowed us to identify FeVI, FeVII, and NiVI in the SW spectrum. As an example Fig. 1 demonstrates that several features are due to FeVI absorption.

Table 1. Preliminary results for the abundances

element	number ratio	predictions
Fe	$(2-3) \cdot 10^{-7}$	$1.5 \cdot 10^{-5}$
Ni	$(1-2) \cdot 10^{-8}$	$2.0 \cdot 10^{-6}$
Ca	$(2-3) \cdot 10^{-8}$	$2.0 \cdot 10^{-6}$
C	$< 3 \cdot 10^{-5}$	$1.8 \cdot 10^{-4}$
N	$< 3 \cdot 10^{-6}$	$2.5 \cdot 10^{-4}$
O	$< 10^{-7}$	$1.3 \cdot 10^{-4}$
Si	$< 10^{-7}$	$< 10^{-8}$

In a second step we calculated synthetic spectra for hydrogen rich atmospheres with only one additional absorber present. This allowed us to confirm that most of the structure in the SW spectrum is indeed produced by iron and nickel absorption. A spectrum for an H+Ca atmosphere revealed that calcium is the third strongest absorber. Besides many features in the SW there are also several lines in the MW spectrum coinciding with the observed spectra. Moreover, there are indications that small amounts of C, N, and O are also present in the atmosphere. Fig. 2 shows model fits for the SW and MW spectrum. The Table summarizes the very preliminary results for the abundances and compares the results to the predictions by Chayer et al. (1995, for $\log g = 7.5$). Note that the error ranges given in the Table are only approximately correct if the stellar parameters are held fixed at $T_{\mathrm{eff}} = 55,600\,\mathrm{K}$ and $\log g = 7.7$ and do not take into account any systematic errors of the models or the observation. Observational uncertainties may result from the fact that the spectrum of PG 1234+482 has not been dithered and some of the structure may be due to fixed pattern noise. Since the flux level of the SW spectrum cannot be reproduced (Fig. 2) we conclude that other elements, not yet included in our calculations, are probably present.

At temperatures as high as $50,000\,\mathrm{K}$ NLTE effects cannot be excluded. We therefore made some test calculations with a NLTE code, which is also able to account for a large number of spectral lines by opacity sampling (Dreizler & Werner 1993). We restricted our test to the iron group elements only. The result was that the iron lines were somewhat weaker in the case of NLTE models, which means that the iron abundance would be a factor of 5 higher compared to the LTE result. We plan to continue with such investigations in the future; in order to be sure that the differences are indeed NLTE effects and not differences in other parts of the code the NLTE program will also be run with artificially high collision rates in order to simulate LTE conditions. There is a slight indication that NLTE effects may be important since the features of Fe VII, responsible for some of the features at $\lambda < 125$ Å, are somewhat too weak in an LTE model of 55,600 K and are better reproduced in strength with a model with $T_{\rm eff} \approx$ 60,000 K. Such a high temperature would enlarge the flux discrepancy in the SW fit significantly and would be also at variance with the optical $T_{\rm eff}$ determination. If metal line blanketing would be taken into account for the optical temperature determination, the result would be even lower than 55,600 K. Therefore it is possible that overionization due to NLTE effects could make the relative strength of the Fe VI and Fe VII features more consistent.

The flux distribution in the MW spectrum is relatively well reproduced under the assumption of an interstellar hydrogen column density of $N_{\rm H} = 10^{19}$ cm^{-2}. Modelling of the He I autoionization transition, first detected in the spectrum of GD 246 by Vennes et al. (1993), we estimated $N_{\rm He} = 10^{18}$ cm^{-2}, meaning that the ISM is mostly neutral. A careful determination of the helium ionization fraction will be performed when a final model for PG 1234+482 is found.

References

Barstow M.A., Fleming T.A., Diamond C.J., et al., 1993, MNRAS 264, 16

Chayer P., Fontaine G., Wesemael F., 1995, submitted to ApJ

Dreizler S., Werner K., 1993, A&A 278, 199

Green R.F., Schmidt M., Liebert J., 1986, ApJS 61, 305

Holberg J. B., Barstow M.A., Buckley D.A.H. et al. 1993, ApJ 416, 806

Holberg J. B., Hubeny I., Barstow M.A., et al. 1994, ApJ 425, L105

Jordan S., 1993, In: Advances in Space Research Vol. 13, No. 12, Perg.Press, p. 319

Jordan S., Koester D., Wulf-Mathies C., Brunner H., 1987, A&A 185, 253

Jordan S., Heber U., Weidemann V. 1991, in White Dwarfs, eds. G. Vauclair and E.M. Sion, p. 121

Jordan S., Wolff B., Koester D., Napiwotzki R., 1994, A&A 290, 834

Kurucz R.L., 1991, in Stellar Atmospheres: Beyond Classical Models, NATO ASI Series, 341, p.441

Morrison R., MacCammon D., 1983, ApJ 270, 119

Napiwotzki R., Barstow M.A., Fleming T.A., Holweger H., Jordan S., Werner K., 1993, A&A 278, 478

Pounds K.A., Allan D.J., Barber C. et al., 1993, MNRAS 260, 77

Rumph T., Bowyer S., Vennes S., 1994, AJ 107, 2108

Seaton M.J., Zeippen C.J., Tully J.A., et al., 1992, Rev.Mexicana Astron.Af. 23, 19

Vennes S., Chayer P., Fontaine G., Wesemael F. 1989, ApJ 336, L25

Vennes S., Chayer P., Thorstensen J. R., Bowyer S., Shipman H. L., 1992, ApJ 392, L27

Vennes S., Dupuis J., Rumph T. et al., 1993, ApJ 410, L119

Vidal-Madjar A., Allard N.F., Koester D. et al., 1994, A&A, 287, 175

Wolff B., Jordan S., Bade N., Reimers D., 1994, A&A, in press

ORFEUS and EUVE Observations of the Cool DO HD 149499 B

R. Napiwotzki[1], S. Jordan[2], D. Koester[2], V. Weidemann[2], S. Bowyer[3], M. Hurwitz[3]

[1] Dr. Remeis-Sternwarte, Sternwartstr. 7, 96049 Bamberg, Germany
[2] Institut für Astronomie und Astrophysik der Universität, D-24098 Kiel, Germany
[3] Center for EUV Astrophysics, 2150 Kittredge Street, University of California, Berkeley, CA 94720, USA

We present an analysis of FUV and EUV observations of the cool DO white dwarf HD 149499 B. It is by far the brightest star ($V \approx 11.7$) of this class. However, it is a secondary in a binary system, separated from a K0V primary 3^m brighter in V by only $1\farcs5$. This makes optical observations of the white dwarf nearly impossible. However, in the FUV/EUV region the flux of the hot white dwarf is virtually undisturbed by the cool companion.

HD 149499 B was observed with the Berkeley EUV/FUV spectrometer of the ORFEUS (Orbiting Retrievable Far and Extreme Ultraviolet Spectrograph) experiment during the September 1993 mission of the Space Shuttle Discovery. The spectrometer covers the wavelength range from 390 Å to 1170 Å with a spectral resolution $\lambda/\Delta\lambda \approx 3000$. Since the interstellar hydrogen column density is relatively high (see below) only the FUV range ($\lambda > 912$ Å) is usable. This data is supplemented by an EUV spectrogram of HD 149499 B taken with the EUVE satellite ($\lambda/\Delta\lambda \approx 300$). Detectable flux is observed in the 240...380 Å region. The first lines of the He II resonance series are clearly visible.

We carried out NLTE and LTE analyses of the FUV and EUV spectrograms with atmospheres containing helium and hydrogen. NLTE models were calculated with the ALI code developed by Werner (1986; A&A 161, 177), for details see Napiwotzki & Rauch (1994; A&A 285, 603). LTE calculations were performed with the program of Koester.

We determined T_{eff} and g simultaneous by a fit of the He II 2-n/H I Lyman series observed by ORFEUS. If He/H is kept fixed at He/H=100 the best solution with the NLTE program can be achieved at $T_{\text{eff}} = 50000 \pm 2000$ K and $\log g = 7.8 \pm 0.4$. The best fit with LTE models was obtained for $\log g = 8.0$, $T_{\text{eff}} = 50300$ K, $\log$ H/He $= -1.07 \pm 0.2$, meaning that the Lyman line profiles are much better reproduced if photospheric and not only interstellar hydrogen is present.

Assuming $\log g = 8$ and He/H=10 the best fit to the EUVE spectrogram with a LTE model can be achieved at an effective temperature of 50000^{+4000}_{-1000} K and an interstellar column density of $7 \pm 0.4 \cdot 10^{18}$ cm^{-2}. The best fits with the NLTE models tend to yield T_{eff} higher by 1000...2000 K.

An independent check of the interstellar hydrogen column density can be performed with the Lyman lines in the ORFEUS spectrogram. Our preliminary analysis yielded $N_{\text{H}} \approx 1 \cdot 10^{19}$.

ROSAT Studies of DA White Dwarfs and the Calibration of the PSPC Detector

Burkhard Wolff, Stefan Jordan, and Detlev Koester

Institut für Astronomie und Astrophysik der Universität, D-24098 Kiel, Germany

We have analyzed ROSAT pointed observations (together with EXOSAT and the WFC all-sky survey measurements) of 49 hot ($T_{eff} \gtrsim 20000$ K) DA white dwarfs with model atmospheres containing hydrogen and helium, either homogeneously mixed or chemically stratified. For 34 stars PSPC data with and without Boron filter were extracted from the ROSAT archive.

The figure compares the effective temperatures from the X-ray analysis with the values derived from optical spectra under the assumption of pure hydrogen atmospheres. At $T_{eff} \lesssim 38000$ K both temperatures agree within the error ranges. At higher effective temperatures, however, most of the objects can no longer be explained by pure hydrogen atmospheres.

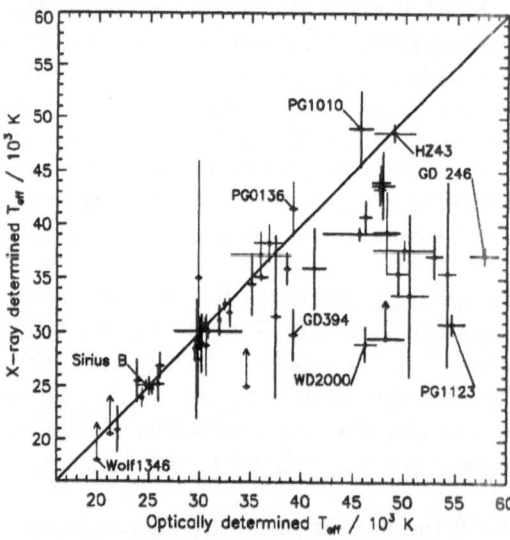

The limited energy resolution of the ROSAT/EXOSAT observations does not allow to determine which elements are responsible for the absorption in the EUV/soft X-ray region. However, in some cases we can exclude helium from being the only absorber, which is consistent with results from other observations (e.g. HST, EUVE) and diffusion calculations showing that metals play the dominant role as absorbers in hot DA white dwarfs.

In addition to these results we have found evidence that the PSPC detector is a factor 2.0 ± 0.2 (1.9 ± 0.2 with Boron filter) more sensitive to very soft X-ray photons ($\lesssim 0.2$ keV, measured at channel energies up to 0.4 keV) compared to the standard calibration. Details of our analysis can be found in Jordan, Wolff, Koester, & Napiwotzki (1994, A&A 290, 834) and Wolff, Jordan, Bade & Reimers (1994, A&A, in press).

List of Participants

N. ACHILLEOS, University College London Gower Street, London WC1E 6BT, UK,
nick@astro.umontreal.ca

M.A. BARSTOW, Dept. of Physics and Astronomy, University of Leicester, Leicester LE1 7RH, UK,
mab@star.le.ac.uk

A. BEAUCHAMP, Departement de Physique, Universite de Montreal, P.O. Box 6128 Succ. A, Montreal P.Q. H3C 3J7, CANADA,
beauchamp@astro.umontreal.ca

T. BLÖCKER, Astrophysikalisches Institut Potsdam, Telegrafenberg, D-14473 Potsdam, GERMANY,
tbloecker@aip.de

H.E. BOND, Space Telescope Science Institute, 3700 San Martin Drive, Home-wood Campus, Baltimore, MD 21218, USA,
bond@stsci.edu

P.A. BRADLEY, Los Alamos National Laboratory, X-2 Division, MS B220, Los Alamos NM 87545, USA,
pbradley@lanl.gov

A. BRAGAGLIA, Osserv. Astron. Bologna, via Zamboni 33, I-40126 Bologna, ITALY,
angela@alma02.cineca.it

I. BUES, Dr. Remeis-Sternwarte Bamberg, Sternwartstrasse 7, D-96049 Bam-berg, GERMANY,
bues@sternwarte.uni-erlangen.d400.de

M. BURLEIGH, Dept. of Physics and Astronomy, University of Leicester, Le-icester LE1 7RH, UK,
mbu@star.le.ac.uk

M. CHEVRETON, Observatoire de Paris-Meudon, DAEC, F-92195 Meudon, FRANCE,
chevreton@mesiob.obspm.fr

C. CLAVER, University of Texas, Department of Astronomy and McDonald Observatory, Austin, TX 78712, USA,
cfc@stimpy.as.utexas.edu

J.C. CLEMENS, Iowa State University, Department of Physics and Astronomy, Ames, IA 50211, USA,
cclemens@iastate.edu

F. D'ANTONA, Osservatorio Astronomico di Roma, I-0040 Monte Porzio, Rome, ITALY,
franca@astrmp.astro.it

S. DREIZLER, Dr. Remeis-Sternwarte Bamberg, Sternwartstrasse 7, D-96049 Bamberg, GERMANY,
dreizler@sternwarte.uni-erlangen.d400.de

J. DRILLING, Department of Physics and Astronomy, 266 Nicholson Hall, Lousiana State University, Baton Rouge LA 70803-4001, USA,
drilling@rouge.phys.lsu.edu

J. DUPUIS, Center for EUV Astrophysics, 2150 Kittredge Street, University of California, Berkeley CA 94720, USA,
jdupuis@cea.berkeley.edu

D. ENGELHARDT, Dr. Remeis-Sternwarte Bamberg, Sternwartstrasse 7, D-96049 Bamberg, GERMANY,
engelhardt@sternwarte.uni-erlangen.d400.de

D.S. FINLEY, Center for EUV Astrophysics, 2150 Kittredge Street, University of California, Berkeley CA 94720, USA,
david@cea.berkeley.edu

T.A. FLEMING, Steward Observatory, University of Arizona, Tucson, AZ 85721, USA,
tfleming@as.arizona.edu

B. FREYTAG, Institut für Astronomie und Astrophysik der Universität, D-24098 Kiel, GERMANY,
supas024@astrophysik.uni-kiel.d400.de

B.T. GÄNSICKE, Universitäts-Sternwarte Göttingen, Geismarlandstr. 11, D-37083 Göttingen, GERMANY,
boris@uni-sw.gwdg.de

E. GARCIA-BERRO, Dept. de Física Aplicada, Universidad Politécnica de Cataluña, Jordi Girona Salgado s/n, Módul B-4, Campus Nord, E-08034 Barcelona, SPAIN,
garcia@etseccpb.upc.es

A. GEMMO, European Southern Observatory, Karl-Schwarzschild-Str. 2, D-85748 Garching, GERMANY,
agemmo@eso.org

S. HAAS Dr. Remeis-Sternwarte Bamberg, Sternwartstrasse 7, D-96049 Bamberg, GERMANY,
haas@sternwarte.uni-erlangen.d400.de

W.-R. HAMANN, Institut für Astronomie und Astrophysik der Universität, D-24098 Kiel, GERMANY,
supas081@astrophysik.uni-kiel.d400.de

U. HEBER, Dr. Remeis-Sternwarte Bamberg, Sternwartstrasse 7, D-96049 Bamberg, GERMANY,
heber@sternwarte.uni-erlangen.d400.de

M. HERNANZ, Centre d'Estudis Avancats Blanes, Cami de Sta. Barbara, E-17300 Blanes (Girona), SPAIN,
marga@ceab.es

T. VON HIPPEL, Institute of Astronomy, Madingley Road, Cambridge CB3 0HA, UK,
ted@mail.ast.cam.ac.uk

J.B. HOLBERG, Lunar and Planetary Laboratory, Gould-Simpson Building, University of Arizona, Tucson AZ 85721, USA,
holberg@looney.lpl.arizona.edu

I. HUBENY, NASA/GSFC Code 681, Greenbelt, MD 20771, USA,
hubeny@stars.gsfc.nasa.gov

I. IBEN, JR., Astronomy Building, Univ. of Illinois, Urbana, IL 61801, USA,
icko@sirius.astro.uiuc.edu

J. ISERN, Centre d'Estudis Avancats de Blanes, Cami de Santa Barbara, E-17300 Blanes (Girona), SPAIN,
jordi@ceab.es

S. JORDAN, Institut für Astronomie und Astrophysik der Universität, D-24098 Kiel, GERMANY,
jordan@astrophysik.uni-kiel.d400.de

L. KARL-DIETZE, Dr. Remeis-Sternwarte Bamberg, Sternwartstrasse 7, D-96049 Bamberg, GERMANY,
karl@sternwarte.uni-erlangen.d400.de

D. KOESTER, Institut für Astronomie und Astrophysik der Universität, D-24098 Kiel, GERMANY,
koester@astrophysik.uni-kiel.d400.de

L. KOESTERKE, Institut für Astronomie und Astrophysik der Universität, D-24098 Kiel, GERMANY,
supas019@astrophysik.uni-kiel.d400.de

J. KUBAT, Astronomický ústav, Akademie věd České republiky, 251 65 Ondřejov, Czech Republic,
kubat@sunstel.asu.cas.cz

W. LANDSMAN, NASA/GSFC Code 681, Greenbelt, MD 20771, USA,
landsman@stars.gsfc.nasa.gov

U. LEUENHAGEN, Institut für Astronomie und Astrophysik der Universität, D-24098 Kiel, GERMANY,
supas023@astrophysik.uni-kiel.d400.de

J. LIEBERT, Steward Observatory, University of Arizona, Tucson, AZ 85721, USA,
liebert@as.arizona.edu

H.-G. LUDWIG, Max-Planck-Institut für Astrophysik, D-85740 Garching, GERMANY,
hgl@mpa-garching.mpg.de

J. MADEJ, Warsaw University Observatory, Al. Ujazdowskie 4, Warszawa, POLAND,
jm@carina.astrouw.edu.pl

M.C. MARSH, Dept. of Physics and Astronomy, University of Leicester, Leicester LE1 7RH, UK,
mcm@star.le.ac.uk

C. MASSACAND, University of Tromsø, Auroral Observatory, N-9037 Tromsø, NORWAY,
xtophe@lie.uit.no

I. MAZZITELLI, Istituto di Astrofisica Spaziale C.N.R., c.p.67 I-00044 Frascati, ITALY,
aton@hyperion.ias.cnr.fra.it

E.G. MEIŠTAS, Institute of Theoretical Physics and Astronomy, Goštauto 12, LT-2600 Vilnius, LITHUANIA,
meistas@itpa.fi.lt

N. MERANI, Ruhr-Universität Bochum, Lehrstuhl für Theoretische Physik I, D-44780 Bochum, GERMANY,
merani@tp1.ruhr-uni-bochum.de

S. MOEHLER, Landessternwarte Königsstuhl, D-69117 Heidelberg, GERMANY,
smoehler@mail.lsw.uni-heidelberg.de

R. NAPIWOTZKI, Dr. Remeis-Sternwarte Bamberg, Sternwartstrasse 7, D-96049 Bamberg, GERMANY,
napiwotzki@sternwarte.uni-erlangen.d400.de

D. O'DONOGHUE, UCT Astronomy Dept., Cape Town, SOUTH AFRICA,
dod@uctvax.uct.ac.za

E. ØSTGAARD, Fysisk Institutt, AVH, Universitetet i Trondheim, N-7055 Dragvoll, NORWAY,
erlend.oestgaard@avh.unit.no

T.D. OSWALT, Dept. Physics and Space Sciences, Florida Institute of Technology, 150. W. University Blvd., Melbourne, FL 32901, USA,
oswalt@tycho.pss.fit.edu

B. PFEIFFER, Observatiore de Midi-Pyrenees, 14 av. E. Belin, F-31400 Toulouse, FRANCE,
pfeiffer@srvdec.obs-mip.fr

J. PROVENCAL, Department of Physics and Astronomy, University of Delaware, Newark, DE 19716, USA,
jlp@chopin.udel.edu

A. PUTNEY, Palomar Observatory, California Institute of Technology, Pasadena, CA 91101, USA,
axp@fido.caltech.edu

T. RAUCH, Institut für Astronomie und Astrophysik der Universität, D-24098 Kiel, GERMANY,
rauch@astrophysik.uni-kiel.d400.de

M.T. RUIZ, Department of Astronomy, Universidad de Chile, Casilla 36-D, Santiago, CHILE,
cuca@das.uchile.cl

R.A. SAFFER, Space Telescope Science Institute, 3700 San Martin Drive, Homewood Campus, Baltimore MD 21218, USA,
saffer@stsci.edu

D. SCHÖNBERNER, Astrophysikalisches Institut Potsdam, Telegrafenberg, D-14473 Potsdam, GERMANY,
deschoenberner@aip.de

T. SCHÖNING, Institut für Astronomie und Astrophysik der Universität, Scheinerstr. 1, D-81679 München, GERMANY,
schoening@usm.uni-muenchen.de

H.L. SHIPMAN, Physics Department, University of Delaware, Sharp Lab., Newark DE 19716, USA,
harrys@strauss.udel.edu

E. SION, Dept. of Astronomy, and Astrophysics, Villanova University, Villanova, PA 19085, USA,
emsion@ucis.vill.edu

N. SKJEI, Fysisk Institutt, AVH, Universitetet i Trondheim, N-7055 Dragvoll, NORWAY,
norunn.skjei@avh.unit.no

J.E. SOLHEIM, University of Tromsø, Auroral Observatory, N-9037 Tromsø, NORWAY,
janerik@mack.uit.no

W. STOLZMANN, Institut für Astronomie und Astrophysik der Universität, D-24098 Kiel, GERMANY,
supas051@astrophysik.uni-kiel.d400.de

A. THEISSEN, Sternwarte Bonn, Auf dem Hügel 71, D-53121 Bonn, GERMANY,
theissen@astro.uni-bonn.de

P. THEJLL, Niels Bohr Institute, Blegdamsvej 17, DK-2100 Copenhagen, DENMARK,
thejll@nordita.dk

A. ULLA, Niels Bohr Institute, Blegdamsvej 17, DK-2100 Copenhagen, DENMARK,
ulla@nordita.dk

K. UNGLAUB, Dr. Remeis-Sternwarte Bamberg, Sternwartstrasse 7, D-96049 Bamberg, GERMANY

H. VÄTH, Institut für Astronomie und Astrophysik der Universität, D-24098 Kiel, GERMANY,
supas097@astrophysik.uni-kiel.d400.de

G. VAUCLAIR, Observatiore de Midi-Pyrenees, 14 av. E. Belin, F-31400 Toulouse, FRANCE,
gerardv@obs-mip.fr

S. VENNES, Center for EUV Astrophysics, 2150 Kittredge Street, University of California, Berkeley CA 94720, USA,
vennes@cea.berkeley.edu

V. WEIDEMANN, Institut für Astronomie und Astrophysik der Universität, D-24098 Kiel, GERMANY,
supas058@astrophysik.uni-kiel.d400.de

K. WERNER, Institut für Astronomie und Astrophysik der Universität, D-24098 Kiel, GERMANY,
werner@astrophysik.uni-kiel.d400.de

F. WESEMAEL, Departement de Physique, Universite de Montreal, P.O. Box
6128, Succ. A, Montreal P.Q. H3C 3J7, CANADA,
wesemael@astro.umontreal.ca

D.T. WICKRAMASINGHE, Astrophysical Theory Centre, Australian National
University, Canberra ACT 2601, AUSTRALIA,
dayal@cygnus.anu.edu.au

B. WOLFF, Institut für Astronomie und Astrophysik der Universität, D-24098
Kiel, GERMANY,
supas089@astrophysik.uni-kiel.d400.de

M. WOOD, Dept. Physics and Space Sciences, Florida Institute of Technology,
150. W. University Blvd., Melbourne, FL 32901-6988, USA,
wood@kepler.pss.fit.edu

Author Index

Achilleos N.   129
Allard N.   196, 197
Andersen J.   203
Aslan T.   259
Barstow M.A.   302, 318, 328
Bauer F.   264, 271
Beauchamp A.   108, 200
Belmonte J.A.   274
Bergeron P.   12, 108, 191, 200
Beuermann K.   263
Blöcker T.   68, 83
Boer K.S. de   268, 271
Bond H.E.   267
Bowyer S.   337
Bradley P.A.   96, 284
Bragaglia A.   238
Bruhweiler F.C.   203
Buckley D.A.   328
Bues I.   118, 123, 201, 259
Burkert A.   19
Burleigh M.R.   318, 328
Chevreton M.   274
Claver C.F.   78
Clemens J.C.   294
Collins J.   202
D'Antona F.   93
Dolez N.   274, 299
Dreizler S.   160, 171, 204, 243, 332
Dupuis J.   323
Engelhardt D.   123
Finley D.S.   150, 332
Fleming T.A.   328
Freytag B.   88
Gänsicke B.T.   263
Garcia-Berro E.   19, 36, 73
Gautschy A.   295
Gilmore G.   95
Grauer A.D.   274
Haas S.   243
Hamann W.-R.   198
Hansen-Ruiz C.S.   272

Heber U.   160, 171, 243, 266, 268, 271
Hernanz M.   19, 36, 73
Hintzen P.   199
Hippel T. von   95
Holberg J.B.   138, 202, 203, 302, 328
Hubeny I.   98
Hurwitz M.   337
Iben I. Jr.   48
Isern J.   19, 36, 73
Jimenez A.   274
Jones D.H.P.   95
Jordan S.   134, 135, 332, 337, 338
Kügler L.   266
Karl-Dietze L.   201
Koesterke L.   198
Koester D.   196, 197, 302, 328, 332, 337, 338
Kubát J.   133
Kunze D.   264
Landsman W.   191, 199
Lannes A.   299
Lanz T.   98
Leuenhagen U.   205
Liebert J.   12, 108, 200, 221, 264
Ludwig H.-G.   88, 128, 295
MacDonald J.   48, 272
Madej J.   130
Main J.   131
Maréchal P.   299
Marsh M.C.   328
Martino D. de   263
Massacand C.   270
Mazzitelli I.   58, 93
Meištas E.G.   296
Meier T.   243
Merani N.   131, 134
Mochkovitch R.   19, 36, 73
Moehler S.   268, 271
Napiwotzki R.   132, 176, 337
Nousek J.A.   302
O'Donoghue D.   297, 328

Oswalt T.D.	24
Pfeiffer B.	274, 299
Provencal J.L.	254, 298
Putney A.	135
Rasilla J.L.	272
Rauch T.	171, 186
Roques S.	299
Ruiz M.T.	46
Saffer R.A.	108, 221, 264, 267
Schöning T.	113
Schaefer K.G.	267
Schmidt J.H.K.	271
Serre B.	299
Shipman H.L.	248, 264, 298
Simon T.	191
Sion E.M.	208, 267
Smith J.A.	24
Solheim J.E.	269, 270
Stauffer J.R.	267
Stecher T.	199
Steffen M.	88, 128
Stolzmann W.	83
Theissen A.	271, 272
Thejll P.	264, 272
Thorstensen J.R.	313
Tweedy R.W.	202
Ulla A.	272
Unglaub K.	118
Väth H.	136
Vauclair G.	196, 274, 299
Vennes S.	313, 323
Vidal I.	274
Weidemann V.	1, 337
Werner K.	160, 171, 186, 204, 243, 302, 332
Wesemael F.	108, 200
Wickramasinghe D.T.	129, 232
Winget D.E.	78
Wolff B.	204, 338
Wood M.A.	41
Wunner G.	131